Florin Gorunescu

D1281400

Intelligent Systems Reference Library, Volume 12

Editors-in-Chief

Prof. Janusz Kacprzyk
Systems Research Institute
Polish Academy of Sciences
ul. Newelska 6
01-447 Warsaw
Poland
E-mail: kacprzyk@ibspan.waw.pl

Prof. Lakhmi C. Jain
University of South Australia
Adelaide
Mawson Lakes Campus
South Australia 5095
Australia
E-mail: Lakhmi.jain@unisa.edu.au

Further volumes of this series can be found on our homepage: springer.com

Vol. 1. Christine L. Mumford and Lakhmi C. Jain (Eds.)
Computational Intelligence: Collaboration, Fusion and Emergence, 2009
ISBN 978-3-642-01798-8

Vol. 2. Yuehui Chen and Ajith Abraham
Tree-Structure Based Hybrid Computational Intelligence, 2009
ISBN 978-3-642-04738-1

Vol. 3. Anthony Finn and Steve Scheding
Developments and Challenges for Autonomous Unmanned Vehicles, 2010
ISBN 978-3-642-10703-0

Vol. 4. Lakhmi C. Jain and Chee Peng Lim (Eds.)
Handbook on Decision Making: Techniques and Applications, 2010
ISBN 978-3-642-13638-2

Vol. 5. George A. Anastassiou
Intelligent Mathematics: Computational Analysis, 2010
ISBN 978-3-642-17097-3

Vol. 6. Ludmila Dymowa
Soft Computing in Economics and Finance, 2011
ISBN 978-3-642-17718-7

Vol. 7. Gerasimos G. Rigatos
Modelling and Control for Intelligent Industrial Systems, 2011
ISBN 978-3-642-17874-0

Vol. 8. Edward H.Y. Lim, James N.K. Liu, and Raymond S.T. Lee
Knowledge Seeker – Ontology Modelling for Information Search and Management, 2011
ISBN 978-3-642-17915-0

Vol. 9. Menahem Friedman and Abraham Kandel
Calculus Light, 2011
ISBN 978-3-642-17847-4

Vol. 10. Andreas Tolk and Lakhmi C. Jain
Intelligence-Based Systems Engineering, 2011
ISBN 978-3-642-17930-3

Vol. 11. Samuli Niiranen and Andre Ribeiro (Eds.)
Information Processing and Biological Systems, 2011
ISBN 978-3-642-19620-1

Vol. 12. Florin Gorunescu
Data Mining, 2011
ISBN 978-3-642-19720-8

Florin Gorunescu

Data Mining

Concepts, Models and Techniques

 Springer

Prof. Florin Gorunescu
Chair of Mathematics
Biostatistics and Informatics University of
Medicine and Pharmacy of Craiova
Professor associated to the Department of
Computer Science
Faculty of Mathematics and Computer Science
University of Craiova
Romania
E-mail: gorun@umfcv.ro

ISBN 978-3-642-19720-8   e-ISBN 978-3-642-19721-5

DOI 10.1007/978-3-642-19721-5

Intelligent Systems Reference Library   ISSN 1868-4394

Library of Congress Control Number: 2011923211

Typeset & *Cover Design:* Scientific Publishing Services Pvt. Ltd., Chennai, India.

Printed on acid-free paper

9 8 7 6 5 4 3 2 1

springer.com

To my family

Preface

Data Mining represents a complex of technologies that are rooted in many disciplines: mathematics, statistics, computer science, physics, engineering, biology, etc., and with diverse applications in a large variety of different domains: business, health care, science and engineering, etc. Basically, data mining can be seen as the science of exploring large datasets for extracting implicit, previously unknown and potentially useful information.

My aim in writing this book was to provide a friendly and comprehensive guide for those interested in exploring this vast and fascinating domain. Accordingly, my hope is that after reading this book, the reader will feel the need to deepen each chapter to learn more details.

This book aims to review the main techniques used in data mining, the material presented being supported with various examples, suggestively illustrating each method.

The book is aimed at those wishing to be initiated in data mining and to apply its techniques to practical applications. It is also intended to be used as an introductory text for advanced undergraduate-level or graduate-level courses in computer science, engineering, or other fields. In this regard, the book is intended to be largely self-contained, although it is assumed that the potential reader has a quite good knowledge of mathematics, statistics and computer science.

The book consists of six chapters, organized as follows:

- The first chapter introduces and explains fundamental aspects about data mining used throughout the book. These are related to: what is data mining, why to use data mining, how to mine data? Data mining solvable problems, issues concerning the modeling process and models, main data mining applications, methodology and terminology used in data mining are also discussed.
- Chapter 2 is dedicated to a short review regarding some important issues concerning data: definition of data, types of data, data quality, and types of data attributes.

- Chapter 3 deals with the problem of data analysis. Having in mind that data mining is an analytic process designed to explore large amounts of data in search of consistent and valuable hidden knowledge, the first step consists in an initial data exploration and data preparation. Then, depending on the nature of the problem to be solved, it can involve anything from simple descriptive statistics to regression models, time series, multivariate exploratory techniques, etc. The aim of this chapter is therefore to provide an overview of the main topics concerning exploratory data analysis.
- Chapter 4 presents a short overview concerning the main steps in building and applying classification and decision trees in real-life problems.
- Chapter 5 summarizes some well-known data mining techniques and models, such as: Bayesian and rule-based classifiers, artificial neural networks, k-nearest neighbors, rough sets, clustering algorithms, and genetic algorithms.
- The final chapter discusses the problem of evaluating the performance of different classification (and decision) models.

An extensive bibliography is included, which is intended to provide the reader with useful information covering all the topics approached in this book.

The organization of the book is fairly flexible, the selection of the topics to be approached being determined by the reader himself (herself), although my hope is that the book will be read entirely.

Finally, I wish this book to be considered just as a "compass" helping the interested reader to sail in the rough sea representing the current information vortex.

December 2010   Florin Gorunescu
                                                                        Craiova

Contents

Chapter 1
Introduction to Data Mining

Abstract. It is the purpose of this chapter to introduce and explain fundamental aspects about data mining used throughout the present book. These are related to: what is data mining, why to use data mining, how to mine data? There are also discussed: data mining solvable problems, issues concerning the modeling process and models, main data mining applications, methodology and terminology used in data mining.

1.1 What Is and What Is Not Data Mining?

Since 1990s, the notion of *data mining*, usually seen as the process of "mining" the data, has emerged in many environments, from the academic field to the business or medical activities, in particular. As a research area with not such a long history, and thus not exceeding the stage of 'adolescence' yet, data mining is still disputed by some scientific fields. Thus, Daryl Pregibons allegation: "data mining is a blend of Statistics, Artificial Intelligence, and database research" still stands up (Daryl Pregibon, *Data Mining*, Statistical Computing & Graphics Newsletter, December 1996, 8).

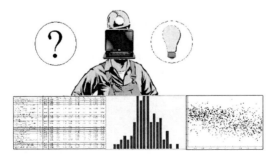

Fig. 1.1 Data 'miner'

F. Gorunescu: Data Mining: Concepts, Models and Techniques, ISRL 12, pp. 1–43.
springerlink.com   © Springer-Verlag Berlin Heidelberg 2011

Despite its "youth", data mining is "projected to be a multi-billion dollar industry by the year 2000", while, at the same time, it has been considered by some researchers as a "dirty word in Statistics" (*idem*). Most likely, they were statisticians and they have not considered data mining as something interesting enough for them at that time.

In this first chapter, we review the fundamental issues related to this subject, such as:

- What is (and what is not) data mining?
- Why data mining?
- How to 'mine' in data?
- Problems solved with data mining methods.
- About modeling and models.
- Data mining applications.
- Data mining terminology.
- Data confidentiality.

However, before attempting a definition of data mining, let us emphasize some aspects of its genesis. Data mining, also known as "*knowledge-discovery in databases*" (KDD), has three generic roots, from which it borrowed the techniques and terminology (see Fig. 1.2):

- *Statistics* -its oldest root, without which data mining would not have existed. The classical Statistics brings well-defined techniques that we can summarize in what is commonly known as *Exploratory Data Analysis* (EDA), used to identify systematic relationships between different variables, when there is no sufficient

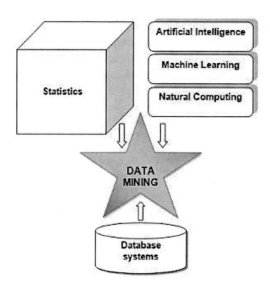

Fig. 1.2 Data mining roots

information about their nature. Among EDA classical techniques used in DM, we can mention:

- *Computational methods*: descriptive statistics (distributions, classical statistical parameters (mean, median, standard deviation, etc.), correlation, multiple frequency tables, multivariate exploratory techniques (cluster analysis, factor analysis, principal components & classification analysis, canonical analysis, discriminant analysis, classification trees, correspondence analysis), advanced linear/non-linear models (linear/non-linear regression, time series/forecasting, etc.);
- *Data visualization* aims to represent information in a visual form, and can be regarded as one of the most powerful and, at the same time, attractive methods of data exploration. Among the most common visualization techniques, we can find: histograms of all kinds (column, cylinders, cone, pyramid, pie, bar, etc.), box plots, scatter plots, contour plots, matrix plots, icon plots, etc. For those interested in deepening EDA techniques, we refer, for instance, to (386), (395), or (251).

- *Artificial Intelligence* (AI) that, unlike Statistics, is built on heuristics. Thus, AI contributes with information processing techniques, based on human reasoning model, towards data mining development. Closely related to AI, *Machine Learning* (ML) represents an extremely important scientific discipline in the development of data mining, using techniques that allow the computer to learn with 'training'. In this context, we can also consider *Natural Computing* (NC) as a solid additional root for data mining.
- *Database systems* (DBS) are considered the third root of data mining, providing information to be 'mined' using the methods mentioned above.

The necessity of 'mining' the data can be thus summarized, seen in the light of important real-life areas in need of such investigative techniques:

- *Economics* (business-finance) - there is a huge amount of data already collected in various areas such as: Web data, e-commerce, super/hypermarkets data, financial and banking transactions, etc., ready for analyzing in order to take optimal decisions;
- *Health care* - there are currently many and different databases in the health care domain (medical and pharmaceutical), which were only partially analyzed, especially with specific medical means, containing a large information yet not explored sufficiently;
- *Scientific research* - there are huge databases gathered over the years in various fields (astronomy, meteorology, biology, linguistics, etc.), which cannot be explored with traditional means.

Given the fact that, on the one hand, there is a huge amount of data systematically unexplored yet and, on the other hand, both computing power and computer science have grown exponentially, the pressure of using new methods for revealing information 'hidden' in data increased. It is worth pointing out that there is a lot of information in data, almost impossible to detect by traditional means and using only the human analytic ability.

Let us try now to define what data mining is. It is difficult to opt for a unique definition providing a picture as complete as possible of the phenomenon. Therefore, we will present some approaches more or less similar, which will outline clearly enough, hopefully, what data mining is. So, by data mining we mean (equivalent approaches):

- The automatic search of patterns in huge databases, using computational techniques from statistics, machine learning and pattern recognition;
- The non-trivial extraction of implicit, previously unknown and potentially useful information from data;
- The science of extracting useful information from large datasets or databases;
- The automatic or semi-automatic exploration and analysis of large quantities of data, in order to discover meaningful patterns;
- The automatic discovery process of information. The identification of patterns and relationships 'hidden' in data.

Metaphorically speaking, by data mining we understand the proverbial "finding the needle in a haystack", using a metal sensor just to speed up the search, 'automating' the corresponding process.

We saw above what data mining means. In this context, it is interesting to see what data mining is not. We present below four different concrete situations which eloquently illustrates what data mining is not compared with what it could be.

- *What is not data mining*: Searching for particular information on Internet (e.g., about cooking on Google).
  What data mining could be: Grouping together similar information in a certain context (e.g., about French cuisine, Italian cuisine, etc., found on Google).

- *What is not data mining*: A physician seeking a medical register for analyzing the record of a patient with a certain disease.
  What data mining could be: Medical researchers finding a way of grouping patients with the same disease, based on a certain number of specific symptoms.

- *What is not data mining*: Looking up *spa* resorts in a list of place names.
  What data mining could be: Grouping together *spa* resorts that are more relevant for curing certain diseases (gastrointestinal, urology, etc.).

- *What is not data mining*: The analysis of figures in a financial report of a trade company.
  What data mining could be: Using the trade company database concerning sales, to identify the customers' main profiles.

A good example, to highlight even more the difference between what is usually a search in a database and data mining, is: "Someone may be interested in the difference between the number of purchases of a particular kind (e.g., appliances) from a supermarket compared to a hypermarket, or possibly from two supermarkets in different regions". In this case, it already takes into account *a priori* the assumption that there are differences between a supermarket and a hypermarket, or the sales between the two regions. On the contrary, in the data mining case, the problem may consist for instance in identifying factors that influence sales volume, without relying on any *a priori* hypothesis. To conclude, the data mining methods seek to identify patterns and hidden relationships that are not always obvious (and therefore easily identifiable) under the circumstances of certain assumptions.

As it is seen from the above examples, we cannot equate a particular search (research) of an individual object (of any kind) and data mining research. In the latter case, the research does not seek individualities, but sets of individualities, which, in one way or another, can be grouped by certain criteria. Metaphorically speaking once more, the difference between a simple search and a data mining process is that of looking for a specific tree and the identification of a forest (hence the well-known proverb *"Can't see the forest for the trees"* used when the research is not sufficiently lax regarding constraints).

Let us list below two data mining goals to distinguish more clearly its area of application (108):

- *Predictive objectives* (e.g., classification, regression, anomalies/outliers detection), achieved by using a part of the variables to predict one or more of the other variables;
- *Descriptive objectives* (e.g., clustering, association rule discovery, sequential pattern discovery), achieved by the identification of patterns that describe data and that can be easily understood by the user.

1.2 Why Data Mining?

At first glance one may think it is easy to answer such a question without a prior presentation of the data mining techniques and especially its applications. We believe that the presentation of three completely different situations in which data mining was successfully used would be more suggestive. First, let us mention a situation, as dramatic as it is true, concerning the possible role of data mining in solving a fundamental nowadays problem that concerns, unfortunately, all of us. According to Wikinews (http://www.wikinews.org/) (408), data mining has been cited as the method by which an U.S. Army intelligence unit supposedly had identified the 9/11 attack leader and three other hijackers as possible members of an Al-Qaeda cell operating in the U.S. more than a year before the attack. Unfortunately, it seems that this information was not taken into account by the authorities. Secondly, it is the case of a funny story, however unpleasant for the person in question. Thus, Ramon C. Barquin -*The Data Warehousing Institute Series* (Prentice Hall) Editor- narrates in "Foreward" to (157) that he received a call from his telephone provider telling him that they had reason to believe his calling card had been stolen. Thus, although the day before he spent all the time in Cincinnati, it seemed he phoned from Kennedy Airport, New York to La Paz, Bolivia, and to Lagos, Nigeria. Concretely, these calls and three others were placed using his calling card and PIN number, facts that do not fit his usual calling patterns. Fortunately, the phone company had been able to early detect this fraudulent action, thanks to their data mining program. In the context of fraudulent use of different electronic tools (credit cards, charge cards, etc.) involving money, the situation is much more dramatic. Industry experts say, that even if a huge number of credit card frauds are reported each year, the fact remains that credit card fraud has actually been decreasing. Thus, improved systems to detect bogus transactions have produced a decade-long decline in fraud as a percentage of overall dollar transactions. Besides the traditional advice concerning the constant vigilance of card issuers, the companies also are seeking sophisticated software solutions, which use high-powered data mining techniques to alert issuers to potential instances of fraud ("The truth about credit-card fraud", *BusinessWeek*, June 21, 2005). Third, let us mention the urban legend concerning the well-known "couple" *beer and diapers*. Briefly, a number of store clerks noticed that men often bought beer at the same time they bought diapers. The store mined its receipts and proved the clerks' observations were correct. Therefore, the store began stocking diapers next to the beer coolers, and sales skyrocketed. The story is a myth, but it shows how data mining seeks to understand the relationship between different actions, (172).

Last but not least, recall that "Knowledge is power" ("*Scientia potentia est*" - F. Bacon, 1597) and also recall that knowledge discovery is often considered as synonymous with data mining -*quod erat demonstrandum*.

These are only three very strong reasons to seriously consider this domain, fascinating and complex at the same time, regarding the discovery of information when human knowledge is not of much use.

There are currently many companies focused on data mining (consulting, training and products for various fields) -for details see KDnuggets^TM for instance (http://www.kdnuggets.com/companies/index.html). This is due mainly to the growing demand for services provided by the data mining applications to the economic and financial market (e.g., *Business intelligence* (BI), *Business performance management* (BPM), *Customer relationship management* (CRM), etc.), health care field (e.g., *Health Informatics*, e-Health, etc.), without neglecting other important areas of interest, such as telecommunications, meteorology, biology, etc.

Starting from the marketing forecast for large transnational companies and passing through the trend analysis of shares trading on the main Stock Exchanges, identification of the loyal customer profile, modeling demand for pharmaceuticals, automation of cancer diagnosis, bank fraud detection, hurricanes tracking, classification of stars and galaxies, etc., we notice a various range of areas where data mining techniques are effectively used, thus giving a clear answer to the question: "*Why Data Mining?*"

On the other hand, we must not consider that data mining can solve any problem focused on finding useful information in data. Like in the original mining, it is possible for data mining to dig the 'mine' of data without eventually discovering the lode containing the "gold nugget" of knowledge. Knowledge/useful information discovery depends on many factors, starting with the 'mine' of data and ending with the used data mining 'tools' and the mastery of the 'miner'. Thus, if there is no gold nugget in the mine, there is nothing to dig for. On the other hand, the 'lode' containing the 'gold nugget', if any, should be identified and correctly assessed and then, if it is worth to be explored, this operation must be carried out with appropriate 'mining tools'.

1.3 How to Mine the Data?

Let us see now what the process of 'mining' the data means. Schematically, we can identify three characteristic steps of the data mining process:

1. *Exploring data*, consisting of data 'cleansing', data transformation, dimensionality reduction, feature subset selection, etc.;
2. *Building the model and its validation*, referring to the analysis of various models and choosing the one who has the best performance of forecast -*competitive evaluation of models*;
3. *Applying* the model to new data to produce correct forecasts/estimates for the problems investigated.

According to (157), (378) we can identify five main stages of the process of 'mining' the data:

- *Data preparation/data pre-processing*. Before using whatever data mining technique to 'mine' the data, it is absolutely necessary to prepare the raw data. There are several aspects of the initial preparation of data before processing them using data mining techniques. First, we have to handle the problem concerning the quality of data. Thus, working with raw data we can find noise, outliers/anomalies, missing values, duplicate data, incorrectly recorded data, expired data, etc. Accordingly, depending on quality problems detected in data, we proceed to solve them with specific methods. For instance, in the case of noise existence (i.e., distortions of the true values (measurements) produced by random disturbances), different filtering techniques are used to remove/reduce the effect of distortion. Thus, in case of signal processing we can mention, besides the electronic (hard) filters, the 'mathematical' (soft) filters consisting of mathematical algorithms used to change the harmonic component of the signal (e.g., moving average filter, Fourier filter, etc.). In case of extreme values, i.e., values that deviate significantly from the average value of data, we can proceed either to their removal or to the alternative use of parameters (statistics) that are not so sensitive to these extreme values (e.g., median instead of mean, which is very sensitive to outliers). The case of missing values is common in data mining practice and has many causes. In this situation we can use different methods, such as: elimination of data objects with missing values, estimation of missing values, their substitution with other available values (e.g., mean/median, possibly weighted), ignoring them during analysis, if possible, etc. In case of duplicate data (e.g., a person with multiple e-mail addresses), the deletion of duplicates may be considered. Once the data quality issue is solved, we proceed to their proper pre-processing, which consists, in principle, of the following procedures:

- *Aggregation* consists in combining two or more attributes (or objects) into a single attribute (or object), aiming to reduce the number of attributes (or objects), in order to obtain more 'stable' data, with less variability (e.g., cities aggregated into regions, states, countries, daily sales aggregated into weekly, monthly, yearly sales, etc.).
- *Sampling* is the main method of selecting data, representing the process of drawing a representative sample from the entire dataset. Methods of creating samples form a classic field of Statistics and we will not go further into technical details (see, for instance (10),(63), (380)). We mention however the problem concerning the sample size, which is important in the balance between the effectiveness of the data mining process (obtained by reducing the amount of data being processed) and the significant loss of information due to a low volume of data. This problem belongs to the "*power analysis and sample size calculation*" domain in Statistics, and is approached by taking into account specific techniques (e.g., one mean *t*-test, two means *t*-test, two proportions *z*-test, etc.), which depend on the problem being solved.
- *Dimensionality reduction*. It is known among the mining practitioners that when the data size (i.e., number of attributes) increases, the spread of data also increases. Consequently, further data processing will be difficult due to the need of increased memory, meaning a lower computation speed. In data mining this situation is called, more than suggestive, the "*curse of dimensionality*". The 'antidote' to this 'curse' is represented by dimensionality reduction. Thus, we obtain a reduced amount of time and memory required by data processing, better visualization, elimination of irrelevant features and possible noise reduction. As techniques for dimensionality reduction, we can mention typical multivariate exploratory techniques such as factor analysis, principal components analysis, multidimensional scaling, cluster analysis, canonical correlation, etc.
- *Feature selection* is used to eliminate irrelevant and redundant features, possibly causing confusion, by using specific methods (e.g., brute-force approach, embedded approach, filter approach, wrapper approach, embedded methods -see, for instance, (241),(163),(378).
- *Feature creation* refers to the process of creating new (artificial) attributes, which can better capture important information in data than the original ones. As methods of creating new features, recall feature extraction, mapping data to a new space, feature construction, (242), (378).
- *Discretization and binarization*, that is, in short, the transition from continuous data to discrete (categorical) data (e.g., switch from real values to integer values), and convert multiple values into binary values (e.g., similar to converting a 256-color image into a black-and-white image), transition from several categories to only two categories, etc.) -see (240), (378).

- *Attribute transformation*, that is, in principle, the conversion of old attributes into new ones, using a certain transformation (mathematical functions (e.g., e^x, log x, sin x, x^n, etc.), normalization $x \rightarrow \frac{x}{\|x\|}$, etc.), transformation that improves the data mining process -see, (235), (236), (378).

- *Defining the study* (*research*) is the second step in the data mining process after the pre-processing phase. Note that, in the whole data mining process, the data processing stage will be repeated whenever necessary. First of all, since it represents a process of data analysis (mining the data), we have to focus on the data to be analyzed, i.e. the 'mine' where it will 'dig' looking for hidden information. Once the data to be 'mined' have been chosen, we should decide how to sample the data, since we usually do not work with the entire database. Let us mention here an important aspect of the data mining process, i.e., the way the selected data will be analyzed. Note that the entire research will be influenced by the chosen methodology. In this context, we will review in a few words two machine learning techniques used extensively in data mining, namely the supervised/unsupervised learning. In brief, the term *supervised learning* means the process of establishing a correspondence (function) using a training dataset, seen as a 'past experience' of the model. The purpose of supervised learning is to predict the value (output) of the function for any new object (input) after completion of the training process. A classical example of the supervised learning technique is represented by the classification process (predictive method). Unlike supervised learning, in *unsupervised learning* the model is adapted to observations, being distinguished by the fact that there is no *a priori* output (the learner is fed with only unlabeled objects). A classical example of the unsupervised learning technique is represented by the clustering process (descriptive method). In the case of using supervised learning methods, the definition of the study refers both to the identification of a dependent variable (attribute), which will be considered as output, and to the choice of other variables which 'explain' the output variable (predictor variables/attributes). For example, in a medical study we are interested to understand the way the onset or progression of certain diseases (e.g., myocardial infarction) is influenced by certain "risk factors" (e.g., weight, age, smoking, heredity, etc.). Conversely, when using unsupervised learning methods, the general purpose of a model is to group similar objects or to identify exceptions in data. For instance, we may wish to identify customers with the same behavior regarding the purchase of certain types of goods, and also the process of identifying exceptions in data may be considered for fraud detection (the example given above in connection with the fraudulent use of phone card is suggestive). Once the data to be analyzed were determined, we focus on defining the purpose of the data mining process. In this respect, we present below some general details, (157):

  - *Understanding limits* refers to a set of problems that a user of data mining techniques has to face, starting from the basic idea that data mining cannot perform miracles, and there are limits about expectations of the results of its application. The first problem concerns the choice of the study purpose: "it is or it is not necessary to *a priori* consider a particular purpose, or we can mine 'blindly' in data for the hidden gold nugget?" A wise answer to this question is that, however, we must set up a certain goal or some general objectives of the study, in order to work properly with the available data. We still face the eternal controversy in this matter -how important is to *a priori* define the study targets? As mentioned above, we must always define a goal more or less precisely, when we start a data mining study (a clever search for a "needle in a haystack"). This approach will save much effort and computation time, through a good design of the study, starting with the selection and preparation of data and ending with the identification of potential beneficiaries. A second problem relates to the way one must proceed in case of inadequate data. In this respect, we can apply the idea that a better understanding of available data (even of doubtful quality) can result in a better use of them. Furthermore, once the design and the use of a model are done, questions not cease but rather multiply (e.g., "can the model be applied in other contexts?", "are there other ways to get similar results?"). Finally, it is possible that after the completion of the study, we may not obtain anything new, relevant or useful. However, this result must not stop us to use data mining techniques. Even if we obtain a result which we expected, especially if the problem is already well-known, we still gain because that result was once again confirmed by data mining. Moreover, using the model that repeatedly confirmed only known information on new data, it is possible at some point to get results different from what we had expected, thus indicating changes in patterns or trends requiring further investigation of the data.

– *Choosing an appropriate study* to address a particular problem refers to the 'natural' way the chosen study is linked to the sought solution. An example of a correctly chosen study is the identification of the patient 'standard' profile for a given disease in order to improve the treatment of that disease. Conversely, an inadequate study aims to understand the profile of viewers who like football, in order to optimize the romantic movies program on a TV channel (!).

– *Types of studies* refer to the goals taken into account when using data mining techniques. For instance, we can mention: identification of smokers profile in relation to non-smokers based on medical/behavioral data, discovery of characteristics of different types of celestial bodies based on data from telescopes (*Sky Survey Cataloging*) in order to classify new ones, segmentation of customers in different categories in order to become 'target customers' for certain products sale, etc.

– *Selection of elements for analysis* is again a problem neither fully resolved, nor that it could be, since it depends on many factors. Thus, it is one to consider for the first time a particular set of data and something else that prior experience already exist in this regard. In this respect, a beginner will choose all the available data, while an experienced researcher will focus only on specific relevant issues. On the other hand, a very important role is played by the type of study we want to perform: classification (e.g., classification/decision trees, neural networks), clustering (e.g., k-means, two-way joining), regression analysis (linear/non-linear, logistic regression), etc. The goal of the study is also important when selecting the items for analysis, especially if we deal with a heteroclite set of data. Thus, if for example we are interested in a certain customer profile, buying a specific consumer goods mix, in order to optimize the arrangement of goods in a supermarket, we must select the relevant features in this respect (e.g., job, annual income, gender, age, hobbies, etc.), ignoring other elements such as the health condition, for instance, not so important for the purpose. These types of information that can be selected from a database are known as *dimensions*, because they can be considered as dimensions of the profile of an individual, profile to be tailored by data mining techniques taking into account a particular purpose. Thus, it must be emphasized that one of the main advantages of data mining compared with other methods is that, in principle, we should not arbitrarily limit the number of elements that we observe, because by its own nature data mining possesses means to filter the information. Obviously, we do not need to use all available information as long as an elementary logic might exclude some parts of it. However, beginners or those who deal with a completely unknown area should not exclude anything that might lead to the discovery of useful knowledge.

- *Issue of sampling* is somehow linked to the previous one and concerns the relevance of the chosen sample, seen in the light of reaching the intended purpose. If it were just the statistical component of the data mining process then things would be much simpler, as we noted above (see "*Sampling*"), because there are clear statistical methods to calculate the sample size given the type of analysis chosen. In the data mining case, given the specific nature of the process, the rules are more relaxed, since the purpose of the study is just looking for useful information in very large datasets, information otherwise difficult if not impossible to discover with other classical methods. Yet in this case, to streamline the process (higher speed/lower computational effort) one can build the model starting from a smaller volume of data, obtained by sampling, and then proceeding to validate it on other available data.

- *Reading the data and building the model.* After making the previous steps of the data mining 'roadmap', we arrive at the moment when we use available data to achieve the intended purpose. The first thing to do at this point is 'reading data' from the existing dataset. Essentially, by reading the data we understand the process of accessing data (e.g., extraction data from a text file and placing them in a matrix form where lines are cases and columns are variables, in order to cluster them (getting similar cases, e.g., synonyms); reading data from an Excel file for processing them with a statistical software package (e.g., SAS, Statistica, IBM-SPSS, etc.). It is worth to know that each data mining product has a mechanism that can 'read' data. Once data read, we pass to build the data mining model. Any model will extract various indicators from the amount of available data, useful in understanding the data (e.g., frequencies of certain values, weights of certain characteristics, correlated attributes (and not considered separately) that explain certain behaviors, etc.). Whatever the considered model, we have to take into account some important features:

- *Model accuracy* refers to the power of that model to provide correct and reliable information, when used in real-world situations. We will discuss the matter at length throughout the book, here we only emphasize that the actual accuracy is measured on new data and not on training data, where the model can perform very well (see the case of overfitting).
- *Model intelligibility* refers to its characteristic of being easily understood by different people with different degrees/types of training, starting with the way of connecting inputs (data entered into the 'mining machinery') with outputs (corresponding conclusions) and finishing with the manner in which the forecast accuracy is presented. Although there are 'hermetic' models (e.g., artificial neural networks) on the one hand, which are similar to 'black boxes' in that few know what happens, and 'open' models (e.g., regressive statistical models or decision trees) on the other hand, very 'understandable' for many people, it is preferable to build and, above all, to present a model so to be easily understood, even if not with all the technical details, by a user without any specialized training. Do not forget, in this context, that data mining was created and grew so strong because of the business, health care, trade, etc. demands, which do not involve a specialized training of the potential customer.
- *The performance* of a data mining model is defined by both the time needed to be built and its speed of processing data in order to provide a prediction. Concerning the latter point, the processing speed on using large or very large databases is very important (e.g., when using probabilistic neural networks, the processing speed drops dramatically when the database size increases, because they use the whole baggage of "training data" when predicting).
- *Noise* in data is a 'perfidious enemy' in building an effective model of data mining, because it cannot be fully removed (filtered). Each model has a threshold of tolerance to noise and this is one of the reasons for an initial data pre-processing stage.

- *Understanding the model* (see also "model intelligibility") refers to the moment when, after the database was mined (studied/analyzed/interpreted), a data mining model was created based on the analysis of these data, being ready to provide useful information about them. In short, the following elements have to be considered at this time, regardless of the chosen model:

- *Model summarization*, as the name suggests, can be regarded as a concise and dense report, emphasizing the most important information (e.g., frequencies, weights, correlations, etc.) explaining the results obtained from data (e.g., model describing the patient recovery from severe diseases based on patients information, model forecasting hypertension likelihood based on some risk factors, etc.).

- *Specific information* provided by a model refers to those causal factors (inputs) that are significant to some effect, as opposed to those that are not relevant. For example, if we aim to identify the type of customers in a supermarket that are likely to frequent the cosmetics compartment, then the criterion (input) which is particularly relevant is the customer sex, always appearing in data (in particular - [women]), unlike the professional occupation that is not too relevant in this case. To conclude, it is very important to identify those factors naturally explaining the data (in terms of a particular purpose) and to exclude the irrelevant information to the analysis.

- *Data distribution*, just as in Statistics, regarding the statistical sampling process, is very important for the accuracy (reliability) of a data mining approach. As well as there, we first need a sufficiently large volume of data and, secondly, these data should be representative for analysis. Unlike Statistics, where the issue relates to finding a lower limit for the sample size so that results can be extrapolated with sufficient margin of confidence to the entire population (*statistical inference*), in this case it is supposed to 'dig' in an appreciable amount of data. However, we need to ensure that the data volume is large enough and diverse in its structure to be relevant for the wider use (e.g., the profile of trusted client for the banking system should be flexible enough for banks in general, and not just for a particular bank -if the study was not commissioned by a particular bank, obviously). Secondly, as we saw above, the data must have a 'fair' distribution for all categories considered in the study (e.g., if the 'sex' attribute is included in the analysis, then the two sexes must be represented correctly in the database: such a correct distribution would be, in general, 51% and 49% (female/male) versus 98% and 2% -completely unbalanced)

- *Differentiation* refers to the property of a predictive variable (input) to produce significant differentiation between two results (outputs) of the model. For example, if young people like to listen to both folk and rock music, this shows that this age group does not distinguish between the two categories of music. Instead, if the girls enjoy listening to folk music (20 against 1, for instance), then sex is important in differentiating the two musical genres. As we can see, it is very important to identify those attributes of data which could create differentiation, especially in studies of building some profiles, e.g., marketing studies.

- *Validation* is the process of evaluating the prediction accuracy of a model. Validation refers to obtaining predictions using the existing model, and then comparing these results with results already known, representing perhaps the most important step in the process of building a model. The use of a model that does not match the data cannot produce correct results to appropriately respond to the intended goal of the study. It is therefore understood that there is a whole methodology to validate a model based on existing data (e.g., *holdout*, *random sub-sampling*, *cross-validation*, *stratified sampling*, *bootstrap*, etc.). Finally, in the understanding of the model it is important to identify the factors that lead both to obtaining 'success' as well as 'failure' in the prediction provided by the model.

- *Prediction/forecast* of a model relates to its ability to predict the best response (output), the closest to reality, based on input data. Thus, the smaller the difference between what is expected to happen (*expected outcome*) and what actually happens (*observed outcome*), the better the prediction. As classic examples of predictions let us mention: the weather forecast (e.g., for 24 or 48 hours) produced by a data mining model based on complex meteorological observations, or the diagnosis for a particular disease given to a certain patient, based on his (her) medical data. Note that in the process of prediction some models provide, in addition to the forecast, the way of obtaining it (*white-box*), while others provide only the result itself, not how to obtain it (*black-box*). Another matter concerning it refers to the competitor predictions of the best one. Since no prediction is 'infallible', we need to know, besides the most probable one, its competitors (challenger predictions) in descending hierarchical order, just to have a complete picture of all possibilities. In this context, if possible, it is preferable to know the difference between the winning prediction and the second 'in race'. It is clear that, the larger the difference between the first two competitors, the less doubt we have concerning the best choice. We conclude this short presentation on the prediction of a data mining model, underlining that some areas such as: software reliability, natural disasters (e.g., earthquakes, floods, landslides, etc.), pandemics, demography (population dynamics), meteorology, etc., are known to have great difficulties to be forecasted.

1.4   Problems Solvable with Data Mining

The core process of data mining consists in building a particular model to represent the dataset that is 'mined' in order to solve some concrete problems of real-life. We will briefly review some of the most important issues that require the application of data mining methods, methods underlying the construction of the model.

In principle, when we use data mining methods to solve concrete problems, we have in mind their typology, which can be synthetically summarized in two broad categories, already referred to as the objectives of data mining:

- *Predictive methods* which use some existing variables to predict future values (unknown yet) of other variables (e.g., classification, regression, biases/anomalies detection, etc.);
- *Descriptive methods* that reveal patterns in data, easily interpreted by the user (e.g., clustering, association rules, sequential patterns, etc.).

We briefly present some problems facing the field of data mining and how they can be solved to illustrate in a suggestive manner its application field.

1.4.1 Classification

The idea that the human mind organizes its knowledge using the natural process of classification is widespread. But when it comes to classification, we speak about taxonomy. *Taxonomy* (gr. *tassein* = classify + *nomos* = science, law) appeared first as the science of classifying living organisms (*alpha taxonomy*), but then it developed as the science of classification in general, including here the principles of classification (taxonomic schemes) too. Thus, the (taxonomic) classification is the process of placing a specific object (concept) in a set of categories, based on the respective object (concept) properties. Note in this respect, as a pioneering reference, the work <<Fisher R.A. (1936) *"The use of multiple measurements in taxonomic problems"* -Annals of Eugenics, 7, Part II, pp.179-188>>, in which the famous *Iris* plant classification appears, already a classic in the field. Modern classification has its origins in the work of the botanist, zoologist and Swedish doctor Carl von Linne (Carolus Linnaeus) - XVIIIth century, who classified species based on their physical characteristics and is considered the "father of modern taxonomy".

The process of classification is based on four fundamental components:

- *Class* -the dependent variable of the model- which is a categorical variable representing the 'label' put on the object after its classification. Examples of such classes are: presence of myocardial infarction, customer loyalty, class of stars (galaxies), class of an earthquake (hurricane), etc.
- *Predictors* -the independent variables of the model- represented by the characteristics (attributes) of the data to be classified and based on which classification is made. Examples of such predictors are: smoking, alcohol consumption, blood pressure, frequency of purchase, marital status, characteristics of (satellite) images, specific geological records, wind and speed direction, season, location of phenomenon occurrence, etc.
- *Training dataset* -which is the set of data containing values for the two previous components, and is used for 'training' the model to recognize the appropriate class, based on available predictors. Examples of such sets are: groups of patients tested on heart attacks, groups of customers of a supermarket (investigated by internal polls), databases containing images for telescopic monitoring and tracking astronomical objects (e.g., Palomar Observatory (Caltech), San Diego County, California, USA, http://www.astro.caltech.edu/palomar/), database on hurricanes (e.g., centers of data collection and forecast of type National Hurricane Center, USA, http://www.nhc.noaa.gov/), databases on earthquake research (e.g., centers of data collection and forecast of type National Earthquake Information Center-NEIC, http://earthquake.usgs.gov/regional/neic/).
- *Testing dataset*, containing new data that will be classified by the (classifier) model constructed above, and the classification accuracy (model performance) can be thus evaluated.

The terminology of the classification process includes the following words:

- The dataset of *records/tuples/vectors/instances/objects/samples* forming the *training* set;

- Each record/tuple/vector/instance/object/sample contains a set of *attributes* (i.e., components/features) of which one is the *class* (label);
- The *classification model* (the *classifier*) which, in mathematical terms, is a function whose variables (arguments) are the values of the attributes (predictive/independent), and its value is the corresponding class;
- The *testing* dataset, containing data of the same nature as the dataset of training and on which the model's accuracy is tested.

We recall that in machine learning, the supervised learning represents the technique used for deducing a function from training data. The purpose of supervised learning is to predict the value (*output*) of the function for any new object/sample (*input*) after the completion of the training process. The classification technique, as a predictive method, is such an example of supervised machine learning technique, assuming the existence of a group of labeled instances for each category of objects.

Summarizing, a classification process is characterized by:

- *Input*: a training dataset containing objects with attributes, of which one is the class label;
- *Output*: a model (classifier) that assigns a specific label for each object (classifies the object in one category), based on the other attributes;
- The classifier is used to predict the class of new, unknown objects. A testing dataset is also used to determine the accuracy of the model.

We illustrated in Fig. 1.3, graphically, the design stages of building a classification model for the type of car that can be bought by different people. It is what one would call the construction of a car buyer profile.

Summarizing, we see from the drawing above that in the first phase we build the classification model (using the corresponding algorithm), by training the model on the training set. Basically, at this stage the chosen model adjusts its parameters, starting from the correspondence between input data (age and monthly income) and corresponding known output (type of car). Once the classification function identified, we verify the accuracy of the classification using the testing set by comparing the expected (forecasted) output with that observed in order to validate the model or not (accuracy rate = % of items in the testing set correctly classified).

Once a classification model built, it will be compared with others in order to choose the best one. Regarding the comparison of classifiers (classification models), we list below some key elements which need to be taken into account.

- *Predictive accuracy*, referring to the model's ability to correctly classify every new, unknown object;
- *Speed*, which refers to how quickly the model can process data;
- *Robustness*, illustrating the model's ability to make accurate predictions even in the presence of 'noise' in data;
- *Scalability*, referring mainly to the model's ability to process increasingly larger volume of data; secondly, it might refer to the ability of processing data from different fields;

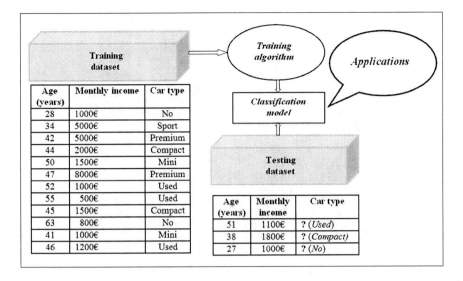

Fig. 1.3 Stages of building a classification model (cars retailer)

- *Interpretability*, illustrating the feature of the model to be easily understood, interpreted;
- *Simplicity*, which relates to the model's ability to be not too complicated, despite its effectiveness (e.g., size of a classification/decision tree, rules 'compactness', etc.). In principle, we choose the simplest model that can effectively solve a specific problem - just as in Mathematics, where the most elegant demonstration is the simplest one.

Among the most popular classification models (methods), we could mention, although they are used, obviously, for other purposes too:

- *Decision/classification trees;*
- *Bayesian classifiers/Naive Bayes classifiers;*
- *Neural networks;*
- *Statistical analysis;*
- *Genetic algorithms;*
- *Rough sets;*
- *k-nearest neighbor classifier;*
- *Rule-based methods;*
- *Memory based reasoning;*
- *Support vector machines.*

Regarding the range of the classification applicability, we believe that a brief overview of the most popular applications will be more than suggestive.

- *Identification* of the customer profile (Fig. 1.4) for a given product (or a complex of goods). The purpose of such a classification model lies in the supply

Fig. 1.4 Supermarket customer

optimization of certain products and a better management of stocks. For example, we would like to build the standard profile of a buyer of washing machines. For this we have to study the available data on this issue (e.g., best selling types of washing machines, how to purchase (cash/credit), average monthly income, duration of use of such property, type of housing (apartment block/house with courtyard) -relative to the possibility of drying laundry, family status (married or not, total of persons in the family, little children, etc.), occupation, time available for housework, etc.). Besides all these information = *input variables*, we add the categorical variable (the label = *output variable*) representing the category of buyer (i.e., buy/not buy). Once these data/information were collected, they are used in the learning phase (training) of the selected model, possibly keeping a part of them as a test set for use in model validation phase (if there are no new data available for this purpose).

Remark 1.1. An extension of the marketing research regarding the customer profile is one that aims to create the profile of the 'basket of goods' purchased by a certain type of customer. This information will enable the retailer to understand the buyer's needs and reorganize the store's layout accordingly, to maximize the sales, or even to lure new buyers. In this way it arrives at an optimization of the sale of goods, the customer being attracted to buy adjacent goods to those already bought. For instance, goods with a common specific use can be put together (e.g., the shelf where hammers are sold put nearby the shelves with tongs and nails; the shelf with deodorants nearby the shelves with soaps and bath gels, etc.). But here are also other data mining models involved, apart from the usual classifiers (e.g., clustering, discovery of association rules, etc.).

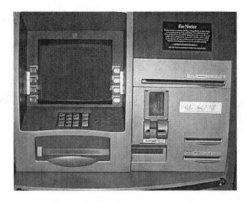

Fig. 1.5 Automated teller machine (ATM)

- *Fraud detection* (e.g., in credit card transactions) is used to avoid as much as possible fraudulent use of bank cards in commercial transactions (Fig. 1.5). For this purpose, available information on the use of cards of different customers is collected (e.g., typical purchases by using the card, how often, location, etc.). The 'labels' illustrating the way the card is used (illegally, fair) are added, to complete the dataset for training. After training the model to distinguish between the two types of users, the next step concerns its validation and, finally, its application to real-world data. Thus, the issuing bank can track, for instance, the fairness of transactions with cards by tracing the evolution of a particular account. Thus, it is often usual that many traders, especially small ones, avoid payment by card, preferring cash, just for fear of being tricked. Recently, there are attempts to build classification models (fair/fraud), starting from how to enter the PIN code, identifying, by analyzing the individual keystroke dynamics, if the cardholder is legitimate or not. Thus, it was observed in this case a similar behavior to that when using the lie detector machine (polygraph) - a dishonest person has a distinct reaction (distorted keystroke dynamics) when entering the PIN.

- *Classification of galaxies.* In the 1920s, the famous American astronomer Edwin Hubble (1889-1953) began a difficult work regarding the classification of galaxies (328). Initially he considered as their main attributes the color and size, but later decided that the most important characteristic is represented by their form (galaxy morphology -Edwin Hubble, 1936). Thus, started the discipline of cataloging galaxies (e.g., lenticular galaxies, barred spiral galaxies, ring galaxies, etc., see pictures below (Fig. 1.6), NASA, ESA, and The Hubble Heritage Team (STScI/AURA)).

1.4.2 Cluster Analysis

By *clustering* we mean the method to divide a set of data (records/tuples/ vectors/instances/objects/sample) into several groups (clusters), based on certain

Fig. 1.6 Galaxy types

predetermined similarities. Let us remember that the idea of partitioning a set of objects into distinct groups, based on their similarity, first appeared in Aristotle and Theophrastus (about fourth century BC), but the scientific methodology and the term 'cluster analysis' appeared for the first time, it seems, in <<C. Tryon (1939), *Cluster Analysis*, Ann Arbor, MI: Edwards Brothers>>. We can therefore consider the method of clustering as a 'classification' process of similar objects into subsets whose elements have some common characteristics (it is said that we partition/divide a lot of objects into subsets of similar elements in relation to a predetermined criterion). Let us mention that, besides the term *data clustering (clustering)*, there are a number of terms with similar meanings, including *cluster analysis, automatic classification, numerical taxonomy, botryology, typological analysis*, etc. We must not confuse the classification process, described in the preceding subsection, with the clustering process. Thus, while in classification we are dealing with an action on an object that receives a 'label' of belonging to a particular class, in clustering the action takes place on the entire set of objects which is partitioned into well defined subgroups. Examples of clusters are very noticeable in real life: in a supermarket different types of products are placed in separate departments (e.g., cheese, meat products, appliances, etc.), people who gather together in groups (clusters) at a meeting based on common affinities, division of animals or plants into well defined groups (species, genus, etc.).

In principle, given a set of objects, each of them characterized by a set of attributes, and having provided a measure of similarity, the question that arises is how to divide them into groups (clusters) such that:

- Objects belonging to a cluster are more similar to one another;
- Objects in different clusters are less similar to one another.

The clustering process will be a successful one if both the intra-cluster similarity and inter-clusters dissimilarity will be maximized (see Fig. 1.7).

To investigate the similarity between two objects, measures of similarity are used, chosen depending on the nature of the data and intended purpose. We present below, for information, some of the most popular such measures:

- *Minkowski* distance (e.g., *Manhattan* (city block/taxicab), *Euclidean*, *Chebychev*);

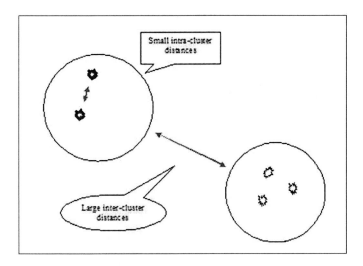

Fig. 1.7 Example of successful clustering

- *Tanimoto* measure;
- *Pearson's r* measure;
- *Mahalanobis* measure.

Graphically, the clustering process may be illustrated as in the Fig. 1.8 below.

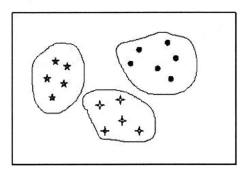

Fig. 1.8 Clustering process

Regarding the area of clustering applications, we give a brief overview of some suggestive examples.

- *Market segmentation*, which aims to divide customers into distinct groups (clusters), based on similarity in terms of purchases usually made. Once these groups established, they will be considered as market target to be reached with a distinct marketing mix or services. Fig. 1.9 illustrates such an example, related to the cars retailers.

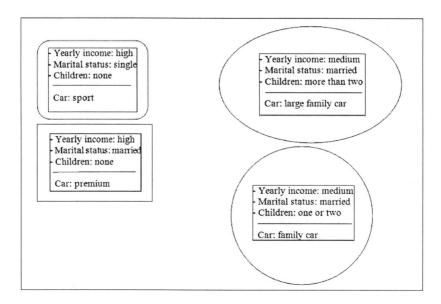

Fig. 1.9 Market segmentation (car retailer)

- *Document clustering*, which aims to find groups of documents that are similar to each other based on the important terms appearing in them, based on their similarity, usually determined using the frequency with which certain basic terms appear in text (financial, sports, politics, entertainment, etc.).
- *Diseases classification*, aiming to gather together symptoms or similar treatments.
- In Biology, the clustering process has important applications (see Wikipedia, http://en.wikipedia.org/wiki/Cluster_analysis) in *computational biology* and *bioinformatics*, for instance:
  - In *transcriptomics*, clustering is used to build groups of genes with related expression patterns;
  - In *sequence analysis*, clustering is used to group homologous sequences into gene families.
- Grouping companies on the stock exchange, based on the fluctuation analysis of their actions (increase/decrease -UP/DOWN).

Finally, let us mention a very important issue in the clustering process - the science to choose the optimal number of clusters (groups) of objects. To solve this problem the *elbow criterion* is usually used (Fig. 1.10). It basically says that we should choose a number of clusters so that adding another cluster does not add sufficient information to continue the process. Practically, the analysis of variance is used to 'measure' how well the data segmentation has been performed in order to obtain a small intra-cluster variability/variance and a large inter-cluster variability/variance, according to the number of chosen clusters. The figure below illustrates

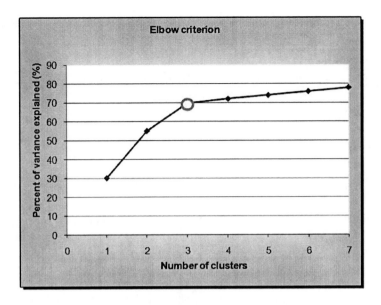

Fig. 1.10 Elbow criterion illustration

this
fact - the graph of the percentage of variance explained by clusters and depending on the number of clusters. Technically, the percentage of variance explained is the ratio of the between-group variance to the total variance.

It is easy to see that if the number of clusters is larger than three, the gained information insignificantly increased, the curve having an "elbow" in point 3, and thus we will choose three as the optimum number of clusters in this case.

Among other criteria for choosing the optimal number of clusters, let us mention BIC (*Schwarz Bayesian Criterion*) and AIC (*Akaike Information Criterion*).

1.4.3 Association Rule Discovery

In principle, by the *association rule discovery/association rule learner* we understand the process of identifying the rules of dependence between different groups of phenomena. Thus, let us suppose we have a collection of sets each containing a number of objects/items. We aim to find those rules which connect (associate) these objects and so, based on these rules, to be able to predict the occurrence of an object/item, based on occurrences of others. To understand this process, we appeal to the famous example of the combination < beer - diaper > , based on tracking the behavior of buyers in a supermarket. Just as a funny story, let us briefly recall this well-known myth. Thus, except for a number of convenience store clerks, the story goes noticed that men often bought beer at the same time they bought diapers. The store mined its receipts and proved the clerks' observations correct. So, the store began stocking diapers next to the beer coolers, and sales skyrocketed (see,

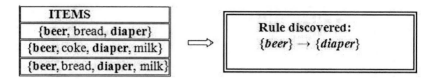

Fig. 1.11 Beer-diaper scheme

for instance http://www.govexec.com/dailyfed/0103/013103h1.htm). We illustrated below (Fig. 1.11) this "myth" by a simple and suggestive example.

The applications field of this method is large; there will be here only a brief overview of some suggestive examples.

- Supermarket shelf/department management, which is, simply, the way of setting the shelves/departments with goods so that, based on the data regarding how the customers make their shopping, goods that are usually bought together, are placed on neighboring shelves (sold in neighboring departments). Technically, this is done based on data collected using barcode scanners. From the database constructed this way, where the goods that were bought at the same time occur, the association rules between them can be discovered. In a similar way as above, we can obtain a rule that associates, for example, beer with diaper, so beer will be found next to diapers, to validate the story.
- Mining the Web has as starting point the way of searching on web for various products, services, companies, etc. This helps companies that trade goods online to effectively manage their Web page based on the URLs accessed by customers on a single visit to the server. Thus, using association rules we can conclude that, for example, 35% of customers who accessed the Web page with URL: http//company-name.com/products/product_A/html have also accessed the Web page with URL: http//company-name.com/products/product_C/html; 45% of customers who accessed the Web page: http//company-name.com/announcements/special-offer.html have accessed in the same session the Web page: http//company-name.com/products/product_C/html, etc.
- Management of equipment and tools necessary for interventions realized by a customer service company (e.g., service vehicles helping drivers whose cars break down on the road, plumbers repairing sinks and toilets at home, etc.). In the first case, for instance, the idea is to equip these intervention vehicles with equipment and devices that are frequently used in different types of interventions, so that, when there is a new application for a particular intervention, the utility vehicle is properly equipped for intervention, saving time and fuel needed to 'repair' the lack of resource management. In this case, association rules are identified by processing the data referring to the type of devices and parts used in previous interventions in order to address various issues arising on the spot. Note that a similar situation can be identified for emergency medical assistance; the problem here is to optimally equip the ambulance so that a first-aid service with timely and maximum efficiency would be assured.

1.4.4 Sequential Pattern Discovery

In many applications such as: computational biology (e.g., DNA or protein sequences), Web access (e.g., navigation routes through Web pages - sequences of accessed Web pages), analysis of connections (logins) when using a system (e.g., logging into various portals, webmail, etc.), data are naturally in the form of sequences. Synthetically speaking, the question in this context is the following: given a sequence of discrete events (with time constraints) of the form $<<...$ $ABACDACEBABC...>>$, by processing them we wish to discover patterns that are frequently repeated (e.g., A followed by B, A followed by C, etc.). Given a sequence of the form: "Time#1 (Temperature = 28°C) $\rightarrow$ Time#2 (Humidity = 67%, Pressure = 756mm/Hg)", consisting of items (attribute/value) and/or sets of items, we have to discover patterns, the occurrence of events in these patterns being governed by time restrictions. Let us enumerate some real-life situations when techniques of discovery sequential patterns are used:

- A good example in this respect refers to the analysis of large databases in which sequences of data are recorded regarding various commercial transactions in a supermarket (e.g., the customer ID -when using payment cards, the date on which the transaction was made, the goods traded -using the barcode technology, etc.), to streamlining the sale.
- In medicine, when diagnosing a disease, symptoms records are analyzed in real time to discover sequential patterns in them, significant for that disease, such as: "The first three days with unpleasant headache and cough, followed by another two days of high fever of 38-39 degrees Celsius, etc."
- In Meteorology -at a general scale- discovering patterns in global climate change (see global warming, for instance), or particularly, discovering the occurrence moment of hurricanes, tsunamis, etc., based on previous sequences of events.

1.4.5 Regression

Regression analysis (regression) as well as *correlation* have their origin in the work of the famous geneticist Sir Francis Galton (1822-1911), which launched at the end of the nineteenth century the notion of "regression towards the mean" -principle according to whom, given two dependent measurements, the estimated value for the second measurement is closer to the mean than the observed value of the first measurement (e.g., taller fathers have shorter children and, conversely, shorter fathers have taller children -the children height regresses to the average height).

In Statistics, regression analysis means the mathematical model which establishes (concretely, by the regression equation) the connection between the values of a given variable (response/outcome/dependent variable) and the values of other variables (predictor/independent variables). The best known example of regression is perhaps the identification of the relationship between a person's height and weight, displayed in tables obtained by using the regression equation, thereby evaluating an ideal weight for a specified height. Regression analysis relates in principle to:

- Determination of a quantitative relationship among multiple variables;
- Forecasting the values of a variable according to the values of other variables (determining the effect of the "predictor variables" on the "response variable").

Applications of this statistical method in data mining are multiple, we mention here the following:

- Commerce: predicting sales amounts of new product based on advertising expenditure;
- Meteorology: predicting wind velocities and directions as a function of temperature, humidity, air pressure, etc.;
- Stock exchange: time series prediction of stock market indices (trend estimation);
- Medicine: effect of parental birth weight/height on infant birth weight/height, for instance.

1.4.6 Deviation/Anomaly Detection

The detection of deviations/anomalies/outliers, as its name suggests, deals with the discovery of significant deviations from 'normal behavior'. Fig. 1.12 below suggestively illustrates the existence of anomalies in data.

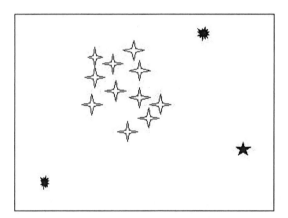

Fig. 1.12 Anomalies in data (outliers)

1.5 About Modeling and Models

In the two preceding subsections, when presenting the way of processing the data, we highlighted some aspects of the main techniques used in data mining models, as well as the common problems addressed with these methods. In this section we make some more general considerations on both the modeling process and models, with the stated purpose of illustrating the complexity of such an approach, and also its fascination exerted on the researcher. In principle, we briefly review the main aspects of the process of building a model, together with problems and solutions related to this complex issue, specifically customized for data mining.

At any age, starting with the serene years of the childhood and ending with the difficult years of the old age, and in any circumstances we might be, we strongly need models. We have almost always the need to understand and model certain phenomena, such as different aspects of personal economic (e.g., planning a family budget as good as possible and adjusted to the surrounding reality), specific activities at workplace (e.g., economic forecasts, designs of different models: architecture, industrial design, automotive industry, 'mining' the data for discovery of useful patterns -as in our case, informatics systems and computer networks, medical and pharmaceutical research, weather forecasting, etc.). Thus, on the one hand we will better know their specific characteristics and, on the other hand, we can use this knowledge to go forward in the research field.

It is more than obvious that in almost all cases, the real phenomena tackled in the study, which are the prototypes for our models, are either directly unapproachable (e.g., the study of the hurricanes movements, the modeling of stars and galaxies evolution), or too complicated on the whole (e.g., the motion analysis of insects to create industrial robots by analogy), or too dangerous (e.g., modeling of processes related to high temperatures, toxic environments, etc.). It is then preferable and more economical at the same time to study the characteristics of the corresponding models and simulations of "actual use", seen as substitutes more or less similar to the original 'prototype'.

It is therefore natural that, in the above scenario, Mathematics and Computer Science will have a crucial role in modeling techniques, regardless of the domain of the prototype to be modeled (economy, industry, medicine, sociology, biology, meteorology, etc.). In this context, mathematical concepts are used to represent the different components constituting the phenomena to be modeled, and then, using different equations, the interconnections between these components can be represented. After the "assembly" of the model using all the characteristic components connected by equations was completed, the second step, consisting in the implementation of the mathematical model by building and running the corresponding software, will end the building process. Afterward, the "outputs" of the model are thoroughly analyzed, changing continuously the model parameters until the desired accuracy in 'imitating' the reality by the proposed model is accomplished - the computerized simulation.

Using the experience gained in the modeling field, one concludes that any serious endeavor in this area must necessarily run through the following steps:

- *Identification.* It is the first step in finding the appropriate model of a concrete situation. In principle, there is no beaten track in this regard; instead, there are many opportunities to identify the best model. However, we can show two extreme approaches to the problem, which can then be easily mixed. First, it is about the conceptual approach concerning the choice of the model from an abstract (rational) point of view, based on an *a priori* knowledge and information about the analyzed situation, and without taking into account specific dates of the prototype. In the conceptual identification stage, data are ignored, the person that designs the model takes into account ideas, concepts, expertise in the field and a lot of references. The modeling process depends on the respective situation, varying from one problem to another, often naturally making the identification, based on classical models in the field. Even if there is not a ready-built model already, which with small changes could be used, however, based on extrapolation and multiple mixing, it is often likely to obtain a correct identification. Secondly, it comes to empirical identification, in which there are considered only the data and the relations between them, without making any reference to their meaning or how they result. Thus, deliberately ignoring any *a priori* model, one wonders just what data want "to tell" us. One can easily observe that this is the situation, in principle, regarding the process of 'mining' the data. It is indeed very difficult to foresee any scheme by just "reading" the data; instead, more experience is needed in their processing, but together with the other method, the first rudiments of the desired model will not delay to appear. Finally, we clearly conclude that a proper identification of the model needs a "fine" combination of the two methods.

- *Estimation and fitting*. After we passed the first step, that is the identification of a suitable (abstract) model for the given prototype, we follow this stage up with the process of "customizing" it with numerical data to obtain a concrete model. Now, abstract parameters designated only by words (e.g., A, B, a, b, α, β, etc.) are no longer useful for us, but concrete data have to be entered in the model. This phase of transition from the general form of the selected model to the numerical form, ready to be used in practice, is called "*fitting the model to the data*" (or, adjusting the model to the data). The process by which numerical values are assigned to the model parameters is called *estimation*.

- *Testing*. This notion that we talked about previously, the chosen term being suggestive by itself, actually means the consideration of the practical value of the proposed model, its effectiveness proved on new data, other than those that were used to build it. Testing is the last step before "the launch of the model on the market", and it is, perhaps, the most important stage in the process of building a model. Depending on the way the model will respond to the 'challenge' of application to new, unknown data (generalization feature of the model), it will receive or not the OK to be used in practice.

- *Practical application (facing the reality)*. We must not forget that the objective of any modeling process is represented by the finding of an appropriate model, designed to solve specific real-world problems. So, the process itself of finding the models is not too important here, although this action has its importance and a special charm for connoisseurs, but finding 'natural' models, that match as close as possible with a given prototype. This activity is indeed fascinating and extraordinary, having its own history connected to different branches of science, such as: mathematics, physics, biology, engineering, economics, psychology, medicine, etc., the models being applied in various concrete situations.

- *Iteration*. When constructing a specific (physical) mechanism, the manufacturer has to consider a very rigorous plan, which should be fulfilled point by point and in the strict order of the points in the program. Obviously, we are not talking about inventors, lying within the same sphere with the creators, artists, etc., all having the chance of a more "liberal" program in the conception process. Although we presented above, in a relative order, the modeling generic stages, this order however should not be considered as "letter of the law." Modeling involves frequent returns to previous stages, changes in the model design, discovery of issues that were initially ignored, but which are essential in a deeper thinking, etc. This repetition of stages, this constant re-thinking of the model is called *iteration* in the modeling process. To conclude, the first model is not the MODEL, the sole and the ultimate, but is only the beginning of a series of iterations of the steps mentioned above, with the sole purpose of finding the most appropriate model for a particular given situation.

When we intend to model a particular phenomenon, situation, etc., it is natural to be interested about the available references concerning that field, to obtain necessary information. The problem consists in how we get the required information and the criteria to decide what is important or not, what best suits or not for the given situation. In what follows, we review some of the most important forms of preliminary information used in modeling.

- *Information about variables.* When conceiving a model we have in mind many variables, which, in one way or another, could enter in the 'recipe' of the model. The choice of variables which are indeed essential for the model is the most important and sensitive issue at the same time, since the neglect of important variables is more "dangerous" than the inclusion of one which is not important. At this stage of modeling, the researcher should draw upon the expert in the specific domain, to support him (her) to establish clearly the constituent variables and their appropriate hierarchy. It would be ideal to work 'in team' in the modeling process, to have the opportunity to choose the optimal variant at any time. Immediately after the selection of the constituent variables of the model, we should identify the domains in which these variables take values (given by 'constraints' imposed to the variables). For instance, such constraints could be: variable X is integer and negative, variable Y is continuous and positive, etc. It is also important to establish relations between variables (e.g., $X < Y$).
- *Information about data.* First, it is necessary that the chosen model is suitable for the volume of available data. There are models more or less sophisticated, e.g., weather forecast models, which require a sufficient large number of data in order that the theoretical model can be adjusted to data (fitting model to data). Secondly, it is about the separate analysis of data (disaggregation), or their simultaneous analysis (aggregation), this fact depending on each case, usually working simultaneously with the two types of analysis. Thirdly, we should consider reliable data only, otherwise the proposed model has no practical value. In this respect, there is a whole literature on how to collect data according to the field under consideration and the proposed objectives. Finally, the database must be sufficiently rich, because for subsequent corrections of the model more data are needed. This is, by the nature of the facts, true in data mining, since it is

a priori assumed that we are 'mining' huge databases. It should be remembered that there are ways to overcome the impediment of a relatively small database, when there is not the possibility of its expansion, e.g., randomization, which implies, in principle, the use of the same data, randomly reordered.

- *Knowledge about errors and variability.* Besides the above information about data and variables related to the model to be studied, we need information about the sources of error that can enter the modeling process, about the data acquisition, and also about the random variability to be modeled.

- *Knowledge about models.* First, note that when we face the problem of choosing an appropriate model for the examined phenomenon, we meet the following situation: on the one hand there is an impressive set of ready built models, available through various books and magazines, and, on the other hand, when we try to apply one of them to real data, we face the situation that our model does not fit satisfactorily, resulting in an insufficiently precise approximation. Secondly, we must choose models to fit data (discrete or continuous, categorical or numerical, uniformly covering the entire field approached by the model, or clustering only around certain values, etc.). In this respect, particularly for those with limited experience in the field or even novices, a special attention to the nature of the data must be paid, because there are dedicated programs which have no included warnings about the nature of input data, and thus the results are compromised.

- *Knowledge about model parameters.* Knowing the intimate nature of the modeled phenomenon, assumptions on the parameters of the proposed model can be made. If, for example, we have a dependent variable Y linked with the explanatory variable X by a linear relationship of parameter b, e.g., $Y = bX + \varepsilon$ (i.e., simple linear regression), and they have the same evolutionary trend, then we choose the parameter $b > 0$ from the beginning. Generally speaking, when a model is built to be used by laymen, for instance a machine learning model concerning the medical diagnosis used by doctors not trained in computer-aided diagnosis, it must have, as far as possible, a practical interpretation clearly presented to be understood by the user. This applies to the model variables and also to its parameters.

- *Knowledge about the domain of applications judging criteria.* It is obvious that a model is not designed for the purpose of making '*art for art's sake*', but to be used in a practical context. Therefore, when we intend to build a model, we must have clear information on its application's field. This information is useful both in the choice of the model and in the way it will fit the data, and concerning its validation. We are then interested in the criteria for deciding its effectiveness, if it is viewed as a whole, applied to a single case, or is part of a more comprehensive modeling process. In the latter case, we are interested in the compatibility with the other components of the overall model too. Finally, we are interested if it will be applied to a given situation in a particular context, or it comes to implementing it to a stable situation in time. We have to mention here the scalability feature of the model too, seen in terms of the area of different situations in which it may be applied.

A pretty complex problem in the modeling process is precisely the way to choose the model. The problem becomes complicated when, often, we have to choose between two or more models. In this situation two ways are to be followed:

1. *Choosing the model from a class of models.* This is the situation in which, on the basis of certain knowledge (experience) and previous data analysis, we choose a particular type of model. In this case the problem is reduced to the use of general techniques, corresponding to the class of chosen models;
2. *Free choice of the model.* This is the case when the selection of the possible types is made from different classes of models, based on the researcher's "free will".

Regardless of the procedure for choosing a model, the referred models will be compared based on a set of criteria, such as:

- The measurement of errors in the phase of estimating the model parameters (*training phase*);
- The measurement of errors in the phase of testing/validating the model (*testing/validation phase*);
- Residual diagnoses and goodness-of-fit tests;
- Qualitative considerations.

Depending on the result of the comparison, we decide which model is optimal for the prototype. Next, we briefly present some general considerations regarding the manner of choosing a model.

- *Hierarchical models.* This case is about choosing the model from a well defined class of models, so that each model represents a special case of a general class. Basically, it is about two extreme approaches in choosing a model from a class. On the one hand, one can act in a 'top-down' manner, i.e. first consider the general model, sometimes called the *saturated model* and, by simplification, taking into account the specific context, it arrives to the sought model. Despite many computing difficulties, generally related to the model complexity, if it is well chosen, the proposed model will certainly be appropriate to the given situation. On the other hand, one can go in reverse order to identify the model within a class of models, namely in a 'bottom-up' manner. Taking into account the principle of simplicity, one starts with the simplest version, one that emphasizes the basic characteristic of the examined phenomenon, and as the need requires, one begins to increase the level of complexity. Whichever method is chosen, the process stops when the reduction or the increase of the model no longer significantly influence its correlation with the actual data.
- *Free models.* Unlike the case of hierarchical models, whose well-defined structure allows the comparison of several models of the same class, for the final selection of the best, in the case of free models, because of the lack of structure of the set they belong to, the choice is made, basically, between two variants. Using certain formal techniques, one can measure the quality of a model regarding the fit to data, in order to find the most suitable one.

An interesting procedure to choose models is the *Bayesian selection*. The Bayesian method of choosing the most suitable model is based on the conditional probability of choosing a model in comparison with the available data, and on the famous formula of Thomas Bayes (1763) (25):

$$P\{A|D\}P\{D\} = P\{D|A\}P\{A\}, \tag{1.1}$$

which establishes the connection between the direct conditional probability and the reverse one. Thus, suppose that event A represents either the choice of the model M_1 or of the model M_2. Denote:

$$P\{A := M_1\} = P\{M_1\}, P\{A := M_2\} = P\{M_2\}, \tag{1.2}$$

which we consider as the *prior probabilities* of the two models and, each model being known, we can calculate the conditional probabilities $P\{D|M_i\}$, $i = 1, 2$, i.e., the probabilities that data, represented here by D, are consistent with the choice made regarding the model (*fitting data to model*). But what we are interested in is how we choose the model based on data, i.e., the reverse conditional probability (*fitting model to data*).

Using Bayes' formula, we have:

$$P\{M_1|D\} = \frac{P\{D|M_1\}P\{M_1\}}{P\{D\}}, \tag{1.3}$$

$$P\{M_2|D\} = \frac{P\{D|M_2\}P\{M_2\}}{P\{D\}}, \tag{1.4}$$

called *posterior probabilities*. We can calculate these conditional probabilities knowing that (*total probability formula*):

$$P\{D\} = P\{D|M_1\}P\{M_1\} + P\{D|M_2\}P\{M_2\}. \tag{1.5}$$

We thus choose the model so that it obtains the largest posterior probability. Generalizing, suppose we have available a complete set of models M_i, $i = 1, 2, ..., k$, from which we can choose one, knowing the corresponding prior probabilities:

$$P\{M_i\}, \quad \sum_{i=1}^{k} P\{M_i\} = 1. \tag{1.6}$$

Suppose we have symbolized the observed (available) data by D. Then, using the Bayes' formula, we obtain the posterior probabilities of each model knowing the dataset D, through the probabilities:

$$P\{M_i|D\} = \frac{P\{D|M_i\}P\{M_i\}}{P\{D\}}, i = 1, 2, ..., k, \tag{1.7}$$

where:

$$P\{D\} = \sum_{i=1}^{k} P\{D|M_i\}P\{M_i\}. \tag{1.8}$$

Thus, we will choose the model for which we obtain the larger posterior probability; for details, see (124).

Finally, it is worth to remember the maxim of A. Einstein *"everything should be made as simple as possible, but no simpler"*, when choosing a model, in other words, the K.I.S.S. concept (*Keep It Simple Series*) applied in this circumstance: "between two models, comparable in performance, the simplest one will be chosen, which is probably closest to the truth, and is more easily accepted by others".

Suppose we have already chosen a model that seems to be appropriate for the prototype (i.e., to the real problem). It remains now to adjust the model to observed data. We will mention three criteria, known as *adjustment (fitness) criteria*, under-lying the assessment of 'fitting the model to data', and based on which we will consider different methods of adjustment (fitting).

- *Establishing the equality between the characteristics of the model form and the characteristics of the data form;*
- *Measuring the deviation between model and data;*
- *Establishing the extent on which the model is justified by the data.*

Once the model fitted (adjusted) to data, it remains to be validated before its success-ful application to solve real-life problems. The notion of validating a given model covers a wide enough range of issues to be considered. Thus, by "validation" we understand the level of the practical value of the model in explaining a particular phenomenon, assessing both its similarity to the prototype and its effectiveness in a straight application to a given situation. We speak of validating the model:

- in the model development, the adjusted model revealing new aspects concerning data;
- in the testing phase, when new data are collected and used to optimize the pro-posed model;
- in the practical application stage, when procedures for "monitoring" are intro-duced to check whether the originally proposed model is effective or not in *"live"* conditions.

Without insisting on each issue separately, we list below the elements considered in the process of validating a model:

- *Analysis of the prototype*, now made *post festum* to reveal new aspects concerning the compatibility of the prototype with the proposed model and initial data, or data collected after the identification process;
- *Analysis of applications*, made to reveal the extent to which the proposed model is effective in solving practical problems for which it was built;
- *Analysis of the model form*, seen as a reconfirmation of the initial choice made in the identification stage;

- *Analysis of the model's behavior* when compared to the prototype, related to the existing data;
- *Analysis of the model sensitivity*, related to the modifications of the data. In this respect, the model has to 'respond' to certain changes of the parameters, corresponding to the modeled situation.

Finally, we mention that even in the model validation stage, the technical procedures from the identification phase are kept. Thus, we speak about the *conceptual* model validation, about the *empirical* validation of the model form and parameters and, finally, about the *eclectic* model validation. Any validation process will be ended in the spirit of "*finis coronat opus*", i.e., its validation in 'real running conditions', in analyzing its effectiveness when applied to real situations for which it has been designed, or to possibly new ones.

The last step and the most important one in building a model is its application in practice. It is clear that when one starts to design and build a model, the purpose of such a work lies in solving a practical problem by using that model. The problem to be solved will lead to the choice of the appropriate means to build the model, so that it can be subsequently applied to solve the initial problem or other similar problems. In what follows we will present a number of issues related to the application-modeling ratio, in order to reveal the connection between them.

Applications of the descriptive type

Whenever we start building a model, we have in view that it represents a more or less accurate description of the prototype, i.e., of the real situation or phenomenon that we want to model. Having in mind this idea, we first need an acceptable description of the main features of the prototype, which must be kept by the 'copy' (i.e. the model) too. At this stage, the model can be used from an applicative point of view only as a simple summarized (overall) description of the actual situation. In this context, a number of models will 'remain' in this primary stage, as descriptive models only. In more complex cases (e.g., dynamic models) the description is much more detailed, depicting each of the different components of the prototype, and picturing all the inter-relationships between components. It is therefore both a static description of the components, and a dynamic one of the relationship between components. More complex models are therefore used in applications, not only for a mere illustrative description of a particular phenomenon, but for verifying and validating (or invalidating) certain theories proposed to explain or solve important issues. A well-built model in relation to the principles of a certain theory can lead to its acceptance or rejection, ultimately being the only means of validating the theory (e.g., models for checking the theory of relativity). This type of use of a model is also included in the class of descriptive applications.

Applications of the exploratory type

When we study a particular real situation, the so-called prototype, one of the most important questions that can be asked is: "what happens if something changes in the prototype data, i.e., what is the influence of some changes on its 'operation'?" The problem here is that we cannot verify what happens using just the prototype, since it is either too complicated, or impossible, or we can even damage it. To solve this problem, we could use a model on which we can practice different modifications to see what changes occur and thus to be able to extrapolate the knowledge obtained for the prototype. This is what is called a *"What-if analysis"*.

Applications of the predictive type

By application of the predictive type we understand that application of the model which attempts to 'predict' the value that a certain variable may take, given what we know at present. There are many examples of such predictive applications, a classic one being the prediction of the occurrence of a hurricane of a certain category (e.g., SaffirSimpson Hurricane Scale), in the following time interval, using a meteorological model. Other models are applied in the framework of the queuing theory, in predicting the mean number of customers waiting in the 'queue', the mean waiting time, the server load, etc., at Stock Exchange in predicting the increasing/decreasing trend of the price of the shares, in medicine in predicting the evolution of a particular disease under a given treatment, etc.

Applications in decision making

Obviously, a built model based on a given prototype can be used in making certain decisions. In this context, we must note that if we dispose of a certain model, no matter for what purpose it was built, it can be used to make decisions. For example, for a weather forecasting model it is obvious that we use the prediction of the hurricane occurrence to make some important decisions for the affected community. Just remember the consequences of Hurricane Katrina (final category 5, on August 28, 2005), along the Gulf of Mexico coast from central Florida to Texas, with its dramatic consequences. However, let us highlight the case where the model is built exclusively for the purpose of making decisions. For instance, when building a queuing model regarding the services provided by a particular company, it will be used primarily by the company staff for taking decisions on how to implement the service (e.g., number of servers required and their degree of use, the mean waiting time, mean number of customers in queue, etc.). When applying the decision models in medicine, it is worth to mention the diagnosis-oriented software, designed exclusively to assist the medical doctor in choosing the best diagnosis, and, implicitly, the best corresponding treatment.

Finally, let us review some problems concerning the risk of separating the model from application, which can occur when neglecting the relationship between the modeling process, i.e., the proposed model, and the application of the model to actual conditions:

- *Inappropriateness of the model to the application.* In this case it is about either the profoundness of detailing the model in relation to the application requirements, or the incompatibility between the assumptions required by the model form and the actual conditions, or the inadequate presentation of the model to the user, resulting in improper applications, etc.;

- *Inappropriateness of the parameters estimation to the practical application.* Here it is about the methodology chosen for establishing the criteria concerning the estimation of the model parameters in relation to the actual situation, especially regarding the assessment of the errors of the estimates;
- *Overestimation of the model* refers especially to the psychological side of the problem, consisting in the fact that the model should obey the reality and not *vice versa*. We should never forget that a model is just a model and nothing more, and, if it is well built, it will reasonably resemble the prototype, helping to better understand the latter, and therefore we should not make it an "idol". Thus, we must not forget the two assertions of the famous statistician G.E.P. Box: *"Essentially, all models are wrong...but some are useful"* and *"Statisticians, like artists, have the bad habit of falling in love with their models"*, true for any scientist, indeed.

Returning to the data mining field, let us outline that the modeling process can be briefly summarized by the following three points:

- Data exploration (data preparation, choice of predictors, exploratory analysis, determination of the nature and/or the complexity of the models to be chosen, etc.);
- Building the model and testing/validating it (select the best model based on its predictive performance - assessing the competitiveness of models);
- Applying the model in practice, evaluating thus its effectiveness.

Regarding the implementation of various data mining models, we present below a list of various software systems based on them. This ever growing list includes the following software products:

- *Statistical packages*:

  - SAS (comprehensive statistical package - http://www.sas.com/)
  - IBM-SPSS (comprehensive statistical package - http://www.spss.com/)
  - Statgraphics (general statistics package - http://www.statgraphics.com/)
  - STATISTICA (comprehensive statistical package -http://www.statsoft.com/)
  - GenStat (general statistics package - http://www.vsni.co.uk/software/genstat/
  - JMP (general statistics package - http://www.jmp.com/)
  - NCSS (general statistics package - http://www.ncss.com/)
  - STATA (comprehensive statistics package - http://www.stata.com/)
  - SYSTAT (general statistics package - http://www.systat.com/)
  - Maplesoft (programming language with statistical features - http://www.maplesoft.com/)
  - MATLAB (programming language with statistical features - http://www.mathworks.com/products/matlab/)

- *Neural networks packages*:

  - STATISTICA (*STATISTICA* Neural Networks (SNN) package - http://www.st
    atsoft.com/; *STATISTICA* Automated Neural Networks -
    http://www.statsoft.co.za/products/stat_nn.html)
  - IBM-SPSS Neural Networks
    (http://www.spss.com/software/statistics/neural-networks/)
  - SAS Enterprise Miner (Neural Networks - http://www.sas.com/technologies/
    analytics/datamining/miner/neuralnet/index.html)
  - MATLAB (Neural Network Toolbox -
    http://www.mathworks.com/products/neuralnet/)
  - NeuroShell Predictor (Ward Systems Group -
    http://www.wardsystems.com/predictor.asp)
  - NeuralTools (Palisade - http://www.palisade.com/neuraltools/)

- *Classification/decision/regression trees*:

  - Data Mining Tools See5 and C5.0 (RuleQuest Research -
    http://www.rulequest.com/see5-info.html)
  - STATISTICA (Classification and Regression Trees module -
    http://www.statsoft.com/textbook/classification-and-regression-trees/)
  - MATLAB (Statistics Toolbox/Decision tree -
    http://www.mathworks.com/products/statistics/)
  - C4.5 (Release 8 - http://www.rulequest.com/Personal/)
  - IBM-SPSS Decision Trees
    (http://www.spss.com/software/statistics/decision-trees/)
  - DTREG (Classification and decision trees -
    http://www.dtreg.com/classregress.htm)
  - CART 5.0 (http://salford-systems.com/cart.php)

- *Evolutionary/Genetic algorithms*:

  - MATLAB (Genetic Algorithm and Direct Search Toolbox - http://www.math
    works.com/products/gads/)
  - GeneHunter (Ward Systems Group -
    http://www.wardsystems.com/genehunter.asp)
  - Jaga - Java Genetic Algorithm Package (GAUL - http://gaul.sourceforge.net/)

- *Nonlinear regression methods*:

  - MATLAB (Statistics Toolbox -Nonlinear Regression)
  - STATISTICA (Statsoft)
  - IBM-SPSS Regression (IBM-SPSS)
  - S-PLUS for Windows (TIBCO Software Inc.)
  - NLREG (Nonlinear regression)
  - STATA (Nonlinear regression)

- *Subject-oriented analytical systems*:

  - MetaStock (Equis International)
  - TradeStation (TradeStation Securities, Inc.)

In the previous section we mentioned some of the most popular models of data mining and some applications related to them. As it was well observed, there is a very wide field both concerning the models/techniques considered in the study and the problems that can be solved with the data mining methodology. What is exciting to data mining is just this opening, quite rare in other areas of research, in terms of both area of applicability and field of used techniques.

1.6   Data Mining Applications

So far we tried to describe, for each method separately, various successful applications of the data mining techniques in real-life situations. Next, we will briefly recall some areas of great interest for application of the data mining techniques.

- The banking and financial services domain is one of the first and most important areas for data mining applications. Thus, in banking, data mining methods were intensively used (and are still successfully used) in:

  - modeling and forecasting credit fraud;
  - risk assessment;
  - trend analysis;
  - profitability analysis;
  - support for direct marketing campaigns.

- In the financial area, we find data mining applications in:

  - stock price forecasting;
  - trading option;
  - portfolio management;
  - forecasting the price of goods;
  - mergers and acquisitions (M&A) of companies;
  - forecasting financial disaster, etc. Unfortunately, the latest global financial 'hurricane' (starting on September 2008) has not been forecasted.

- Sales policy in retail and the supermarket (hypermarket) sales strategy have taken full advantage of using data mining techniques:

  - data warehousing;
  - direct mail campaign;
  - customers segmentation, identification of customer profile;
  - price evaluation of specific products (antiques, used cars, art, etc.).

- Health care is also one of the first important areas of activity that boosted the intensive development of the data mining methods, starting from visualization techniques, predicting health care costs and ending with computer-aided diagnosis.
- Telecommunications, especially in recent years, have taken full advantage of access to data mining technology. Due to the fierce competition currently known in this area, problems of identifying customer profile, to create and maintain their loyalty, strategies for selling new products, are vital to the companies operating in this area. Some problems that can be solved by data mining techniques in this area are the following:

  - fraud prediction in mobile telephony;
  - identifying loyal/profitable customer profile;
  - identifying factors influencing customer behavior concerning the type of phone calls;
  - identifying risks regarding new investments in leading-edge technologies (e.g., optic fiber, nano-technologies, semiconductors, etc.);
  - identifying differences in products and services between competitors.

Regarding the companies that sell data mining products, and their list is very large, below (Fig. 1.13) we only show just a small 'sample' of it (company/product), as it can be found in a simple Web search.

*IBM-SPSS (Clementine) http://www.ibm.com/ http://www.spss.com/ *IBM (Data Warehouse) http://www.ibm.com/ *MICROSOFT (SQL Server) www.microsoft.com *STATISTICA (Statistica Data miner) http://www.statsoft.com/ *MATLAB (ARMADA Data Mining Tool) http://www.mathworks.com/ *SAS (SAS Data mining-Enterprise Miner) http://www.sas.com/	* ORACLE (Oracle Data Mining) http://www.oracle.com/ *Angoss software (Knowledge Seeker) http://www.angoss.com *XpertRule Software Ltd (Knowledge Builder Rules Authoring Studio) http://www.xpertrule.com/ *SAP BusinessObjects (Business intelligence solutions) http://www.sap.com/ *Salford Systems (CART 6.0 ProEX) http://salford-systems.com/ *Analytics1305(Cloud1305) http://www.analytics1305.com/ *Data Applied (Data mining tools) http://www.data-applied.com/

Fig. 1.13 Sample of company/products dealing with data mining technologies

In the past years, since we have been experiencing an exponential growth of data mining products, it was necessary to create standards for these products. In this respect we mention the DMG group (Data Mining Group -http://www.dmg.org/index .html), as an independent group of companies developing data mining standards, consisting of the following partners:

A. *Full Members*:

- IBM Corp. Somers, NY
- MicroStrategy Inc., McLean, VA
- SAS, Cary, NC
- SPSS Inc., Chicago, IL

B. *Associate Members*:

- Pervasive Software, Austin, TX
- Zementis Inc., San Diego, CA

C. *Contributing Members*: Equifax, Atlanta, GA; Fair Isaac, Minneapolis, MN; KNIME, Konstanz, Germany; NASA, KSC, FL; National Center for Data Mining, University of Illinois at Chicago; Open Data Group, River Forest, IL; Rapid-I, Dortmund, Germany; Togaware Pty Ltd, Canberra, Australia; Visa, San Francisco, CA.

Also in this respect, we mention that, in recent years, international conferences with the main theme concerning the standardization of the data mining procedures took place. As examples of such recent events, we can mention:

- 2011 European Conference on Machine Learning and Principles and Practice of Knowledge Discovery in Databases, Athens, Greece.
- 2010 -16th ACM SIGKDD Conference on Knowledge Discovery and Data Mining Washington DC, USA.
- 2010 -10th IEEE International Conference on Data Mining, Sydney, Australia.
- 2010 -SIAM International Conference on Data Mining, Columbus, Ohio, USA.
- 2009 -15th ACM SIGKDD international conference on Knowledge discovery and data mining, Paris, France.
- 2009 -SIAM International Conference on Data Mining, Reno-Sparks, Nevada, USA.
- 2008 -14th ACM SIGKDD international conference on Knowledge discovery and data mining, Las Vegas, Nevada, USA.
- 2008 -IEEE International Conference on Data Mining, Pisa, Italy.

Fig. 1.14 (http://www.kdnuggets.com/polls/2008/data-mining-software-tools-used. htm -Copyright ©2008 KDnuggets) displays the poll results concerning the use of Data Mining commercial software (May 2008) -the first 15 used data mining software.

Fig. 1.15 below (http://www.kdnuggets.com/polls/2009/industries-data-mining-applications.htm - Copyright ©2010 KDnuggets) outlines the poll results concerning the data mining applications fields (December 2009).

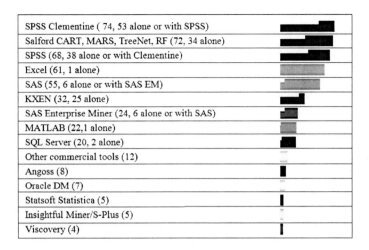

Fig. 1.14 Poll results concerning the use of data mining software (May 2008)

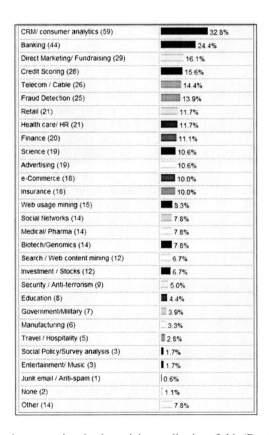

Fig. 1.15 Poll results concerning the data mining applications fields (Dec 2009)

1.7   Data Mining Terminology

In the data mining area there are already fundamental concepts and a specific terminology, even if the field is still not mature enough. As it will be seen below, many
terms are previously known from other well-established research areas, but since
those techniques were naturally borrowed by data mining, they were adopted without any complex. Thus, we could say that it is difficult to speak of a genuine specific
data mining terminology. In conclusion, once borrowing "tools" from neighboring
domains, their names were also borrowed. As noted from the very beginning, data
mining uses any method of data analysis to discover the information hidden there
and, naturally, uses the appropriate terminology too.

Obviously, it is impossible to review even a part of the terms used in data mining.
A brief Internet search will produce a huge list of terms in connection with data
mining, which is not surprising at all. Below it is presented just a small sample of
terms -randomly chosen- with direct reference to data mining applications.

- Algorithms
- Artificial neural network
- Business performance management
- Data clustering
- Data warehouse
- Discovery science
- Forecasting
- Knowledge discovery
- Logit (logistic regression)
- Machine learning
- Nearest neighbor (pattern recognition)
- Preprocessing data
- Regression analysis
- Statistics
- Treatment learning
- Vizualization

- Artificial intelligence
- Business intelligence
- Database
- Data stream mining
- Decision tree
- Document warehouse
- Java Data Mining
- Discriminant analysis
- Frauding card
- Modeling, models
- Pattern recognition
- Principal components analysis
- Relational data mining
- Text mining
- Verification and Validation
- Web mining

A fairly extensive list of a possible explained glossary regarding the data mining
terminology can be found at:

- http://webdocs.cs.ualberta.ca/~zaiane/courses/cmput690/glossary.html
- http://www.twocrows.com/glossary.htm
- http://www.thearling.com/glossary.htm

1.8   Privacy Issues

Since the very early use of data mining techniques, the issues concerning the impli-
cations that may arise concerning the privacy of individuals (privacy issues), related

to the source of the data to be processed, raised many questions and discussions. For example, financial and banking information, necessary to grant credits to individuals or legal entities, can be used in a wrongful way for the individual/company, revealing confidential information, possibly dangerous to their future developments. Also, another 'hot' area in this context is represented by the use of medical databases. The misuse of medical information of a certain individual can seriously harm his/her interests (e.g., when contracting a life insurance, granting of credit, employment, etc.).

By its nature, data mining deals with databases providing information which, often, can be obtained by different (questionable) means. In the context in which these data relate to people, all issues regarding privacy, legality and ethics must be clearly solved. Because billions of data of all kinds are collected yearly, then being processed by data mining means, it is absolutely necessary to develop a legal framework for the protection of the privacy, in order to avoid the emergence of an "*e*-Big Brother". In this regard it is noteworthy, for instance, that the European Union prohibits the use of financial information by banks in the credit field, although in the U.S. this is allowed with certain restrictions (157).

Chapter 2
The "Data-Mine"

Abstract. Data mining deals with data. Basically, a huge amount of data is processed for extracting useful unknown patterns. Accordingly, we need more information concerning the "nugget of knowledge" -data- we are dealing with. This chapter is dedicated to a short review regarding some important issues concerning data: definition of data, types of data, data quality and types of data attributes.

2.1 What Are Data?

The word "data" is the Latin plural of "datum", coming from the verb "dare = to give". Let us mention that Euclid is one of the first to use this term in his writing "Data" -from the first Greek word in the book, *dedomena* <given> ("Euclid." Encyclopdia Britannica. 2010. Encyclopdia Britannica Online. 03 Feb. 2010 <http://www.britannica.com/EBchecked/topic/194880/Euclid>. This short appeal to history shows how old is mankind's concern for collecting and then using the information hidden in data.

"Raw" data, as they were directly obtained by various processes of acquisition, refer to numbers, figures, images, sounds, computer programs (viewed as collections of data interpreted as instructions), etc. These data, once collected, are then processed, thus obtaining information that is stored, used or transmitted further in a 'loop' type process, i.e., with the possibility that some of the processed data will represent 'raw' data for subsequent processes. In the following we consider data collections, regarded as sets of objects/samples/vectors/instances/etc., placed on the rows of a table. Each element of the collection is characterized by a set of features/attributes, placed on the columns of the table, as shown below (Fig. 2.1).

F. Gorunescu: Data Mining: Concepts, Models and Techniques, ISRL 12, pp. 45–56.
springerlink.com   © Springer-Verlag Berlin Heidelberg 2011

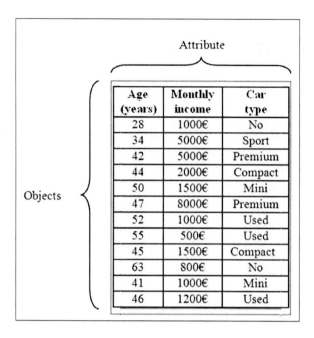

Fig. 2.1 Example of a dataset

When we speak on types of data, in statistical terminology, we actually refer to the attributes of objects from the collection. We considered that it is necessary to make this statement, because, in general, in the statistical literature, e.g., (10), (114), it is about (statistical) data of quantitative, qualitative, categorical, etc. type, the terms actually referring to the attributes of certain objects. This parenthesis being made, henceforth we will consider as data an object/sample/vector, which is characterized by its attributes that can be quantitative, qualitative, categorical, etc.

2.2 Types of Datasets

Let us briefly discuss some best-known types of datasets. Thus, we can mention the following types of datasets often encountered in practical data mining problems (378):

- *Records*:

    - Data Matrix;
    - Document Data;
    - Transaction Data.

- *Graphs*:

    - World Wide Web (WWW);
    - Molecular Structures.

- *Ordered datasets*:

  - Spatial Data;
  - Temporal Data;
  - Sequential Data;
  - Genetic Sequence Data.

Record data consist of collections of records, each record (object/sample) being characterized by a set of attributes. In principle, each object-data has a fixed number of attributes (i.e., constant length of the *tuple*), so that it can be considered as a vector in a multidimensional vector space whose dimension is obviously given by the number of attributes/components of the actual object. Such a data collection can thus be represented as a matrix of the $m \times n$ type, where each of the m rows corresponds to an object, and each of the n columns corresponds to an attribute. A classic example of such a data collection is the hospital patient registry containing patient medical records. Thus, each row of the registry is allocated to a patient (in data mining terms: object/sample/vector) and each column represents different recorded values (numerical or not), representing specific clinical parameters (e.g., age, gender, address, glycemia, cholesterol, the existence or non-existence of certain symptoms, diagnosis, etc.). Below (Table 2.1) a fictitious example of such data is presented in short (we have deliberately ignored the first column of the registry containing the patient's identity).

Table 2.1 Example of record data concerning three hepatic diseases

Diagnosis	GGT(u/l)	Cholesterol (mg/dL)	Albumin (g/dL)	Age (year)	Glycemia (mmol/L)	Sex
Cirrhosis	289	148	3.12	57	0.9	M
Hepatitis	255	258	3.5	65	1.1	M
Hepatitis	32	240	4.83	60	1.14	F
Hepatitis	104	230	4.06	36	0.9	F
Cirrhosis	585	220	2.7	55	1.5	F
Cirrhosis	100	161	3.8	57	1.02	M
Hepatitis	85	188	3.1	48	1.09	M
Cirrhosis	220	138	3.84	58	0.9	M
Cancer	1117	200	2.3	57	2.2	F
Cancer	421	309	3.9	44	1.1	M

Concerning the document data, each document recorded in the database becomes a 'term' vector, each term being an attribute of the vector, so the value assigned to that component means the number of occurrences of that word in the document, as observed in the Table 2.2 below.

Transaction data commonly refer to commercial transactions, each record involving a traded group of goods, so mathematically speaking, represented by a vector whose components stand for the traded goods (Table 2.3).

Table 2.2 Example of document data

Documents	Color	Value	Score	Currency	Party	Recipe	Team
Document A	30	25	17	15	0	2	43
Document B	4	13	2	0	14	2	2
Document C	6	42	0	0	0	123	0
Document D	0	104	0	23	0	0	12
Document E	3	585	0	0	0	60	0

Table 2.3 Example of transaction data

Transaction	Item
1	{bread, cheese, bananas}
2	{beer, coca cola, wine}
3	{meat, sausages, salami}
4	{beer, bread, meat, milk}
5	{coca cola, wine, sausages}

The data in the form of graphs (diagrams), as their name suggests, are charts incorporating information of a certain type (e.g., chemical representations, HTML links, molecular structures, etc.) -see Fig. 2.2 and Fig. 2.3 below.

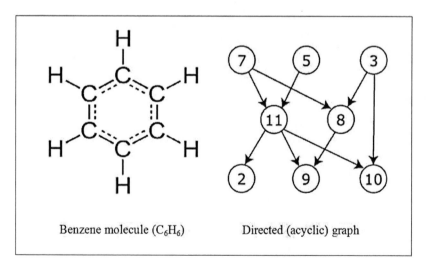

Benzene molecule (C_6H_6)   Directed (acyclic) graph

Fig. 2.2 Examples of graphs (benzene formula and directed graph)

```
<a href = "doc/dm.html">Data Mining</a>
<br />
<ul>
        <li><a href = "doc/mlp.html">Multilayer perceptron</a>
        </li>
        <li><a href = "doc/pnn.html">Probabilist Neural Networks</a>
        </li>
        <li><a href = "doc/rbf.html">Radial Basis Function</a>
        </li>
</ul>
```

HTML link

Fig. 2.3 Example of a graph diagram (HTML link)

Ordered datasets refer to those data collections which are, one way or another, governed by a specific order in which objects appear. Below (Fig. 2.4) we illustrated both ordered data from the genetics field (*left*) -genome sequences (DNA)- and meteorology (*right*) -GISS (Goddard Institute for Space Studies) Surface Temperature Analysis (http://data.giss.nasa.gov/gistemp/).

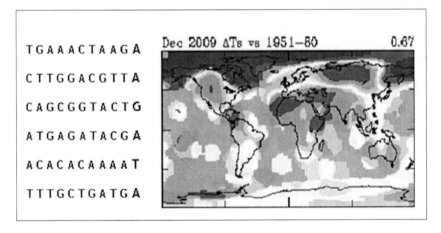

Fig. 2.4 Example of ordered datasets (genome sequences DNA, and meteorology)

Needless to emphasize that genomics problems (i.e., the study of the genomes of organisms -hereditary information of an organism encoded in DNA) or the study of the global weather evolution (e.g., global warming effect) are among the most 'hot' topics in current research. For example, regarding the issue of genomics, there are public databases devoted to this subject (e.g., NCBI -http://www.ncbi.nlm.nih.gov/ mapview; UCSC -http://genome.ucsc.edu; ENSEMBL -http://www.ensembl.org). Regarding the space-time reference data concerning Earth, in general, and global temperature, in particular, we can access sites like that of NASA (http://www.nasa. gov/topics/earth/).

Another example of ordered data concerns the sequences obtained from transactional data, i.e., sequences of collections of items. A typical example of such data refers to data stored in a *customer database*, in which each transaction is a collection of goods purchased by a customer in a single visit to the supermarket (*market basket data*) -see Fig. 2.5.

Another type of such transactional data refers to data concerning the Web usage. We illustrate below examples of such ordered data -see Fig. 2.5

	Commercial transactional data	**Data concerning Web usage**
Data	(*Basket of items*)	(*Visiting History*)
Patterns	Sequence of items e.g. (<A,B>; <D>; <C,E,F>)	Accessing sequence e.g. <1, 3, 5, 6, 9>

Fig. 2.5 Examples of transactional data (commercial and Web usage)

2.3 Data Quality

Data quality refers to their feature to more or less match the use for which they were collected or, alternatively, if they properly reflect the real context from where they originate. As regards the points from which data quality is viewed, we mention the following:

- Integration of the manufactured goods with the corresponding service (compliance with specifications) to meet customer expectation, (210);
- The quality of form, of significance and of use, (305);
- Ontological nature of information systems, (396).

The quality of recorded data is particularly important when they are used in certain specific purposes, such as modeling time (e.g., analysis of survival times -Health care, intra-day transaction price dynamics -Stock Exchange, etc.). The quality of data is strongly connected to the process of:

- Collecting them from the environment (first/original record);
- Measuring objects to obtain values for their attributes;
- Transcribing from the original source (possible second record);
- Feeding the computer software.

In this respect, the process of identifying features that are out-of-range values, or not meeting the assumptions of the methods to be used, and consequently, causing possible difficulties during the data processing, is called *data screening*.

In essence, the main characteristics taken into consideration when discussing data quality are: accuracy, reliability, validity/movement, completeness and relevance. The main problems we face in the data collection process are the following:

- Noise and outliers;
- Missing values;
- Duplicate data.

Thus, noise refers to distortion of original values, due to different interferences mainly occurring in the process of data collecting. Fig. 2.6 illustrates the case of *"white noise"*, i.e., acoustical or electrical noise of which the intensity is the same at all frequencies within a given band, and drawing its name from the analogy with white light (http://www.thefreedictionary.com/white+noise).

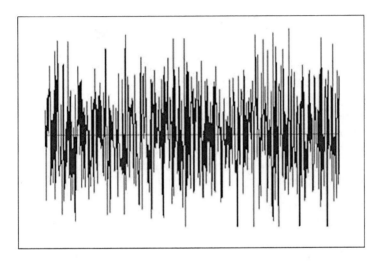

Fig. 2.6 Example of white noise in data

Basically, an *outlier* is an object (observation) that is, in a certain way, distant from the rest of the data. In other words, it represents an 'alien' object in the dataset, with characteristics considerably different to most of the other objects in the dataset.

Outliers are particularly important in any data mining technique because they can have a considerable influence on the results using that method (e.g., in exploratory data analysis, including standard statistical analyses, regression analysis, etc.).

Outliers can occur by chance in any dataset, but they are often generated by measurement errors. Usually, outliers are either discarded from the dataset, or methods that are robust to them are used. It is worth to highlight the fact that an outlier in a safety critical environment (e.g., fraud detection, image analysis, intrusion monitoring, etc.) might be detected by specific data mining methods. Fig. 2.7 illustrates such data.

In cases where there is no value available for some attributes of objects in the database, we speak of *missing values*. Frequently, values are missing essentially at random. For instance, in health care, some clinical records may have been not recorded, or may have been destroyed by error, or lost. Another situation here

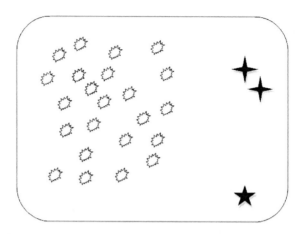

Fig. 2.7 Example of outliers in datasets (black stars)

concerns the possibility to withdraw data from the study because the attribute interest responded inappropriately to the common expectation, (10).

Because missing values are a common occurrence in many real-life situations, as seen above, specific methods have been developed to deal with this problem. Among the main methods used in such cases we mention the following two procedures:

- Attributes whose empty cells (blank) in the database are assigned as *'missing value'* are ignored in all analyses;
- Alternatively, we can use the procedure called *"mean substitution of missing data"* (replacing all missing values of an attribute by the mean of that attribute) in order to eliminate missing values in the dataset. Moreover, we can use the median of that attribute instead of its mean.

Fortunately, modern statistical packages provide the user with different options to deal with such data (e.g., *pairwise deletion of missing data, casewise deletion of missing data*).

Datasets may include data objects that are duplicates (i.e., identical objects occurring repeatedly in the dataset). As examples of duplicate data, we mention: multiple e-mail addresses, duplicate addresses, duplicate contacts (e.g., customers, clients, students, patients, members, subscribers, users, staff, etc.). The solution to this problem is the elimination of duplicates (*data cleaning*). There are a lot of techniques dealing with duplicate data (e.g., macro to delete duplicate items in a list within MS Excel - http://support.microsoft.com/kb/291320, duplicate finding/record remover software - http://www.duplicaterecordremover.com/, etc.).

2.4 Types of Attributes

As stated above, an object from a dataset is determined by its attributes/characteristics /features. We will now talk about attributes, recalling once again that, especially in statistical studies, the concepts of data and attributes are similar to a certain extent. The data mining process is influenced by the type of attributes corresponding to the analyzed objects. For example, when using data visualization, the graphic representation depends on the nature of the observations. Thus, technically, there are different methods when considering the marital status of individuals in comparison with the analysis of their bank accounts, or the investigation whether they smoke or drink alcohol, etc. It is therefore absolutely necessary to detail the problem concerning the nature of data with which data mining operates.

Data, seen in this section as the values taken by attributes, can be divided basically into two major types: numerical data (quantitative) and categorical data (qualitative), although other types of data are (rarely) used. Let us briefly present the main characteristics of each type of data.

- *Numerical data.* Numerical data (quantitative data) are themselves of two kinds: discrete data and continuous data. Discrete data appear in the case of discrete observations (represented by integers) about a particular counting process such as, for instance, the number of children of a family, the pulse of an individual, the number of yearly consultations a patient has undergone, the zip code, the personal numerical code, the PIN code, the number of certain words appearing in a document, binary attributes, etc. Let us mention in this context the four types of discrete data (integers) storage:

  - *byte* -8 bits- numbers between -128 and 127 (i.e., -2^7 and 2^7-1);
  - *short* -16 bits- numbers between -32,768 and 32,767 (i.e., -2^{15} and 2^{15}-1);
  - *int* -32 bits- numbers between -2,147,483,648 and 2,147,483,647 (i.e., -2^{31} and 2^{31}-1);
  - *long* -64 bits- numbers between -9,223,372,036,854,775,808 and 9,223,372,036,854,775,807 (i.e., -2^{63} and 2^{63}-1).

Unlike the discrete data, usually obtained from a counting process, the continuous data are commonly obtained from measurements, e.g., height, weight, blood pressure, cholesterol of a certain individual, temperature, atmospheric pressure, wind speed, value of a bank account or the value of shares traded on the Stock Exchange, etc. These data are usually expressed by real numbers, unlike the discrete data that are restricted to integers. We mention here that, in many analyses, discrete data are treated as continuous data, e.g., the number of heart beats per minute. To avoid a mistreatment in analyzing such data (discrete, but considered as continuous) a sufficient number of different potential values is needed to create the prerequisites concerning their hypothetical continuous nature. Let us mention, in this context, the continuous data (with floating point) that can be stored:

- *float* - numbers between 1.4E-45 and 3.4E+38;
- *double* - numbers between 4.9E-324 and 1.7E+308.

- *Categorical data.* Unlike the numerical data, the categorical (or qualitative) data are those data, as their name says, that divide objects into different categories such as:

  1. male/female;
  2. married/single /divorced;
  3. smoker/non-smoker;
  4. hypertensive/normotensive/hypotensive;
  5. pregnant/not pregnant;
  6. stages in certain diseases (e.g., cancer): I, II, III, IV;
  7. existence of symptoms: YES, NO;
  8. type diagnosis: A, B, C, D, etc.;
  9. alcoholic/non-alcoholic;
  10. employed/unemployed.

  Let us note that discrete data are sometimes treated as categorical data, e.g., the number of children per woman, e.g., 0, 1, 2, 3, 4, divides mothers into categories corresponding to the number of children. It is important to ignore in this case the notion of 'order' or 'numerical parameters', such as average, or median. Conversely, it is inappropriate to interpret the categorical ordered data as numerical data. For example, in certain disease stages, it is wrong to affirm that stage IV is two times worse than stage II, etc. It is good to mention that in the case of the categorical data, we can speak of *nominal* data such as: blood type (A/B/AB/O), eye color (brown, green, blue, black), sex (male, female), etc., and ordinal data, such as: 'degree' of smoking (e.g., non-smoker, former smoker, 'amateur' smoker, 'hard' smoker), the ranking of pain (e.g., small, medium, high), the ranking of height (e.g., short, medium, high), the ranking of minerals hardness, etc. To conclude, in some circumstances it is a good idea to convert the measurements on a group of objects into a rank ordering before analyzing the data, and thus, dealing with ordinal data.

- *Other types of data.* Apart from the two major types of data mentioned above, some miscellaneous other types of data are also used. We briefly describe below the most popular types of such data.

  - *Rank*, representing the place occupied by an object in a hierarchy (e.g., sporting competitions, examinations, physician preference for a particular treatment, customer preference for a particular product, etc.).
  - *Percentage*, as its name suggests, describes a certain proportion (ratio) between two quantities (e.g., percentage of men in a population, relative body weight (the ratio of observed weight to ideal weight), percentage of left-handed in a population, percentage of loyal customers, percentage of objects correctly classified, percentage of missing data, etc.).
  - *Rates and ratios*, related to the observed frequency of a phenomenon or ratios between two values, other than percentages (e.g., relative mortality per

thousand inhabitants, rate of occurrence of a disease depending on sex or geographical area, currency exchange rates, ratio price/quality, etc.).

- *Score*, used when a direct and unambiguous measurement is not possible and a certain value should be however quantified (e.g., *Apgar* score for newborn children immediately after childbirth; severity of an illness quantified as mild, moderate, severe; skin color in certain diseases, etc.).
- *Visual analogue scales*, mainly used in medical studies when, for instance, the subject is asked to indicate on a scale the point that is considered the best level to illustrate his/her degree of pain. Although it is a very subjective representation, almost impossible to quantify numerically, it is however a way to 'measure' a certain phenomenon. It is worth to mention that caution is required in handling such data.

A common situation occurring in exploratory data analysis (i.e., the statistical approach in data mining), especially regarding data originating in the medical field, concerns the case of *censored data*. Thus, there are cases in which a particular observation cannot be well specified. For example, in survival analysis -a classical statistical technique to study the dynamics of survival time after a specific operation or treatment- a part of the subjects included in the study group die during the monitoring period, but another part will survive during this period or will withdraw voluntarily, and thus the time of death cannot be recorded clearly, without any doubt. Another example is when certain measurements are made and the device cannot record lower or higher values, i.e., outside its scale, although they exist in reality -data undetectable by that device. Summarizing, in any situation where there is certain data but, for various reasons, they cannot be stated clearly, we say that we are dealing with censored data.

Also in the context of using statistical methods in data mining, it is useful to discuss briefly both about the data variability and about the probabilistic model for data, with a special reference to the term '*variable*', which is often equivalent to the term '*attribute*' of an object.

Thus, when processing data during statistical analyses within the data mining process, the existence of the so-called data variability is absolutely necessary. By *variability* we mean any change that takes place in a set of data, regardless of their type, i.e., the variability is the opposite of the 'fixedness' of data. As it is well known, Statistics is largely about variability, and consequently, data mining is also interested in variability. The term 'variable' is thus used to denote anything that varies within a set of data. That is why, in Statistics attribute is equivalent to variable, i.e., something that varies somehow. Be aware that we cannot do statistical analysis on variables that are constant. Much of the classical statistical analyses (e.g., the regression analysis) appeals to connections between different data referring to the same 'subjects', studying how changes of some of them affect the change in others (e.g., the connection between height and weight; between the risk factors and the likelihood of a disease trigger, etc.). Or, if a factor from the statistical analysis does not have variability (i.e., it is constant), then it is virtually non-existent in the analysis. The greater the variability, the richer the corresponding statistical analysis is in consistent results.

So far, throughout this section, we have talked about data seen from a statistical point of view only. We have been concerned only about their statistical description, without attempting to define them in a probabilistic context. Since Statistics cannot be broken off from the Probability theory, which provides the theoretical background and the means of investigation, it is imperative to define the data in this context. Suppose we dispose of a certain set of objects/items/samples (i.e., a *statistical population* in the statistical terminology) and we are interested in analyzing their main characteristics, which represent, as it is well known, attributes, or (*statistical*) *data*, or (*statistical*) *characters*, in statistical language. Let us see the way in which the notion of statistical data is defined in terms of probabilities. Generally speaking and from a probabilistic viewpoint, by *data* (or *character*) we mean a function defined on a set, representing the population (objects), and with values in a given set which depends on the nature of the data. For instance, for a population of individuals, their heights may be considered as the values taken by a random variable corresponding to this data (character); for the same population, the color of eyes may be considered as the 'values' of a random variable which does not take numerical values, but categorical. Theoretically, considering a *probability space* (Ω, Σ, P), where Ω is just the considered population, and Σ is a *σ-field* of measurable sets in Ω (if Ω is finite, then Σ coincides with the family of its subsets), the *data X* of the statistical population Ω represents a *random variable* on the probability space (Ω, Σ, P), if data X is numerical.

In Probability Theory, a real-valued measurable function X (i.e., the preimage of every interval of $\mathbf{R}$ is an event, i.e., an element of the *σ-field* Σ), whose domain is the sample space Ω, is called a *random variable* (for details, see for instance, (243), (287)). Such a random variable, seen in terms of Probability, becomes a *statistical variable*, from a statistical viewpoint (more explanations concerning the connection between a (probabilistic) random variable and a statistical variable are presented in subsection 3.2.1). If X does not take numerical values, it can sometimes consider X as a random variable, based on a certain numerical equivalence of these values (e.g., gender equivalence: Male = 0 and Female = 1; numerical equivalence of ordinal data, etc.).

To avoid any vagueness, let us underline that, while in Statistics by (statistical) data we understand the value of an object's attribute, in data mining data represents an object, characterized by a certain number of attributes.

Once these specifications established, we can use the whole probabilistic-statistical arsenal within the data mining framework to process available data.

Chapter 3
Exploratory Data Analysis

Abstract. As we stated in the introductory chapter, data mining originates from many scientific areas, one of them being Statistics. Having in mind that data mining is an analytic process designed to explore large amounts of data in search of consistent and valuable hidden knowledge, the first step made in this fabulous research field consists in an initial data exploration. For building various models and choosing the best one, based on their predictive performance, it is necessary to perform a preliminary exploration of the data to better understand their characteristics. This stage usually starts with data preparation. Then, depending on the nature of the problem to be solved, it can involve anything from simple descriptive statistics to regression models, time series, multivariate exploratory techniques, etc. The aim of this chapter is therefore to provide an overview of the main topics concerning this data analysis.

3.1 What Is Exploratory Data Analysis?

The motivation behind the first step in the data mining process -the exploratory data analysis- is very simple and, at the same time, very serious. First of all, such an analysis makes use of the human ability to recognize patterns based on the previous experience. Based on information and knowledge accumulated over time, people can recognize certain forms, trends, patterns, etc., systematically appearing in data, and that cannot always be emphasized by classical methods of investigation. On the other hand, all the experience gained in a given area can significantly help in choosing the best pre-processing techniques and data analysis. Thus, in order to effectively use these opportunities, we need a data analysis, an exploration of them with well-known statistical means, being then able to choose the optimal data mining methodology for the available data.

Basically, exploratory data analysis (EDA) is the Statistics part which deals with reviewing, communicating and using data in case of a low level of information on them. Unlike the classic case of statistical hypothesis testing, used in Statistics to verify certain *a priori* assumptions (e.g., certain correlations between different attributes/variables, where there is some information concerning a possible

F. Gorunescu: Data Mining: Concepts, Models and Techniques, ISRL 12, pp. 57–157.
springerlink.com

dependency), in EDA different techniques are used to identify systematic relations between certain attributes/variables about which there is no prior information. EDA was created and named by the American statistician John Tukey (in Tukey J., *Exploratory Data Analysis*, Addison-Wesley, 1977). Computational EDA techniques include both basic statistical methods and other advanced exploratory processes (fully using the opportunities opened by the computerized data processing), i.e., *multivariate exploratory techniques*, designed to identify certain patterns hidden in complex datasets. EDA uses various techniques -many of them based on visualization- in order to:

- Maximize the innermost knowledge of the data;
- Reveal underlying structure;
- Extract important variables;
- Detect outliers/anomalies;
- Identify fundamental assumptions to be tested afterwards;
- Develop simple enough models;
- Determine the optimal setting of parameters;
- Suggest some hypotheses concerning the causes of the observed phenomena;
- Suggest appropriate statistical techniques for the available data;
- Provide knowledge for further data collection in support of research or experimentation.

Once EDA techniques are used as a preamble to the data mining process, one raises the question of verifying the results thus obtained. It should be stressed that the exploration of data must be regarded just as a first stage of data analysis, and the results thus obtained will be considered on an experimental basis only, until they are validated alternatively. If, for example, the result of applying EDA suggests a particular model, then it must be validated by applying it to another set of data, and thus testing its predictive quality.

In conclusion, EDA can be considered - in principle - a philosophy about the way in which data are 'dissected', are 'looked at' and, finally, are interpreted. For more details concerning the role and means of EDA in the data mining framework, we may use, among others, the following links:

- Nist-Sematech: http://www.itl.nist.gov/div898/handbook/eda/eda.htm;
- Statgraphics: http://www.statgraphics.com/eda.htm;
- Wikipedia: http://en.wikipedia.org/wiki/Exploratory_data_analysis.

In this presentation we will emphasize the following EDA techniques:

- Descriptive Statistics -numerical summary and graphical representation;
- Analysis of correlation matrix;
- Data visualization;
- Examination of distributions of variables (analysis of symmetry, non-Normality, multi-modal case, etc.);
- Advanced linear and additive models;
- Multivariate exploratory techniques;
- OLAP -Online Analytical Processing.

3.2 Descriptive Statistics

Descriptive statistics gathers together a suite of different methods, aiming to summarize a large number of observations regarding data, highlighting thus their main features. In this respect, there are two main approaches of the statistical description of data:

- Determination of numerical parameters, our interest being focused on their mathematical properties;
- Various simple graphical representations of data, whose interpretation is not difficult, being often very suggestive, even though it is still strictly limited in terms of information.

The subject of the data mining study concerns (large) datasets, consisting of objects characterized by a certain number of attributes. Each such attribute has a value belonging to a certain specified set. For each attribute separately, we consider all taken values, i.e., the set of values corresponding to all objects in the dataset. Thus, a number of sets equaling the number of attributes of the objects, each set containing the values related to a certain attribute, will correspond to a given dataset. From a statistical point of view, we will consider each of these sets of attributes' values as *statistical series*.

To summarize, for a given dataset of attributes, we dispose of a number of statistical series equaling the number of attributes. Under these circumstances, our goal is to perform a statistical description of these statistical series, taken separately or together.

It is a good idea to highlight that descriptive statistics, as a complex of descriptive statistical tools, is distinguished from the inferential statistics, in that descriptive statistics aims to quantitatively summarize a dataset, rather than being used to support inferential statements about the population that the data are thought to represent. To conclude, in what follows we will perform a statistical description, both quantitatively and visually, of the statistical series corresponding to the given dataset of objects.

Remark 3.1. Before we start presenting the main techniques regarding the descriptive statistics of data, we want to point out a possible broader connection between the descriptive statistics and the inferential statistics, seen in the data mining context.

Thus, a fundamental notion in Statistics, especially in inferential statistics, is the concept of "sample". Recall that, given a (statistical) population, a *sample* is a subset of that population. Basically, statistics are computed from samples in order to make (statistical) inferences or extrapolations from the sample to the entire population it belongs to. As we mentioned above, in the data mining context, we might use the notion of sample in a broader sense, with the meaning of 'dataset of objects' to be processed by data mining techniques. Thus, a dataset consisting of objects will be considered, in terms of descriptive statistics, as a sample from a theoretical statistical population. For example, when we want to identify the customer profile of a chain of supermarkets, the available data might be considered as a sample of the theoretical population consisting of all possible clients of all possible chains of supermarkets. In another context, if we want to classify galaxies, then the available data might be considered as a sample of the population of all galaxies in the universe. In all these cases, when using the exploratory data analysis instruments and the dataset is considered as a sample, the main goal will not be that of statistical inference, but just that of the statistical description of a set of data with specific statistical tools. On the other hand, basically, everything which is discovered using these data can be extrapolated to other similar data, regarded as samples of the same population (e.g., supermarket customers, galaxies in the universe). The knowledge gained by statistically evaluating a dataset can then be used in similar situations, thus defining the generalization power of a data mining model. In this way, we may extrapolate knowledge based on some situations to other quite similar cases.

3.2.1 Descriptive Statistics Parameters

Descriptive statistics, viewed as a statistical tool borrowed by EDA from applied Statistics, refers, in principle, to those numerical parameters that give a synthetic image of the data (e.g., mean, median, standard deviation, mode, etc.), summarizing thus the main statistical characteristics of data. In what follows we might think about descriptive statistics as a group of numerical parameters and graphical representations, statistically describing the set of data we are concerned with.

Descriptive statistics aims to summarize a large number of observations concerning a set of objects, using different specific methods, thus highlighting the main features of their attributes. There are two main ways to achieve this goal: either using a simple graphical representation, about which we shall speak later during this chapter, or by using numerical representations containing the main statistical characteristics of data. Whatever the approach, we are concerned about the representation of data variability. This variability may be one with known causes, i.e., a 'deterministic' variability, which is statistically described in order to be better put in evidence and precisely quantified, or may be a variability with only suspected or even unknown causes, i.e., 'random' variability, and which, using statistics, it is hoped to be clarified in terms of causality.

As it is well-known, an object corresponds, from a probabilistic point of view, to a multidimensional random variable (i.e., it is represented mathematically by a random n-dimensional vector), thus becoming subject to a multivariate statistical analysis. Each component of the object, representing a particular attribute, is seen in turn as a random variable. It is then natural to consider in this case the numerical (statistical) parameters that characterize a random variable, parameters very useful to describe its dynamics.

In what follows, we will consider that every attribute x of an object (i.e., a component of an n-dimensional vector denoting the object) is represented mathematically by a random variable X, which may take values corresponding to the nature of that attribute. Although the term *'statistical variable'* is usually considered synonymous with *'random variable'*, we further show the way we refer here to those concepts. Thus, while in Probability Theory a random variable is, simply, a measurable function on a sample space (Ω, Σ), in Statistics a statistical variable refers to measurable attributes. Of the two mathematical concepts, the former refers to measurable functions and the latter to values taken by object's attributes. To be more specific, assume we consider a certain attribute corresponding to the objects belonging to a given dataset (e.g., monthly income, age, eye color, etc.). Theoretically, the values of that attribute (either numerical or categorical) are considered as values of a random variable 'governing' the nature of that attribute. Next, to this parent random variable X there corresponds a statistical variable, denoted naturally by the same letter X, given by the set of the actual values of the attribute, obtained from a particular dataset. Philosophically speaking, a statistical variable from Statistics represents the practical concept, i.e., when working with real objects (data) belonging to a certain dataset, of the abstract (mathematical) notion of random variable from Probability Theory.

For the parent random variable X, we consider the function $F_X : \mathbf{R} \to [0, 1]$, defined by:

$$F_X(x) = P\{X < x\} = P\{\omega \in \Omega; X(\omega) \in (-\infty, x)\}, \tag{3.1}$$

and called *probability distribution* (or simply, *distribution*) of the random variable X, where (Ω, Σ, P) is a probability space. Thus, a random variable is characterized by its distribution function F_X.

As in the case of a random variable, we can define the concept of distribution for a statistical variable too. Thus, by the *distribution* (or *cumulative frequency*, or *empirical distribution*) of a statistical variable X (corresponding to the parent random variable X), related to the statistical series $\{x_i\}$, $i = 1,..., n$, we mean the function $F: \mathbf{R} \to [0, 1]$, given by:

$$F(x) = \frac{f_x}{n}, x \in \mathbf{R}, \tag{3.2}$$

where f_x represents the number of observations x_i strictly less than x. Let us mention that $F(x)$ is also called the *distribution* of the statistical series $\{x_i\}$, $i = 1,..., n$. For instance, if we speak of the monthly income of individuals, then there is a parent random variable "*Monthly income*" only with theoretical role, and the corresponding statistical series $\{x_i\}, i = 1,..., n$, denoting the values of the monthly incomes of those individuals. Another example concerns the clinical parameters (e.g., cholesterol, glycemia, etc.) of patients with a specific disease. In this case we will also speak of random variables and statistical variables/series respectively, corresponding to the above parameters. Below (Fig. 3.1) are illustrated the distributions (cumulative frequencies) corresponding to both the cholesterol and glycemia levels for a group of 299 patients with liver diseases.

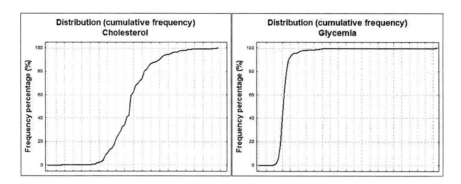

Fig. 3.1 Distribution (cumulative frequency) of cholesterol and glycemia

First, just as in the probabilistic case, we can define the *quantile* of order α, $0 < \alpha < 1$ of the statistical variable X (term first used by Kendall in 1940) the number q_α with the property $F(q_\alpha) = \alpha$. From our point of view related to data exploration, the quantiles involve practically the process of dividing ordered data into $1/\alpha$ equally sized data subsets; thus, the quantiles are the data values marking the boundaries between these consecutive subsets. Let us point out, just for information, that the word "quantile" derives from the Lat. *quantillus = how little, how small?*

It is worth mentioning that in descriptive statistics the quantiles are not always defined, in general, limiting us to specify particular types of quantum that are more appropriate to usual computation. In this context, a *percentile* (Galton, 1885) represents any of the 99 values that divide ordered data into 100 consecutive subsets of equal size. For example the 50th percentile divides the dataset into 50% data above it and 50% data below it. Similarly, the *deciles* represent the 9 values that divide ordered data into 10 consecutive subsets of equal size, and the *quartiles* (Galton, 1882) are the 3 values denoted Q_1, Q_2, Q_3, which divide ordered data into 4 consecutive subsets of equal size (i.e., quantile of type $q_{i/4}$). The most used quantiles are Q_1, Q_2, and Q_3. Thus, for the first quartile Q_1 (the lower quartile - the 25th percentile) there are 25% of data below it and 75% above it, for the second quartile Q_2 (the 50th percentile) there are 50% of data below it and 50% above it, and for the third Q_3 quartile (the upper quartile - the 75th percentile) there are 75% of data below it and 25% above it.

We illustrate below the quantiles values for a dataset corresponding to data collected from a group of 299 patients with different liver diseases (hepatitis C, cirrhosis and hepatic cancer). Thus, Table 3.1 and Table 3.2 show the deciles, while Table 3.3 and Table 3.4 show the quartiles regarding cholesterol and glycemia (blood glucose).

In principle, the typical values corresponding to a data analysis are the following:

- Typical measures of the central tendency (location): *mode, median, mean (average, arithmetic mean), geometric mean* and *harmonic mean*;
- Typical measures of spread: *variance* and *standard deviation*;
- Typical measures concerning the shape of the distribution: *skewness (asymmetry)* and *kurtosis (excess)*.

Table 3.1 The deciles corresponding to cholesterol

$q_{1/10}$	$q_{2/10}$	$q_{3/10}$	$q_{4/10}$	$q_{5/10}$	$q_{6/10}$	$q_{7/10}$	$q_{8/10}$	$q_{9/10}$
144	164	176	190	200	201	220	230	265

Table 3.2 The deciles corresponding to glycemia

$q_{1/10}$	$q_{2/10}$	$q_{3/10}$	$q_{4/10}$	$q_{5/10}$	$q_{6/10}$	$q_{7/10}$	$q_{8/10}$	$q_{9/10}$
0.78	0.81	0.87	0.90	0.96	1.00	1.06	1.12	1.29

Table 3.3 The quartiles corresponding to cholesterol

Q_1	Q_2	Q_3
170	200	227

Table 3.4 The quartiles corresponding to glycemia

Q_1	Q_2	Q_3
0.85	0.96	1.10

The most common parameter measuring the 'central tendency' of a sample is the *mean* (or *average*), which is practically the arithmetic mean of all observations, being given by:

$$\bar{x} = \frac{1}{n} \sum_{i=1}^{n} x_i. \tag{3.3}$$

We must mention that, despite the fact that the mean is the most common measure of the central tendency, being a very suggestive characteristic of the data it represents, it is at the same time very 'sensitive' to the existence of extreme values (outliers), which can seriously perturb its capability of illustrating the data. To verify this observation, consider the following sequence of data that may represent, for instance, the heights (in meters) of certain people:

$\{1.70, 1.67, 1.87, 1.76, 1.79, 1.66, 1.69, 1.85, 1.58, 1.78, 1.80, 1.83, 2.20\}$

It is easy to observe that, in this sequence of heights, the last value of 2.20 is an 'extreme' value (a 'special' height, corresponding more likely to a basketball player). Calculating the mean of the above values, either including or excluding this value, we obtain $m_1 = 1.78$ and $m_2 = 1.74$ respectively, values that are different enough due to the influence of a single data. To avoid such situations it is better to use the median instead of the mean. Thus, the *median* is defined as the real number that divides a given sample into two equal subsets, the observations being considered in ascending order, i.e., the median is just the second quartile Q_2. Formally, the median is given by:

$$P\{X \leq Q_2\} = P\{X > Q_2\} = \frac{1}{2}. \tag{3.4}$$

If the sample size is an odd number $n = 2k + 1$, then the median is the $(k + 1)$th value of the sample, and if the sample size is an even number $n = 2k$, then the median is replaced with the 'median interval' given by the kth and $(k + 1)$th values (the median may be considered as the middle of this interval, i.e., the arithmetic mean of its endpoints). Moreover, we can mention the fact that the median is also useful when there is the possibility that some extreme values of the sample are censored. When there are observations that are under a certain lower threshold or above a certain upper threshold and, for various reasons, these observations are not exactly enough specified, we cannot use the mean. Under these circumstances, we will replace it by the median when we dispose of accurate values for more than half of the observations (the case of physicochemical measurements, when there are values outside the normal scale of the measuring device). We must understand that both measures are equally effective and, although the mean is more commonly used than the median, the latter may be more valuable in certain circumstances. In the example above, concerning the set of data about the heights of certain individuals, the corresponding medians of the two cases (with and without the extreme value) are $med_1 = 1.78$ and $med_2 = 1.77$, respectively, so the median is not significantly influenced by the extreme values.

Another measure of the central tendency presented here is the *mode*, which is simply the value that occurs the most frequently in a dataset or a (probability) distribution, being rarely used for continuous data (representing there the maximum of the probability density distribution), and most often used for categorical data. Note that there are multimodal datasets, i.e., there is more than one value with the same maximum frequency of occurrence (several modes). For instance, the statistical series concerning the values taken by a discrete variable:

$\{9, 9, 12, 12, 12, 12, 13, 15, 15, 15, 15, 17, 19, 20, 20, 25, 26, 29, 29, 30\}$ that

can be summarized as follows (Table 3.5):

Table 3.5 Distribution of a discrete data (variable)

x_i	9	12	13	15	17	19	20	25	26	29	30
n_i	2	4	1	4	1	1	2	1	1	2	1

is bi-modal, since the values 12 and 15, respectively, occur four times. If we consider the clinical parameter 'cholesterol' corresponding to a group of 299 patients with liver diseases as a continuous attribute, Normally distributed, then its mean value represents the mode for this attribute, as shown in the graph below. Technically, the graph illustrates both the distribution of the number of observations of the cholesterol values and the parental Normal (Gaussian) distribution (background continuous line). As it is easy to see (Fig. 3.2), the mean value, very close to the theoretical mean of the parent Normal distribution, represents the mode for this attribute.

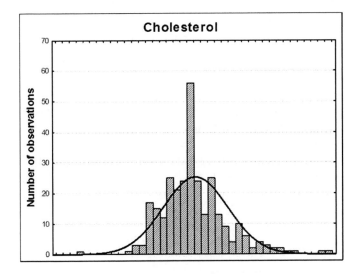

Fig. 3.2 Histogram of the cholesterol (with parent Normal distribution)

Note. If the data are grouped into classes (i.e., the values belong to intervals), we call modal class each class corresponding to a maximum of frequency.

The *geometric mean* is given by:

$$\sqrt[n]{x_1 \cdot x_2 \cdot \ldots \cdot x_n}. \tag{3.5}$$

The geometric mean is used in particular in case of measurements with non-linear scale (e.g., in Psychometrics, where the rate of a stimulus intensity is often a logarithmic function with respect to intensity, in which case the geometric mean is used rather than the arithmetic mean). For instance, for the clinical parameter '*glycemia*' corresponding to a group of 299 patients with liver diseases, the geometric mean is 0.99 and the mean is 1.04, close enough in this case.

The *harmonic mean*, given by:

$$\frac{n}{\sum_{i=1}^{n} 1/x_i}, \tag{3.6}$$

is sometimes used for determining the average frequency. For instance, in the above example, concerning glycemia, the harmonic mean is 0.96.

Summarizing, for the statistical series concerning the hypothetic people's heights, earlier considered, we have the following values for mean, mode, median, geometric mean and harmonic mean (Table 3.6).

Table 3.6 The mean parameters of the central tendency

Mean	Mode	Median	Geometric mean	Harmonic mean
1.78	1.65/1.75	1.78	1.77	1.77

From the table above we see that:

- The mean coincides with the median;
- The (continuous) parental distribution is bi-modal: two modal classes (1.6, 1.7) and (1.7, 1.8) have the same frequency 30.8% -see graph in Fig. 3.3;
- The geometric mean coincides with the harmonic mean;
- The distribution is 'symmetrical' enough, since the parameters of the central tendency are sufficiently close -*idem*.

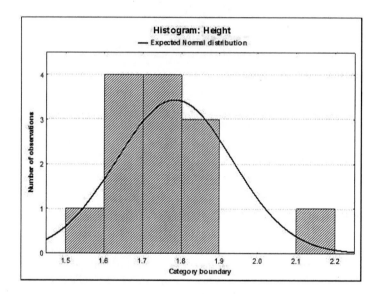

Fig. 3.3 Histogram of people's height

Another approach concerning data analysis focuses on evaluating the spread of data around the mean, i.e., on measuring the overall distance between each value of the statistical series and the mean. Starting from the classical probabilistic notion of *'variance'*, term introduced by R.A. Fisher in 1918, we define the *variance* corresponding to a statistical series $\{x_i\}$, $i = 1,..., n$, using the formula:

$$\sigma^2 = \frac{1}{n} \sum_{i=1}^{n} (x_i - m)^2, \tag{3.7}$$

i.e., the average of the squares of the deviations of the values belonging to the statistical series from their mean, where m is the (assumed known) mean of the statistical parent population from which the statistical series has been selected. Because we usually consider the statistical series, working with as just a sample from the entire population, and thus the theoretical mean m is not *a priori* known, we replace it by the estimated mean $\bar{x}$ of the statistical series, and, accordingly, the estimated variance is given by:

$$\sigma^2 = \frac{1}{n} \sum_{i=1}^{n} (x_i - \bar{x})^2. \tag{3.8}$$

Let us remark here that the statistical software packages use a slightly different formula for variance in this case, namely the *unbiased* sample estimate of the population variance, given by:

$$\sigma^2 = \frac{1}{n-1} \sum_{i=1}^{n} (x_i - \bar{x})^2. \tag{3.9}$$

Let us note that, for large statistical series, the difference between the value given by the formula above and the standard formula (3.8) is negligible.

Often, it is preferable that, instead of using the variance, we should use another parameter which is measured with the same unit as the statistical series, namely the *standard deviation* (SD), denoted by σ and given by:

$$\sigma = \sqrt{\frac{1}{n-1} \sum_{i=1}^{n} (x_i - \bar{x})^2} \tag{3.10}$$

Standard deviation is mainly used in descriptive statistics to define certain intervals which contain most observations. Thus, in case of reasonably symmetric distributions, the vast majority of observations composing the statistical series (approximately 95% of them) fell into the interval defined by *mean* $\pm 2 \times SD$, called the *95% confidence interval*. We stress again that it is absolutely necessary to deal with relatively symmetrical distributions, otherwise what was said above is no longer relevant. If the distribution of the given statistical variable is far from a sufficiently symmetrical distribution, there are methods to statistically describe its variability using symmetrical distributions, for example by considering a mathematical transformation of the original statistical series (e.g., computing the logarithm of the original series).

We illustrated the typical measures of spread for the statistical series related to the height, the corresponding values being displayed in the Table 3.7 below.

Table 3.7 The main parameters of the spread of data

Variance	Standard deviation (SD)	95% confidence interval
0.023	0.151	(1.69, 1.87)

Remark 3.2. Besides the statistical parameters mentioned above, sometimes the following parameters are also used:

- The *range* (of values), represented by the difference between the maximum and the minimum of the values taken by a variable;
- The *interquartile range*, defined by the difference $Q_3 - Q_1$;
- The *average absolute deviation*, given by:

$$AAD(x) = \frac{1}{n} \sum_{i=1}^{n} |x_i - \bar{x}|$$

- The *median absolute deviation*, given by:

$$MAD(x) = med\{|x_1 - \bar{x}|, |x_2 - \bar{x}|, ..., |x_n - \bar{x}|\}$$

3.2.2 Descriptive Statistics of a Couple of Series

In the previous section we presented different ways to describe a series of observations (statistical series/sample), corresponding to a single statistical variable, 'governing' a certain attribute. When we consider two or more series of observations, represented by two or more statistical series, corresponding to two or more different attributes of the objects to be statistically analyzed, in addition to their individual description, as done in the previous subsection, it is indispensable to analyze the relationship that may exist between them. Thus, other statistical techniques (numerical or graphical) are used for this purpose. For instance, we use conditional and marginal parameters, correlation, covariance, regression, plotting the spread, etc.

(a) *Conditional and marginal parameters* Suppose, simplifying, that we dispose of two statistical variables X and Y, corresponding to two attributes of the objects belonging to the database to be analyzed (e.g., cholesterol and glycemia for a database concerning liver diseases). Theoretically speaking, the two statistical variables X and Y correspond to two parent random variables X and Y. In what follows we are interested in their joint probability distribution. Consider a couple of random variables (X, Y) on the same probability space (Ω, Σ, P). The function given by:

$$F_{XY}(x,y) = P\{X < x, Y < y\}, \tag{3.11}$$

represents the so-called *bivariate joint distribution* of the two random variables. The *marginal distribution* of the random variable X is given by:

$$F_X(x) = P\{X < x\} = F_{XY}(x, \infty). \tag{3.12}$$

In this context, the random variables are said to be *independent* if:

$$F_{XY}(x, y) = F_X(x)F_Y(y), \ \forall x, y \tag{3.13}$$

Further:

$$F_{X|Y}(x|y) = \sum_{a \le x} \frac{p_{XY}(a, y)}{p_y(y)}, p_y(y) > 0 \tag{3.14}$$

represents the *conditional distribution* of X given $Y = y$ (the case of discrete data), where p_{XY} is the *joint probability mass function* of the two random variables, and p_Y is the *probability mass function* of Y.

Next:

$$F_{X|Y}(x|y) = \int_{-\infty}^{x} \frac{f_{XY}(t, y)}{f_Y(y)} dt \tag{3.15}$$

represents the *conditional distribution* of X given $Y = y$ (the case of continuous data), where f_{XY} is the *joint density* of X and Y, and f_Y is the *density* (or *probability density function* $\sim$ p.d.f.) of Y.

Example 3.1. In the probability space (Ω, Σ, P) we consider the couple of discrete independent random variables (X, Y), with distributions given by:

Table 3.8 Distribution of X (tabulated)

x_i	1	2	3	4
p_i	0.1	0.2	0.3	0.4

Table 3.9 Distribution of Y (tabulated)

y_i	-1	0	1
p_i	0.25	0.5	0.25

Table 3.10 below shows both their bivariate joint distribution and the corresponding marginal distributions.

Table 3.10 Bivariate joint distribution (X, Y)

$X \backslash Y$	-1	0	1	F_X
1	0.025	0.05	0.025	0.1
2	0.05	0.1	0.05	0.2
3	0.075	0.15	0.075	0.3
4	0.1	0.2	0.1	0.4
F_Y	0.25	0.5	0.25	1

Since, in principle, there may be some connection between the attributes associated with the variables X and Y, we consider the conditional expectations as parameters of their statistical description.

Thus, theoretically speaking, for discrete variables (attributes), the *conditional expectation* of X given $Y = y$ is defined by:

$$E[X|Y=y] = \sum_x \frac{x p_{XY}(x,y)}{p_y(y)}, \ p_y(y) > 0, \tag{3.16}$$

and, then, the *expectation* of X is given by:

$$E[X] = \sum_x x p_X(x) = \sum_x \sum_y x p_{X|Y}(x|y) p_y(y) = E[E[X|Y]]. \tag{3.17}$$

Next, for continuous variables (attributes), the *conditional expectation* of X given $Y = y$ is defined by:

$$E[X|Y=y] = \int_{-\infty}^{\infty} x \frac{f_{XY}(x,y)}{f_Y(y)} dx, \tag{3.18}$$

and, thus, the *expectation* of X is given by:

$$E[X] = \int_{-\infty}^{\infty} x f_X(x) dx = \int_{-\infty}^{\infty} \int_{-\infty}^{\infty} x f_{X|Y}(x|y) f_Y(y) dx dy = E[E[X|Y]]. \tag{3.19}$$

In this context, recall the following computational formula in the statistical series case. Thus, the means of the statistical variables X and Y, forming the couple (X, Y), are given by:

$$m_X = E[X] = \sum_i \sum_j x_i p_{ij}, \tag{3.20}$$

$$m_Y = E[Y] = \sum_i \sum_j y_j p_{ij}, \tag{3.21}$$

where $p_{ij} = P\{X = x_i, Y = y_j\}$ -the discrete case.

In the continuous case, the corresponding formulas are:

$$m_X = E[X] = \int \int x f_{XY}(x,y) dx dy, \tag{3.22}$$

$$m_Y = E[Y] = \int \int y f_{XY}(x,y) dx dy. \tag{3.23}$$

Regarding the variances, they are given by the formulas (discrete case):

$$D^2(X) = \sum_i \sum_j (x_i - m_X)^2 p_{ij}, \tag{3.24}$$

$$D^2(Y) = \sum_i \sum_j (y_j - m_Y)^2 p_{ij}, \tag{3.25}$$

and:

$$D^2(X) = \int \int (x - m_X)^2 f_{XY}(x,y) dx dy, \tag{3.26}$$

$$D^2(Y) = \int \int (y - m_Y)^2 f_{XY}(x,y) dx dy, \tag{3.27}$$

respectively (continuous case).

Example 3.2. Consider a couple (X,Y) of independent discrete (statistical) variables, with the joint distribution given by:

Table 3.11 Bivariate joint distribution (X,Y)

$X \backslash Y$	1	2	3
1	1/18	1/12	1/36
2	1/9	1/6	1/18
3	1/6	1/4	1/12

Thus, the means of the two variables are: $m_x = 2\frac{1}{3}$ and $m_y = 1\frac{5}{6}$, where the point $(2\frac{1}{3}, 1\frac{5}{6})$ is sometimes called the *variance center* of the couple (X,Y); the corresponding variances are: $D^2(X) = \dfrac{5}{9}$ and $D^2(Y) = \dfrac{17}{36}$.

Example 3.3. Assume that a couple of continuous (statistical) variables is governed by a probability law with the joint density given by:

$$f(x,y) = \begin{cases} 1/2 \cdot sin(x+y) \,, (x,y) \in D \\ \qquad 0 \qquad\quad\, , (x,y) \notin D. \end{cases} \tag{3.28}$$

where $D = \{0 \le x \le \pi/2, 0 \le y \le \pi/2\}$.

Then, using the above formulas, we obtain:

$$m_X = \frac{\pi}{4}, m_Y = \frac{\pi}{4}, D^2(X) = \frac{\pi^2 + 8\pi - 32}{16}, D^2(Y) = \frac{\pi^2 + 8\pi - 32}{16}$$

Most studies concerning usual statistical analyses deal with the relationship between two statistical variables (attributes, in the data mining process) corresponding to the same group of objects. The most popular example concerns the relationship that exists between an individual's height and weight, individual belonging to certain population groups. To identify it, we study the relationship between the two characteristics/attributes, measured on the items from a particular dataset. In other words, it is about two statistical series in which the couples of values (x_i, y_i), corresponding to the couple of (statistical) variables (X, Y), are measured on the same object. There are two main reasons to perform such a study:

- The description of the relationship that might exist between the two variables is done by examining the possible connection between the two series of observations. Specifically, we examine whether the upward trend of one implies an upward trend for the other, or a downward trend, or no clear trend;
- Assuming there is a real connection between them, identified in the first instance, it is necessary to enable the value of one variable to be predicted from any known value of the other variable, based on the equation that establishes the connection between them.

As seen from the above, the ultimate goal of such a study is the process of *forecasting*, on condition that it is possible, the two variables being indeed connected. The method by which we analyze the possible association between the values of two statistical variables, taken from the same group of objects, is known as the *correlation method*, and is based on the correlation coefficient. The correlation coefficient can be calculated for any set of data, but in order to undoubtedly ensure its statistical relevance, two major conditions must be satisfied:

1. The two variables must be defined by the same group of objects, the couples of data corresponding to the same object;
2. At least one of the variables should have an approximately Normal distribution, ideally both variables should be Normally distributed.

If the data are not Normally distributed (at least one of the variables), we proceed either to their transformation in order to be thus normalized, or we consider some non-parametric correlation coefficients (e.g., rank correlation, see (10)).

As mentioned in the beginning, for a couple (X, Y) of random variables, we are interested in how we can identify a possible connection between the components of the couple, through their mean and variance. Thus, from a theoretical point of view (i.e., for random variables), the *covariance* between the random variables X and Y is given by:

$$cov(X,Y) = E[(X - E[X])(Y - E[Y])] = E[XY] - E[X]E[Y],\qquad(3.29)$$

and the ratio:

$$r = r(X,Y) = \frac{cov(X,Y)}{\sqrt{D^2(X)D^2(Y)}},\qquad(3.30)$$

is called *Pearson product-moment correlation coefficient r*, or *Pearson's correlation* (concept introduced by Sir F. Galton and subsequently developed by K. Pearson). As its name suggests, the correlation coefficient gives us a 'measure' of the relationship (correlation) between the two variables. It might be considered as the '*intensity*' (or '*strength*') of the (linear) relationship between X and Y.

In terms of applied Statistics, the above formulas will be written in the following form. Consider, therefore, two statistical series $\{x_i\}, i = 1,...,n$ and $\{y_i\}, i = 1,...,n$, corresponding to the couple (X,Y) of the statistical variables X and Y. Then, the covariance between the two variables is given by:

$$cov(X,Y) = \frac{1}{n}\sum_{i=1}^{n} x_i y_i - \overline{xy}.\qquad(3.31)$$

The correlation coefficient r between the two variables is a real number ranging between -1 and 1, being defined by the formula:

$$r = \frac{\sum_{i=1}^{n}(x_i - \overline{x})(y_i - \overline{y})}{\sqrt{\sum_{i=1}^{n}(x_i - \overline{x})^2 \sum_{i=1}^{n}(y_i - \overline{y})^2}},\qquad(3.32)$$

where its value should be regarded as a measure of the relative elongation of the ellipse formed by the cloud of points in the scattering diagram (see the next subsection). For practical calculations, we may use the above formula, written as:

$$r = \frac{\sum x_i y_i - \frac{1}{n}\sum x_i \sum y_i}{\sqrt{\left[\sum x_i^2 - \frac{1}{n}(\sum x_i)^2\right]\left[\sum y_i^2 - \frac{1}{n}(\sum y_i)^2\right]}}.\qquad(3.33)$$

Starting from the fact that the random variable:

$$z = \frac{1}{2}\ln\left(\frac{1+r}{1-r}\right),\qquad(3.34)$$

is Normally distributed, it results that the 95% confidence interval for the variable z has the form (z_1, z_2), where:

$$z_1 = z - \frac{1.96}{\sqrt{n-3}}, \quad z_2 = z + \frac{1.96}{\sqrt{n-3}}.\qquad(3.35)$$

Hence, applying the inverse transformation, the 95% confidence interval for r is given by:

$$\left(\frac{e^{2z_1} - 1}{e^{2z_1} + 1}, \frac{e^{2z_2} - 1}{e^{2z_2} + 1} \right). \tag{3.36}$$

Let us now interpret the correlation between two statistical variables. As we stated before, the (Pearson) correlation coefficient r takes values between -1 and 1, passing through 0, in which case it indicates a nonlinear association between the two variables (i.e., in other words, they are linearly independent). A value of r close to -1 indicates a strong negative correlation, i.e., the tendency (predisposition) of a variable to decrease significantly while the other variable increases; a value of r close to 1 indicates a strong positive correlation, i.e., a significantly increasing trend of a variable while the other variable increases too. Note that there are cases where dependent variables have the correlation coefficient zero. The problem is the way to establish a threshold for the correlation coefficient, from which we can conclude that the two variables are indeed connected (correlated). In this regard it is suggested either a threshold defined by $|r|\sqrt{n-1} \geq 3$, for instance, from which we could consider the connection between the two variables as sufficiently probable, or to use the well-known (statistical) significance level p, associated with the computation of the coefficient r. Let us note that, in the past, when there were no computers and appropriate statistical software, the above threshold had been used. Now, the significance level p is solely used (for details regarding the level p, see for instance (335)).

Remark 3.3. Despite what we pointed out above, we should not overlook the fact that a significant correlation coefficient does not necessarily involve always a natural connection between the attributes that define the two statistical variables. There are cases in health care practices, for example, when high levels of the correlation coefficient, indicating a significant statistical correlation, have no medical relevance and *vice versa*. For instance, the same low value of the correlation coefficient may be important in epidemiology but clinically insignificant, (10). In conclusion, the correlation coefficient is a measure of the linear relationship, simply from an 'arithmetic' point of view, between two variables, connection which may be sometimes by chance, without real relevance. This fact should be considered especially in data mining, where there is not always well structured prior knowledge about the analyzed phenomenon.

Assuming that the relationship between two variables X and Y, underlined by the correlation coefficient r, is not by chance, there are three possible explanations:

1. Variable X influences ('causes') variable Y;
2. Variable Y influences variable X;
3. Both variables X and Y are influenced by one or more other variables (from background)

Note. When there is no supplementary information about the context in which the two variables operate, especially in data mining studies, it is unrealistic to use Statistics to validate one of the three hypotheses, without an alternative analysis.

The next step in the analysis of the relationship between two statistical variables, when they are correlated, is to concretely establish the nature of their linear relationship, describing it by a mathematical equation. The ultimate goal of this approach is forecasting the values of one variable based on the values of the other variable, forecast made using the equation describing the relationship between the two sets of data. The way to present the linear relationship between two variables, when it really exists, is called the *linear regression* (method). To achieve this, one of the variables is considered as an *independent variable* (or *predictor variable*), and the other variable is considered as *dependent variable* (or *outcome variable*). The (linear) relationship between the two variables is described by a (linear) *regression equation*, which geometrically corresponds to the *regression line*. As methodology, the dependent variable is distributed on the y-axis (axis of ordinates), while the independent variable is distributed on the x-axis (axis of abscissas). The equation of the regression line is determined by the "*least squares method*-LSM", which intuitively minimizes the distance between the points represented by pairs of data (*observed values*) and the corresponding points on the regression line (*fitted values*). The distances to be minimized, called *residuals*, are represented by the vertical distances of the observations from the line (see Fig. 3.4 below).

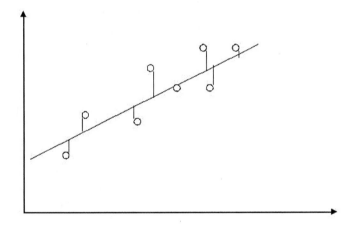

Fig. 3.4 Plot of regression line and residuals

The LSM method produces thus the line that minimizes the sum of the squares of the residuals, also minimizing the variance of the residuals. This variance is called *residual variance* and is used to measure the "*goodness-of-fit*" given by the regression line.

Finally, we obtain the regression equation as:

$$Y = a + b \cdot X, \qquad (3.37)$$

where the constant a is called *intercept* and b *regression coefficient* (or *slope*), the two parameters being obtained from the formulas:

$$b = \frac{\sum_1^n (x_i - \bar{x})(y_i - \bar{y})}{\sum_1^n (x_i - \bar{x})^2}, a = \bar{y} - b\bar{x}. \tag{3.38}$$

From a practical perspective, software for computing the regression line parameters can be easily built using the equivalent formula:

$$\sigma_{XX} = \sum_1^n x_i^2 - \frac{1}{n}\left(\sum_1^n x_i\right)^2, \tag{3.39}$$

$$\sigma_{YY} = \sum_1^n y_i^2 - \frac{1}{n}\left(\sum_1^n y_i\right)^2, \tag{3.40}$$

$$\sigma_{XY} = \sum_1^n x_i y_i - \frac{1}{n}\sum_1^n x_i \sum_1^n y_i, \tag{3.41}$$

from where we obtain:

$$b = \frac{\sigma_{XY}}{\sigma_{XX}}. \tag{3.42}$$

Remark 3.4. When the relationship between the two variables is not linear, but we suspect that there is a certain type of connection between them, we can use a *nonlinear regression* (e.g., polynomial regression). Then, instead of finding the regression line, we find the corresponding regression curve. For more details on this subject, see, for instance, (24), (338), (198).

Note. If we consider pairs of data from two different groups of objects, having the same meaning, we can use the regression lines computed for each group, in order to compare the two groups. If, for example, two regression lines have approximately the same slope b (i.e., they are parallel), then we can consider the difference on the vertical axis (y) as the difference in the means of the Y variable in the two groups, adjusted for any difference in the distribution of the X variable. We end the study by testing the statistical significance of the difference. Such a statistical analysis is part of a wider statistical study which is called *analysis of covariance*.

Example 3.4. Consider data collected from a group of 299 patients with different hepatic diseases (hepatitis C, cirrhosis and hepatic cancer). We are now interested in checking that two clinical parameters, namely cholesterol and gamma-glutamyl-transferase, are correlated and, if so, to find the corresponding regression equation and plot the associated regression line on a graph. Thus, the variables in this regression analysis are cholesterol and gamma-glutamyl-transferase. The correlation coefficient between the two variables is $r = 0.20$, with a statistical significance *level* $p < 0.05$, confirming the fact that they are really correlated. Next, we may consider gamma-glutamyl-transferase as the response (or outcome) variable, and cholesterol as a predictor variable. Fig. 3.5 below shows at the same time the scatterplot of data, the regression line, and the corresponding 95% confidence interval.

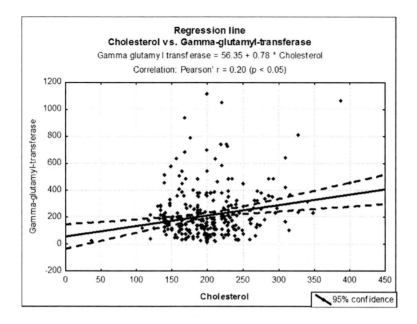

Fig. 3.5 Regression line with the 95% confidence interval

The (least squares) regression equation is given by:

$$Gamma - glutamyl - transferase = 56.35 + 0.78 \times Cholesterol, \qquad (3.43)$$

Fig. 3.6 shows a Normal plot of residuals, which is reasonably straight.

From Figures 3.5 and 3.6, the assumptions of this analysis seem reasonable enough. Thus, the scatter around the regression line is fairly even and symmetric, implying a quite plausible linear relation, while the residuals have a distribution that is not too far from Normal. As it is well-known, the regression line gives us an estimate of average gamma-glutamyl-transferase for a given cholesterol level. Figure 3.5 illustrates both the regression line and the 95% confidence interval for the line. Thus, we can consider this interval as including the true relation between the two variables with 95% probability. As it can be easily seen, the confidence interval is narrowest at the mean gamma-glutamyl-transferase (213.39) and gets wider with increasing distance from the mean.

The consistency of the relationship between the two clinical parameters is indicated by the scatter of the observed data around the fitted line. Thus, the nearer the points are to the line, the narrower the confidence interval will be for that line. To conclude this short analysis, with the present data there is nevertheless considerable scatter, and this is more noticeable if we consider the prediction of gamma-glutamyl-transferase for a new patient with known cholesterol level. However, using the regression equation to predict the gamma-glutamyl-transferase level for a new patient with cholesterol level equaling, for instance, 161, we obtain 181.93 as the

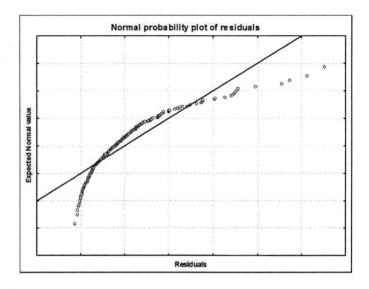

Fig. 3.6 Plot of residuals (Normal probability plot -*Henry's line*)

predicted value for gamma-glutamyl-transferase. In conclusion, a much tighter prediction interval is needed for such a relation to have a consistent clinical value.

Remark 3.5. It is useful to mention here the three assumptions that underlie the linear regression analysis:

1. The values of the dependent variable Y (gamma-glutamyl-transferase in our example) should have a Normal distribution for each value of the independent variable X (cholesterol in our example);
2. The variability of Y should be the same for each value of X;
3. The relation between the two variables should be linear.

If the above assumptions hold, then the residuals should be Normally distributed with mean equaling zero.

Example 3.5. Consider the *Iris* plant dataset, (113), (UCI Machine Learning Repository, http://archive.ics.uci.edu/ml/datasets/Iris). Briefly, it is about a dataset containing 150 objects (*Iris* flowers), equally distributed (class attribute): 50 *Iris* Setosa, 50 *Iris* Versicolour, and 50 *Iris* Virginica. There are four predictive attributes:

1. sepal length (in cm);
2. sepal width (in cm);
3. petal length (in cm);
4. petal width (in cm).

The linear regression performed here refers to establishing the probable relation between petal length and width. The Pearson coefficient r equals 0.96, confirming thus a very high level of correlation between the two main characteristic of a flower, significantly larger than in the first case. The graph illustrating the scatterplot of data, the regression line and the corresponding 95% confidence interval is displayed in Fig. 3.7.

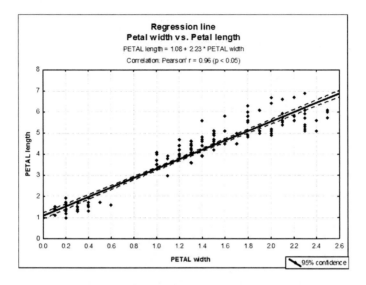

Fig. 3.7 Regression line with 95% confidence interval

As we see from the figure above, in the *Iris* flowers case, the linear relationship between the two characteristics of the flower, namely the petal length and width, is much stronger than in the case of the clinical parameters presented in the previous example. The slope of the regression line is consistent, associated with a high correlation coefficient, near the maximum possible value 1. The scatter of observations

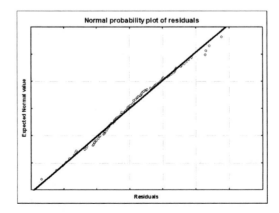

Fig. 3.8 Probability plot of residuals (Normal probability plot -*Henry's line*)

around the regression line is insignificant, demonstrating again a strong connection between the two variables. We also remark a very narrow 95% confidence interval around the line. Thus, we expect a small uncertainty when trying to predict the petal length for a certain flower. Since the assumptions underlying the linear regression analysis can be assessed visually by plotting the distribution of residuals, Fig. 3.8 displayed the Normal probability plot of residuals.

The 'linear' distribution of observations illustrated in Fig. 3.8 is more suggestively pictured in Fig. 3.9 below, showing the histogram of residuals vs. expected Normal distribution.

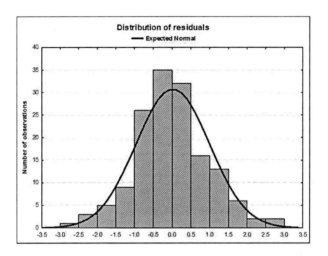

Fig. 3.9 Probability density function of residuals

As it is easy to see, the residuals have a quite Normal distribution, with a mean of zero.

Remark 3.6. The observation of two quantitative variables (attributes) X and Y, which correspond to the same objects from a dataset, leads to the graphical representation of pairs of values (x_i, y_i) relative to object i. Geometrically speaking, the pairs of values (x_i, y_i) represent a *'cloud'* of points in the Cartesian coordinate system xOy. In this graphical representation (see, for instance, the examples above), the choice of proper units for each of the two (connected) variables has its importance, because this choice influences the *'elongation'* of the cloud of points. The difficulty arises when many of the points are very close one to each other and, therefore, the corresponding values are very close. In this case we use to group data, defining p classes for variable X and q classes for variable Y. We then build an array of elements $[a_{jk}]$, where a_{jk} represents the number of couples so that x_j and y_k belong to the classes j and k, obtained from regrouping the initial data of the variables X and Y. Such an array is called *contingency table* (or *correlation table*), being used to define quantitative parameters, as well as for the graphical representation of grouped data, as shown in Fig. 3.10 below. Note that the area of each little circle is proportional to the number of couples corresponding to the groups.

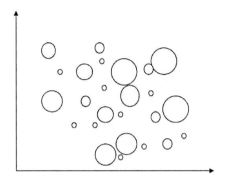

Fig. 3.10 Representation of grouped data

3.2.3 Graphical Representation of a Dataset

The elementary graphical representation of data means the conversion of data into a simple visual or tabular format, so that we can quickly analyze and report both the data characteristics and the relationships between attributes.

Graphical representations related to a dataset (attributes of an object, in our case) depend on the nature of attributes: qualitative or quantitative.

Let us consider first the case of qualitative data. Such data may be plotted using various diagrams, consisting of bi- or tri-dimensional vertical or horizontal bars, circles, ellipses, etc., starting from the partition of the dataset done by the specific attribute. This visualization type depends on the chosen method. Thus, in case of using circles, ellipses, etc., i.e., a circular representation, the whole set of objects is represented by a circle, ellipse, etc.; each attribute of the objects is represented by a circular section whose area is proportional to the number of objects having the corresponding attribute (or corresponding percentage). The other way of representing a qualitative attribute -that with charts with bars- relates to visualization in a system of axes: on the x-axis we represent the attributes and on the y-axis there appear the number of objects with those attributes (or the corresponding percentage). We present below such a visualization manner. Consider, for example, that we deal with a certain population, and we are interested in the eye color of individuals, i.e., a quality attribute. Suppose that the set of eye color, denoted by Eye_color, is given by {black, blue, green, brown}, and from the statistical study resulted that 37% of the population has black eyes, 39% has brown eyes, 8% has blue eyes and 16% has green eyes. We present below (Fig. 3.11 and Fig. 3.12) two types of charts for these data: circular representation (also called '*pie*'), and representation with rectangular bars. In each case we represented (either by circular sectors or by bars/columns) the frequency of occurrence of different types of eye colors in that population.

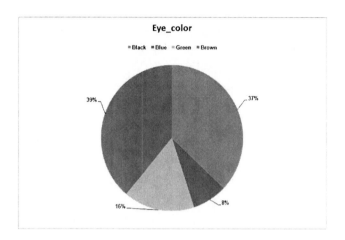

Fig. 3.11 'Pie' chart type representation of Eye_color

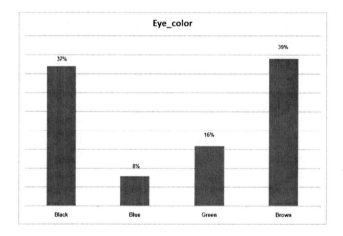

Fig. 3.12 Rectangular bars type representation of Eye_color

Remark 3.7. Let us note in this context that the term "histogram" is sometimes used for these diagrams, although, in principle, the representation by histograms refers to the graphical visualization of tabulated frequencies, with direct reference to the quantitative data (attributes). Mathematically speaking, the *histogram* is the way of displaying the number of observations (data values) belonging to certain intervals (equivalently, frequency of observations belonging to different intervals of values). Visually, by a (bi-dimensional) histogram, for instance, we mean the graphical representation of the distribution of the number of values (frequencies) of a certain numerical attribute, in which each bar (column) represents a particular range of values of the attribute, and its height is proportional to the number (frequency) of values in the respective interval. The term was first used by Pearson -1895 (293).

Now, consider the case of numerical attributes (i.e., quantitative attributes). In the graphical visualization for numerical data case, we meet two modes of handling the representation of data using histograms. Thus, depending on the data type, i.e., discrete data or continuous data, we will choose the corresponding representation mode. For discrete data, the graphical representation is similar to the qualitative data case, although there are conceptual differences highlighted in the remark above. In this case, if we consider diagrams with bars, their length (height) has a precise numerical significance. Specifically, as showed in the histogram below (Fig. 3.13), on the y-axis we represented the values taken by the discrete variable (data), while on the x-axis there lies the relative frequency of occurrence of each value.

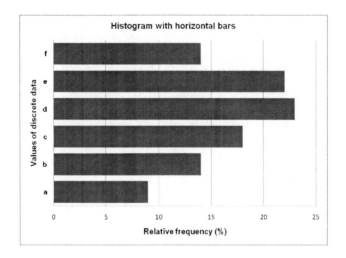

Fig. 3.13 Histogram with horizontal bars

The problem becomes more complicated in the case of continuous numerical data. Here, the division (grouping) of the numerical data in certain classes (usually intervals) is necessary to draw the corresponding histogram. Concretely, to each class (group) represented on one axis, there will correspond the relative frequency of occurrence (or the number of observations) on the other axis. In the histogram below (Fig. 3.14), on the x-axis we represent the classes (i.e., intervals of values I_k) and on the y-axis there lie the corresponding percentage or number of observations.

Let us note that, before drawing histograms, we should calculate all the heights of the bars representing frequencies, in order to identify the optimal scale of values, so that the histogram is really useful from a visual point of view.

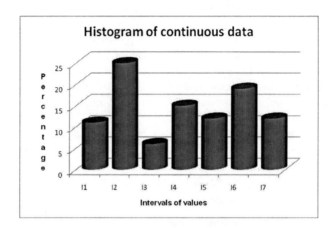

Fig. 3.14 Histogram of continuous data (grouped data)

Example 3.6. Returning to the case of the *Iris* flower and focusing on the petals length, by dividing these continuous numerical data into seven intervals of values, we obtain the following distribution, displayed in Table 3.12 below.

Table 3.12 Distribution of grouped data (*Iris* petal lengths)

Intervals	Count	Frequency (%)
$0 < x \le 1$	1	0.66
$1 < x \le 2$	49	32.68
$2 < x \le 3$	1	0.66
$3 < x \le 4$	15	10.00
$4 < x \le 5$	42	28.00
$5 < x \le 6$	33	22.00
$6 < x \le 7$	9	6.00

In Fig. 3.15 we illustrated by an '*exploded 3D pie*' histogram the continuous variable given by the petal length, grouped as in Table 3.12.

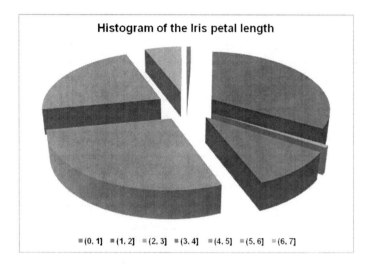

Fig. 3.15 Histogram of continuous data (grouped data)

3.3 Analysis of Correlation Matrix

If for the analysis of a series of couples we have considered couples consisting of two variables (attributes), in the analysis of correlation matrix case we take into account *tuples* of variables $X_1, X_2, ..., X_k, k > 2$. As for couples of variables, we can analyze the correlations between each two pairs of variables, as well as the 'cloud' corresponding to their spread and the subsequent regression lines. The advantage of the latter approach (i.e., the global analysis of the *tuple* $(X_1, X_2, ..., X_k)$) is that the numerical results and graphical representations are presented together (correlation matrix and multiple scattering diagrams, respectively).

We present below a multiple analysis (also known as *multivariate analysis*) in case of data concerning both patients with different liver diseases and *Iris* flower characteristics. In the first case, the attributes taken into account are the following four clinical parameters: age, albumin, cholesterol and glycemia (blood glucose). In the second case we considered the petal and sepal length and width. As mentioned above, the advantage of presenting the relationship between all attributes of objects using the correlation matrix and not the correlations of each pair of attributes is that we have an overview of all connections between the analyzed attributes and not partial connections to be afterwards assembled together. The 'cloud' of scattered points, as well as other common graphical representations (e.g., surfaces, three-dimensional histograms, etc.) are much more suggestively illustrated for collective representation of data than separately (see figures below).

First, Tables 3.13 and 3.14 present the correlation matrices, containing both the Pearson's *r* correlation coefficient and the corresponding significance level p ($p = ns$ means *non-significant*); the significant correlations (i.e., $p < 0.05$) are highlighted.

Table 3.13 Correlation matrix of clinical parameters: Pearson's *r*/level *p* (hepatic diseases)

	Cholesterol	Albumin	Age	Glycemia
Cholesterol	1.00/ns	0.245/**0.00**	-0.010/0.853	-0.091/0.115
Albumin	0.245/**0.00**	1.00/ns	-0.098/0.091	-0.073/0.204
Age	-0.010/0.853	-0.098/0.091	1.00/ns	0.062/0.281
Glycemia	-0.091/0.115	-0.073/0.204	0.062/0.281	1.00/ns

Table 3.14 Correlation matrix of *Iris* flower characteristics: Pearson's *r*/level *p*

	Sepal length	Sepal width	Petal length	Petal width
Sepal length	1.00/ns	-0.117/0.152	0.872/**0.00**	0.818/**0.00**
Sepal width	-0.117/0.152	1.00/ns	-0.428/**0.00**	-0.366/**0.00**
Petal length	0.872/**0.00**	-0.428/**0.00**	1.00/ns	0.963/**0.00**
Petal width	0.818/**0.00**	-0.366/**0.00**	0.963/**0.00**	1.00/ns

Since there are four variables in each *tuple*, they have been plotted in groups of three in the following two figures (Fig. 3.16 and Fig. 3.17).

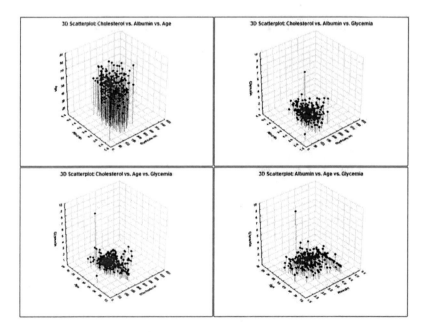

Fig. 3.16 3D scatterplot concerning four clinical parameters

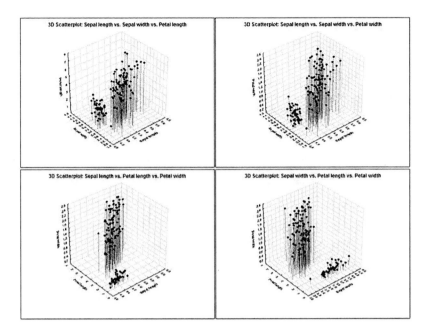

Fig. 3.17 3D scatterplot concerning the *Iris* flower features

Finally, Fig. 3.18 and Fig. 3.19 show the *scatterplot matrices* for the two cases, which can be considered the graphical equivalent of the correlation matrix.

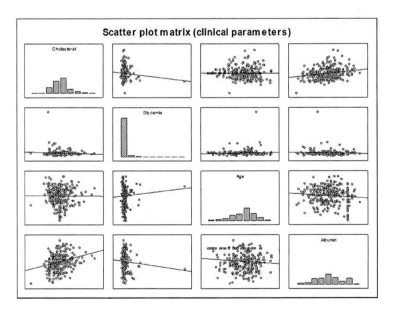

Fig. 3.18 Scatter plot matrix for clinical parameters

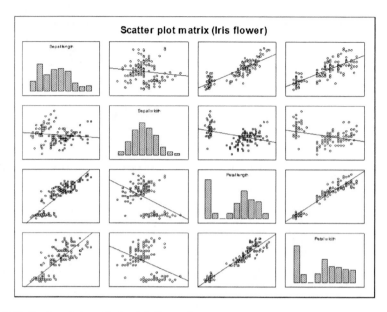

Fig. 3.19 Scatter plot matrix for *Iris* flower features

The correlation matrix is also useful for analyzing the characteristics (attributes) that correspond to objects of different kinds. For example, in the case of medical data regarding different liver diseases (clinical parameters), we may be interested in establishing possible correlations concerning the values of certain important medical parameters, like glycemia and cholesterol, for different types of diseases. Such an analysis could reveal hidden connections that may exist between the same type of attributes, but belonging to different objects. For the above case (i.e., liver diseases), we present such an analysis for cholesterol and glycemia between four classes of diagnosis: liver cirrhosis (LC), liver cancer (HCC), chronic hepatitis C (CH) and healthy persons (HP), seen as control group. Tables 3.15 and 3.16 below show the correlation matrices for cholesterol and glycemia. Note the cholesterol case for healthy people (HP) in study -the same value regardless of the individual- which produces the label "ns" (i.e., non-significant) in the corresponding correlation matrix.

Table 3.15 Correlation matrice for cholesterol -Pearson's r/level p

	LC	HCC	CH	HP
LC	1.00/ns	0.019/0.918	-0.132/0.485	ns/ns
HCC	0.019/0.918	1.00/ns	-0.078/0.679	ns/ns
CH	-0.132/0.485	-0.078/0.679	1.00/ns	ns/ns
HP	ns/ns	ns/ns	ns/ns	1.00/ns

Table 3.16 Correlation matrice for glycemia -Pearson's *r*/level *p*

	LC	HCC	CH	HP
LC	1.00/ns	-0.143/0.450	0.218/0.246	0.145/0.442
HCC	-0.143/0.450	1.00/ns	-0.067/0.724	0.122/0.520
CH	0.218/0.246	-0.067/0.724	1.00/ns	0.128/0.498
HP	0.145/0.442	0.122/0.520	0.128/0.498	1.00/ns

Regarding the graphical visualization of the above correlations (i.e., scatter plot matrix), the following illustration (Fig. 3.20) is very useful, suggestively summarizing, using the regression lines, all numerical information presented so 'unfriendly' in the table.

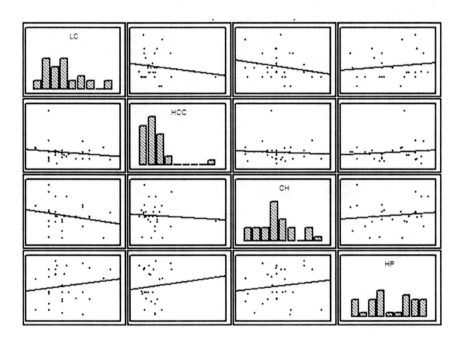

Fig. 3.20 Scatter plot matrix for different liver diseases

3.4 Data Visualization

Data visualization is one of the most powerful and appealing techniques for data exploration, one of the fundamental pillars of the EDA. Unlike the elementary graphical representation that we talked about in the previous subsection, which is practically attached to the statistical description of data, the visualization techniques used in EDA appeal to more sophisticated ways of processing the data. Data visualization is the first contact made with the intimate nature of the information that we try to decipher in the available dataset. Visualization appeals to the power of synthetic grasp and the human capability to decipher and interpret the information hidden in images rather than in 'dry' numbers. Below (Fig. 3.21) we present a suggestive image of the water temperature close to the surface (SST - Sea Surfaces Temperatures), image that shows "at a glance" a lot more than if it had been previously 'digitized' and then tabulated.

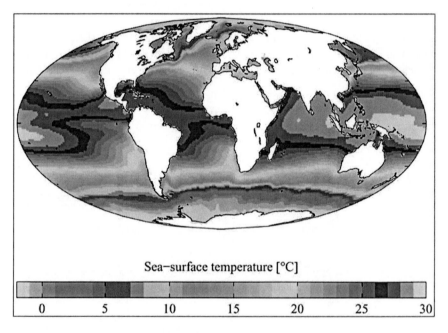

Fig. 3.21 Annual mean sea surface temperature for the World Ocean (data from the World Ocean Atlas 2005, http://en.wikipedia.org/wiki/Sea_surface_temperature)

Another simple visualization technique is represented by the 'rearrangement' of data in an appropriate and intuitive form, being thus much more understandable at the first sight. This rearrangement of data should reveal connections between objects, otherwise hidden or fuzzy. For example, let us consider a set of objects which are analyzed on the basis of a certain criterion that provides the value 0 if the objects do not verify a certain condition and the value 1 if they do. The values provided by the criterion for each pair of objects are represented, as seen below, in a numerical format in two tables 2×2. The objects in the original arrangement are displayed in the left table and the objects reorganized considering the criterion are displayed in the right table. It is easy to see that the rearrangement of data in the right table provides valuable information on different subgroups of objects (clustering the data), information otherwise hidden in the original format (left table) -see Fig. 3.22.

	a	b	c	d	e	f
a	0	1	0	0	0	1
b	1	1	1	1	1	1
c	0	1	0	0	0	1
d	0	1	0	0	0	0
e	1	1	1	1	1	1
f	0	1	1	1	1	0

	f	b	c	e	d	a
a	1	1	0	0	0	0
c	1	1	0	0	0	0
d	0	1	0	0	0	0
f	0	1	1	1	1	0
e	1	1	1	1	1	1
b	1	1	1	1	1	1

Fig. 3.22 Rearrangement of data

Starting from the medical data on liver diseases mentioned above, and focusing on three main clinical parameters, namely cholesterol, glycemia and age, we present below some versions of graphical representations, more often used in data visualization. These graphical representations are common to all software packages in the statistical field, and can be utilized very easily and in a wide range of different ways by users without deep knowledge in computer applications. On the other hand, people with appropriate knowledge of computer graphics could easily design their own software dedicated to this issue, based on the classical statistical formulas.

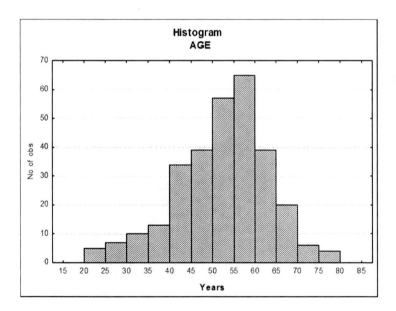

Fig. 3.23 Histogram plot (probability density function)

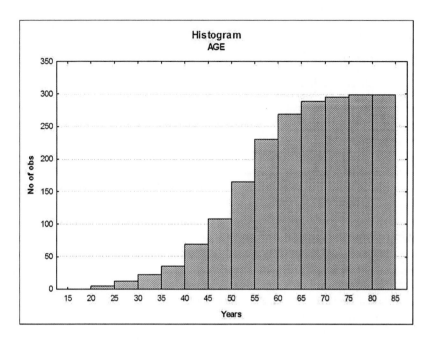

Fig. 3.24 Histogram plot (distribution function)

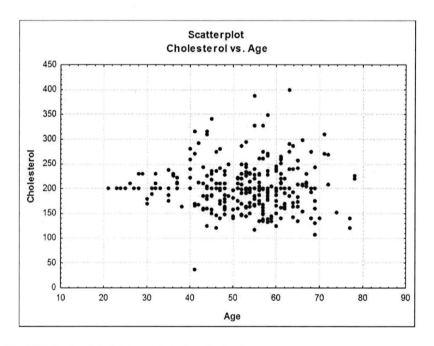

Fig. 3.25 Scatterplot of data points (pairs of points)

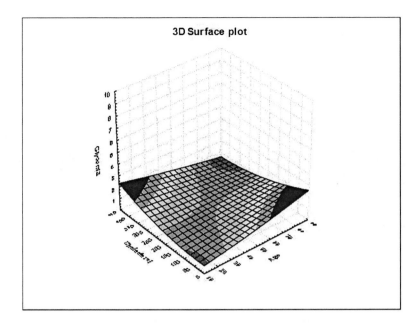

Fig. 3.26 3D surface plot (3-*tuple*)

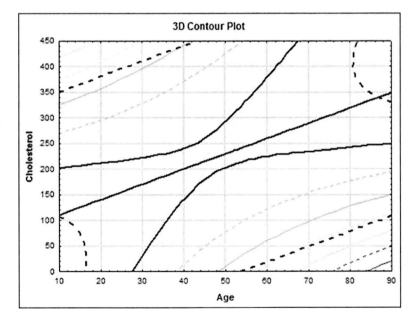

Fig. 3.27 3D contour plot (pairs of variables)

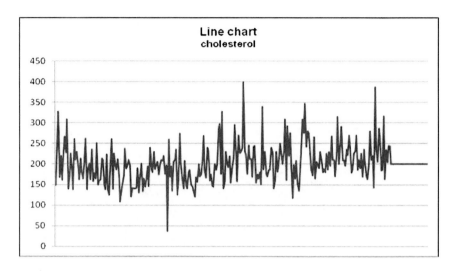

Fig. 3.28 Line chart

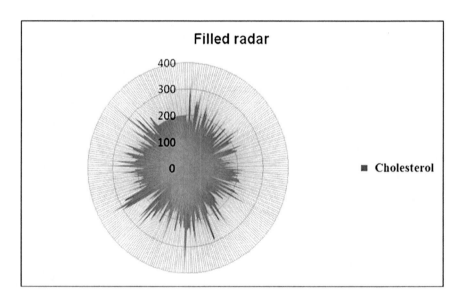

Fig. 3.29 Filled radar plot

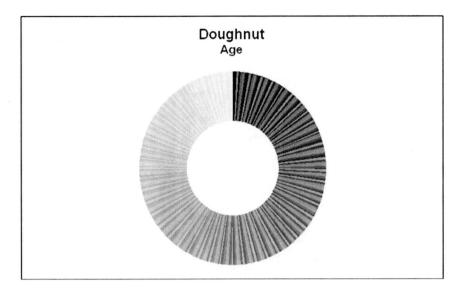

Fig. 3.30 Doughnut plot

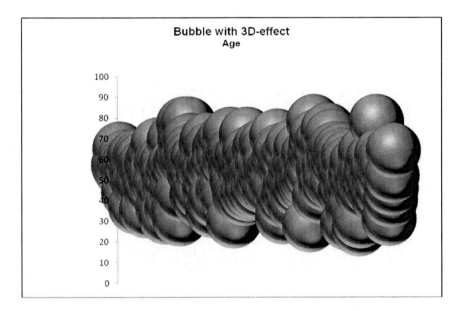

Fig. 3.31 Bubble with 3D-effect

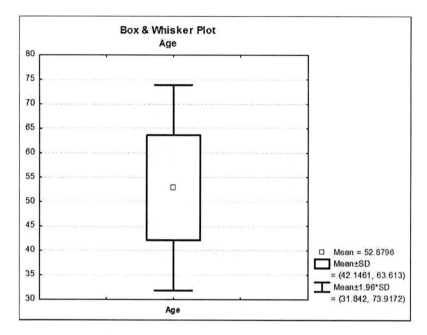

Fig. 3.32 Box & Whisker plot

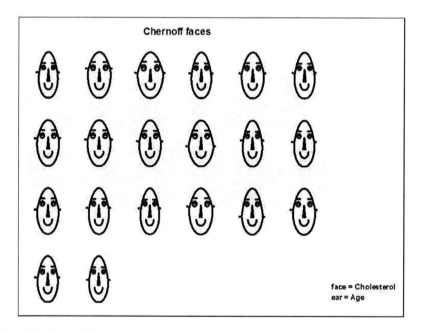

Fig. 3.33 Chernoff faces (pairs of variables)

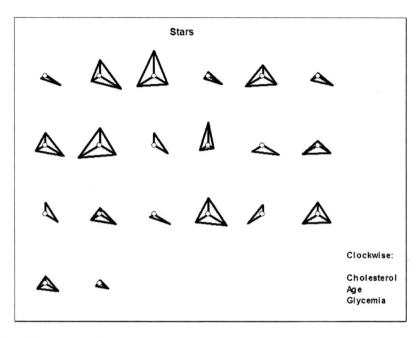

Fig. 3.34 Stars plot (3-*tuple*)

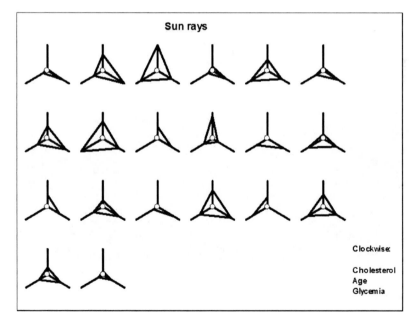

Fig. 3.35 Sun rays plot (3-*tuple*)

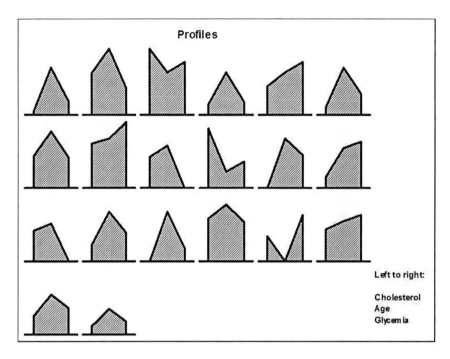

Fig. 3.36 Profiles plot (*3-tuple*)

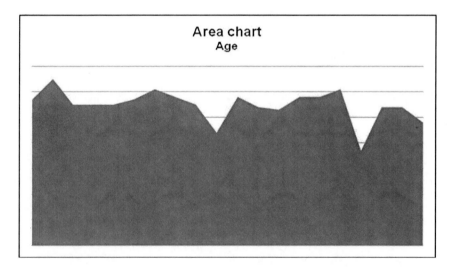

Fig. 3.37 Area chart

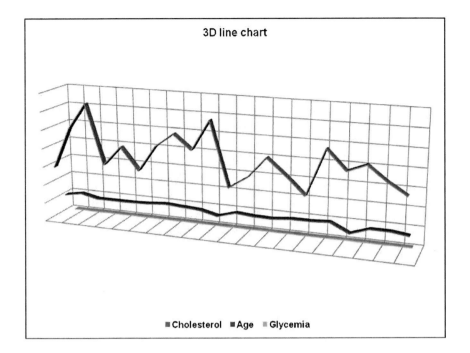

Fig. 3.38 3D line chart (*3-tuple*)

3.5 Examination of Distributions

A distinct approach in data analysis relates to typical measures of the distribution shape - asymmetry (*skewness*) and excess (*kurtosis*) - important to the analysis of the non-Gaussian distributions and multi-modal distributions. Thus, if the data distribution is not symmetric, we must raise the question of deviations from symmetry. In this context, we define the skewness -term first used by Pearson, 1895- as being the measure of the deviation of a given distribution from symmetry. The formula for skewness is given by:

$$Skewness = \frac{n \cdot \sum_1^n (x_i - \bar{x})^3}{(n-1)(n-2)\sigma^3} \tag{3.44}$$

If skewness is different from zero, then the distribution is asymmetrical. By contrast, the Normal (Gaussian) distribution is perfectly symmetrical, representing the generic pattern of symmetry. Fig. 3.39 below shows the "*Gauss (bell) curve*", i.e., the 'most' symmetrical distribution.

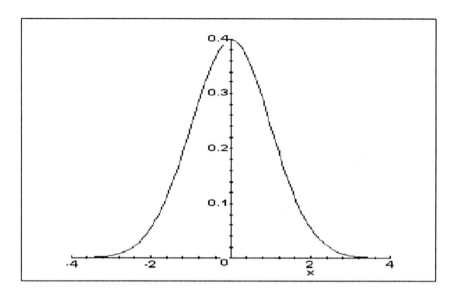

Fig. 3.39 Gauss (bell) curve -standardized Normal distribution $N(0, 1)$

If, by contrast, the distribution is asymmetrical, then it will have a "tail" either to the left or to the right. If the "tail" is to the right, we say that the distribution has a *positive* asymmetry, while in the other case, when the "tail" is to the left, the asymmetry is *negative*. Fig. 3.40 and Fig. 3.41 illustrate these two types of (non-zero) skewness.

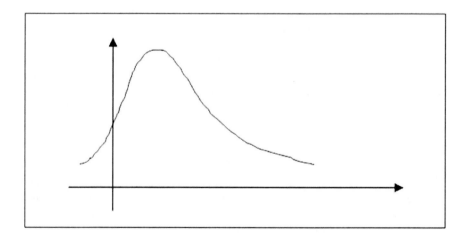

Fig. 3.40 Positive skewness

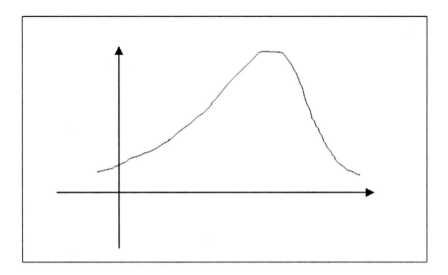

Fig. 3.41 Negative skewness

As regards the other typical measure of the distribution shape, the *kurtosis* (term introduced by Pearson, 1905, (294)), measures the 'peakedness' of a distribution. If the kurtosis is non-zero, then the distribution is either more 'flat', or more 'spiky' than the Normal distribution, which is even zero. The kurtosis is computed using the following formula:

$$Kurtosis = \frac{n \cdot (n+1) \cdot \sum_1^n (x_i - \bar{x})^4 - 3 \cdot (n-1) \left[\sum_1^n (x_i - \bar{x})^2 \right]^2}{(n-1)(n-2)(n-3)\sigma^4} \qquad (3.45)$$

We illustrate in Table 3.17 the two typical measures of the distribution shape for the statistical series concerning the <height>, presented in subsection 3.2.1.

Table 3.17 Skewness and kurtosis

Skewness	Kurtosis
1.73	4.69

Besides the Normal (Gaussian) distribution, there are countless other continuous distributions, more or less far-away from it (see also the skewness feature, described above), corresponding to different types of data. All these distributions are non-Normal (non-Gaussian), characterizing data frequently met in reality.

We present below four such distributions, with broad use in Statistics and data mining.

(a) **Exponential distribution - EXP** (λ), $\lambda > 0$, with the probability density function given by:

$$f_X(x) = \begin{cases} 0 & ,x < 0 \\ \lambda e^{-\lambda x} & ,x \geq 0 \end{cases} \tag{3.46}$$

and graphically represented in Fig. 3.42 below.

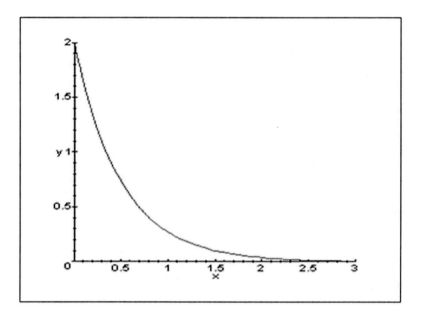

Fig. 3.42 Exponential distribution

(b) **Gamma distribution - GAM** (λ, k), $\lambda > 0$, $k > 0$, with the probability density function given by:

$$f(x) = \begin{cases} \frac{\lambda^k x^{k-1} e^{-\lambda x}}{\Gamma(k)} & ,x \geq 0 \\ 0 & ,x < 0 \end{cases} \tag{3.47}$$

where $\Gamma(k) = \int_0^\infty e^{-x} x^{k-1} dx$ is a *gamma* function of order k. Its density, for different parameters λ and k, is displayed below (Fig. 3.43).

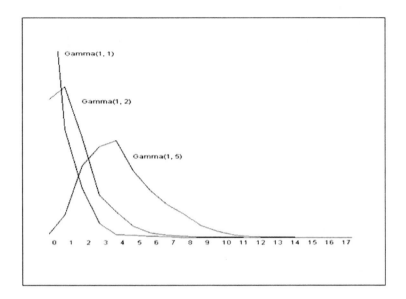

Fig. 3.43 Gamma distribution

(c) **Weibull distribution** - $WEI(\alpha, \beta)$, $\alpha > 0$, $\beta > 0$, with the probability density function given by:

$$f(x) = \begin{cases} \alpha\beta(\alpha x)^{\beta-1}e^{-(\alpha x)^{\beta}} & ,x \geq 0 \\ 0 & ,x < 0 \end{cases} \tag{3.48}$$

and with the graphical representation, for different parameters α and β, showed below (Fig. 3.44).

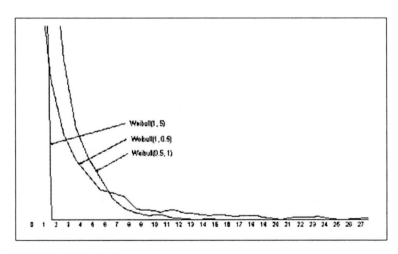

Fig. 3.44 Weibull distribution

(d) **Log-normal distribution** - *LOGN*(μ, σ^2). If log X is Normally distributed, then the random variable X is *log-normal* distributed with the corresponding probability density function given by:

$$f_X(x) = \frac{1}{\sqrt{2\pi}\sigma x} \exp\left[-\frac{(\ln x - \mu)^2}{2\sigma^2}\right], \, x > 0, \tag{3.49}$$

with the graphical representation displayed below (Fig. 3.45).

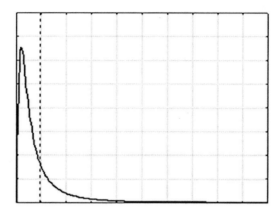

Fig. 3.45 Log-normal distribution

With regard to the multi-modality (i.e., more than one mode) characteristic of data distributions, the visualization of the corresponding histogram emphasizes the existence of certain values with similar occurrence frequency (maximum frequency), as we saw when we analyzed the numerical characteristics of data.
Fig. 3.46 below illustrates a bi-modal distribution.

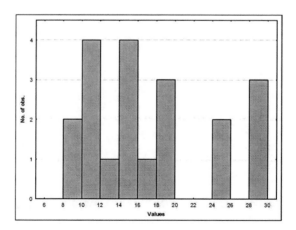

Fig. 3.46 Bi-modal distribution

3.6 Advanced Linear and Additive Models

3.6.1 Multiple Linear Regression

Unlike in the case of the simple linear regression, where we tried to express one (outcome) variable depending on another (explanatory, predictive) variable, now we consider the situation in which we are dealing with at least three variables, one of which is dependent and the others are (independent) predictors. The statistical technique handling such a situation, called *multiple (linear) regression* (the term was first used by Pearson, 1908, (295)), yields a regression model in which the dependent variable is expressed as a (linear) combination of the explanatory variables (or *predictor variables*, or *covariates*). We will present below in brief such a multiple linear regression model. As with analysis of variances and linear regression, the residual variance provides a measure of how well the model fits the data. Mathematically speaking, the way the dependent variable is expressed as a linear combination of the explanatory variables is simply the *multiple regression equation*, given by:

$$Y = a + b_1X_1 + b_2X_2 + + b_kX_k, \tag{3.50}$$

where Y is the dependent (*outcome*) variable, and the variables X_1, ..., X_k are the explanatory (*predictive*) variables. The constants b_1, ..., b_k are the *regression coefficients*, and a is the regression constant or *intercept*.

Among the situations taken into account when applying the multiple linear regression, we mention the following three, considered as more important:

- The necessity to remove the possible effects of other insignificant variables, when studying the relationship of a group of variables;
- Exploration of possible prognostic variables without prior information of which variables are important;
- Development of a certain prognostic index from several explanatory variables for predicting the dependent variable.

Multiple regression is a direct method to be used when we have a clear *a priori* picture of the variables we wish to include in the model. In this case we can proceed with the standard multiple regression method, which represents, basically, a generalization of the simple linear regression. We have to choose a completely different approach when there is no prior knowledge about the real situation we wish to model, and when we face many virtual predictive variables, from which we need to select only those who are really essential to the model. In this case an exploratory analysis is necessary to confirm the chosen model, based on testing its statistical significance.

Remark 3.8. If we include a binary variable in the regression model, taking values 0 and 1 (e.g., indicating situations such as: single/married, sick/healthy, fraud/fair transactions, smoker/non-smoker, etc.) the regression coefficients indicate the average difference in the dependent variable between the groups defined by the binary variable, adjusted for any differences between the groups with respect to the other variables of the model. The explanation is simple: the difference between the codes for the groups is 1, since the values are 0 and 1. In case of dealing with categorical variables having more than two values, we can handle the problem by grouping the variables and, thus, we create new binary variables (e.g., starting from the variable for marital status with three outcomes: 1-married, 2-single and 3-divorced, we can consider two new binary variables: {1-single, 0-otherwise} and {1-divorced, 0-otherwise}). Another problem arises when we deal with ordered categories (e.g., stages of a disease). In this case, they can sometimes be treated as ordered discrete variables, indicating a certain trend. More details concerning this subject are to be found in (10).

As we pointed out above, when we know in advance which variables will be included in the multiple regression analysis, the model can be easily built, the only remaining question concerns the concrete estimation of the regression equation. If the goal of the study also includes the ordering of predictors depending on their importance, then we have to choose from all available variables the essential predictors in order to obtain a clear and simple model. In this case we have to consider the statistical significance level p for each variable in the regression analysis to decide the hierarchy of their importance.

Turning now to the effective building of the multiple linear regression model, in the case of not knowing in advance which variables should be included in the predictive model, we briefly describe the two main algorithms that are commonly used:

- *Forward stepwise regression*;
- *Backward stepwise regression*.

Generally speaking, the *stepwise regression* refers to a model-building technique that finds subsets of predictor variables that most adequately predict responses on a dependent variable by linear (or nonlinear) regression, given the specified criteria for adequacy of model fit (e.g., the significance p-level). The basic procedures involve:

- Identification of an initial model;
- Iteratively 'stepping', i.e., repeatedly changing the model at the previous step by adding or removing a predictor variable in accordance to the 'stepping criteria';
- Terminating the search when stepping is no longer possible given the stepping criteria, or when a specified maximum number of steps has been reached.

More details about this well-known statistical technique are to be found, for instance, in (326), (74), (238).

Now, we briefly present the two stepwise regression techniques.

Basically, in the forward stepwise regression the variables are moved into the model (equation) in successive steps. At each step, the variable with the smallest p-level will be chosen for inclusion in the model. The stepping will terminate when no other variable has a p-level value that is smaller than a pre-specified value.

Next, we present in short the steps of the forward stepwise regression algorithm.

- Forward stepwise regression algorithm:

1. Identify the single explanatory variable that has the strongest association with the dependent variable and enter it into the model;
2. Identify the variable among those not included in the model yet that, when added to the model so far obtained, explains the largest amount of the remaining variability;
3. Repeat step 2) until the addition of an extra variable is not statistically significant at some level *p*.

Remark 3.9. 1. The variable with the strongest association is that with the most significant slope, in other words, with the smallest *p*-level. Concretely, this is the variable that is most correlated with the dependent variable.
2. The second step refers to finding the variable that is so far most correlated with the residuals from the model.
3. Usually, the significance level *p* is chosen as equaling the standard cut-off value 0.05.

- Backward stepwise regression algorithm.

In the backward stepwise regression all variables are moved into the model (equation), and then variables are removed one by one.

At each step, the variable with the largest *p*-level will be removed from the model.

The stepping will terminate when no other variable in the model has a *p*-level larger than a pre-specified value.

As we saw above and its name also implies, with the backward stepwise regression method we approach the problem from the opposite direction. Thus, it is easy to design the corresponding algorithm, by replacing the term "inclusion" with "removing", and "the smallest *p*-level" with "the largest *p*-level".

Example 3.7. We will use the multiple linear regression to predict the index of respiratory muscle strength, expressed by the maximal static expiratory pressure (PEmax -in cm H_2O) using data from a study of 25 patients with cystic fibrosis, (286), (10). The explanatory variables chosen for this statistical study are the following: Age, Sex (categorical variable, coded: 0 = male, 1 = female), Height, Weight, Body Mass Percentage (BMP -%), Forced Expiratory Volume in 1 second (FEV_1), Residual Volume (RV), Functional Residual Capacity (FRC) and Total Lung Capacity (TLC). The dependent variable of the model is, as we pointed above, the index PEmax. Table 3.18 summarizes these data.

Table 3.18 Medical data from 25 patients with cystic fibrosis

Age	Sex	Height	Weight	BMP	FEV_1	RV	FRC	TLC	PEmax
7	0	109	13.1	68	32	258	183	137	95
7	1	112	12.9	65	19	449	245	134	85
8	0	124	14.1	64	22	441	268	147	100
8	1	125	16.2	67	41	234	146	124	85
8	0	127	21.5	93	52	202	131	104	95
9	0	130	17.5	68	44	308	155	118	80
11	1	139	30.7	89	28	305	179	119	65
12	1	150	28.4	69	18	369	198	103	110
12	0	146	25.1	67	24	312	194	128	70
13	1	155	31.5	68	23	413	225	136	95
13	0	156	39.9	89	39	206	142	95	110
14	1	153	42.1	90	26	253	191	121	90
14	0	160	45.6	93	45	174	139	108	100
15	1	158	51.2	93	45	158	124	90	80
16	1	160	35.9	66	31	302	133	101	134
17	1	153	34.8	70	39	204	118	120	134
17	0	174	44.7	70	49	187	104	103	165
17	1	176	60.1	92	29	188	129	130	120
17	0	171	42.6	69	38	172	130	103	130
19	1	156	37.2	72	21	216	119	81	85
19	0	174	54.6	86	37	184	118	101	85
20	0	178	64.0	86	34	225	148	135	160
23	0	180	73.8	97	57	171	108	98	165
23	0	175	51.5	71	33	224	131	113	95
23	0	179	71.5	95	52	225	127	101	195

Table 3.19 presents the correlation matrix of all variables included in the regression model (i.e., the matrix of the correlation coefficients). The multiple correlation analysis is necessary to establish the existence of the relationship between the

variables included in this statistical study. Basically, we are interested in examining the direct relationships between each explanatory variable and the dependent variable of this model, namely PEmax.

Table 3.19 Correlation matrix for PEmax and nine predictors

	Age	Sex	Height	Weight	BMP	FEV_1	RV	FRC	TLC	PEmax
Age	1	-0.17	0.93	0.91	0.38	0.29	-0.55	-0.64	-0.47	0.61
Sex	-0.17	1	-0.17	-0.19	-0.14	-0.53	0.27	0.18	0.02	-0.29
Height	0.93	-0.17	1	0.92	0.44	0.32	-0.57	-0.62	-0.46	0.60
Weight	0.91	-0.19	0.92	1	0.67	0.45	-0.62	-0.62	-0.42	0.64
BMP	0.38	-0.14	0.44	0.67	1	0.55	-0.58	-0.43	-0.36	0.23
FEV_1	0.29	-0.53	0.32	0.45	0.55	1	-0.67	-0.67	-0.44	0.45
RV	-0.55	0.27	-0.57	-0.62	-0.58	-0.67	1	0.91	0.59	-0.32
FRC	-0.64	0.18	-0.62	-0.62	-0.43	-0.67	0.91	1	0.7	-0.42
TLC	-0.47	0.02	-0.46	-0.42	-0.36	-0.44	0.59	0.7	1	-0.18
PEmax	0.61	-0.29	0.60	0.64	0.23	0.45	-0.32	-0.42	-0.18	1

We begin by considering the first variant of the multiple regression methodology, namely the forward stepwise regression. In this case, the first eligible positions, in descending order, for the predictor variables, established by measuring the correlations between each explanatory variable and the dependent variable, are given in Table 3.20.

Table 3.20 Forward stepwise regression

	Coefficient b	Standard error of b	t-test (20)	p-level
Intercept	44.276	2.604		
Weight	1.7761	0.363	4.844	0.0001
BMP	1.770	0.563	3.142	0.0051
FEV_1	-1.336	0.536	-2.503	0.0211

Concerning the application of the other method -backward stepwise regression- the characteristics of the extended model (i.e., with three predictors) are presented in Table 3.21.

Table 3.21 Backward stepwise regression -extended variant

	Coefficient b	Standard error of b	t-test (20)	p-level
Intercept	126.334	34.720		
Weight	1.536	0.364	4.220	0.0004
BMP	-1.465	0.579	2.530	0.019
FEV_1	1.109	0.514	2.160	0.043

We can consider an even 'stricter' model, based on weight only (see Table 3.22), which represents the simple regression case, connecting the dependent variable PE-max with the predictor variable Weight. If we look at the correlation matrix, we see that, indeed, the highest correlation (0.64) is between PEmax and Weight, confirming thus this result and the results above concerning the importance of Weight in this regression analysis.

Table 3.22 Backward stepwise regression -restricted variant

	Coefficient b	Standard error of b	t-test (20)	p-level
Intercept	63.616	12.710		
Weight	1.184	0.301	3.936	0.0006

Remark 3.10. In principle, none of the two variants of the multiple regression model is ideal for sure. Usually, if we want the 'widest' model, we choose the 'forward' model, and if we want the 'narrowest' model, we choose the 'backward' model. However, we notice that the approach based on the significance level p only does not completely resolve the problem.

The regression equations for the three cases are:

$$PEmax = 44.27 + 1.77 \times Weight + 1.77 \times BMP - 1.33 \times FEV_1, \qquad (3.51)$$

$$PEmax = 126.33 + 1.53 \times Weight - 1.46 \times BMP + 1.11 \times FEV_1, \qquad (3.52)$$

$$PEmax = 63.61 + 1.18 \times Weight. \qquad (3.53)$$

We can assess the "*overall goodness-of-fit*" (ANOVA) of the above models (i.e., how well the models 'fit' the data), by considering the corresponding analysis of variances, displayed in the tables below (Tables 3.23, 3.24 and 3.25).

Table 3.23 Analysis of variances (forward stepwise regression)

Effect	Sums of squares	df	Mean squares	F-value	p-level
Regression	17373.18	4	4343.294	9.183	0.0002
Residual	9459.46	20	472.973		
Total	26832.64				

Table 3.24 Analysis of variances (backward stepwise regression -extended)

Effect	Sums of squares	df	Mean squares	F-value	p-level
Regression	15294.46	3	5089.15	9.280	0.0004
Residual	11538.18	21	549.44		
Total	26832.64				

Table 3.25 Analysis of variances (backward stepwise regression -restricted)

Effect	Sums of squares	df	Mean squares	F-value	p-level
Regression	10799.28	1	10799.28	15.492	0.0006
Residual	16033.36	23	697.10		
Total	26832.64				

Technically, we assess the "goodness-of-fit" of the model or, in other words, how well the model predicts the dependent variable, by considering the proportion of the total sum of squares that can be explained by the regression. Thus, for the forward stepwise regression, the sum of squares due to the model is 17373.18, so that the proportion of the variation explained is $17373.18/26832.64 = 0.6474$. This statistic is called R^2 and is often expressed as a percentage, here 64.74%. In the second case, the backward stepwise regression (extended), the R^2 statistics equals 56.99%, while in the third case, the backward stepwise regression (restricted), $R^2 = 40.25\%$.

Note. The smaller the variability of the residual values around the regression line relative to the overall variability, the better is the model prediction. Concretely, in the case of the forward stepwise regression, $R^2 = 64.74\%$ shows that we have explained about 65% of the original variability, and 35% are left to the residual variability. In the second case, $R^2 = 57\%$ implies that we have explained about 57% of the original variability, and 43% are left to the residual variability, while in the last case, $R^2 = 40\%$ that 40% of the original variability has been explained, the rest of 60% are left to the residual variability. In our case, the backward stepwise regression (extended) will be kept in the subsequent regression analysis. Ideally, we would like to explain the most if not all of the original variability. To conclude, the R^2 value is an indicator of how well the model fits the data (more details in (10)).

On the other hand, we can assess the "goodness-of-fit" of the model by analyzing the distribution of residuals. Multiple regression assumes that the residual values (i.e., observed minus predicted values) are Normally distributed, and that the regression function (i.e., the relationship between the independent and dependent variables) is linear in nature. Thus, a good prediction accuracy implies a Normal (Gaussian) distribution of residuals, i.e., placing them along a (straight) line (*Henry's line*). Recall that the Henry's line (*la droite de Henry*) is a graphical method to fit a normal (Gaussian) distribution to that of a series of observations (a continuous numerical variable). Normal probability plot (*Henry's line*) provides a quick way to visually inspect to what extent the pattern of residuals follows a Normal distribution. A deviation from this (linear) arrangement of residuals also indicates the existence of outliers. We present below (Fig. 3.47 and Fig. 3.48) the graphs of the *Henry's lines* for the two regression models (forward stepwise regression and (extended) backward stepwise regression).

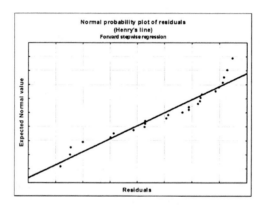

Fig. 3.47 Normal probability plot of residuals (forward stepwise regression)

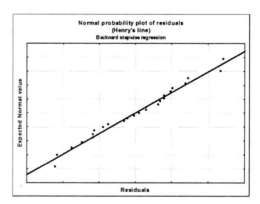

Fig. 3.48 Normal probability plot of residuals (backward stepwise regression)

As it is easy to see from the figures above, both graphs indicate satisfactory validation of the two regressive models.

Next, the graphical representation of the observed values against the expected values is particularly useful for identifying potential clusters of cases that are not well forecasted. From the graphical representations below we also deduce a good forecast of the dependent variable, regardless of the regression model.

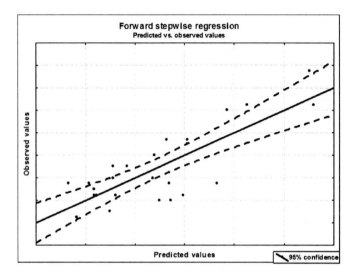

Fig. 3.49 Predicted vs. observed values (forward stepwise regression)

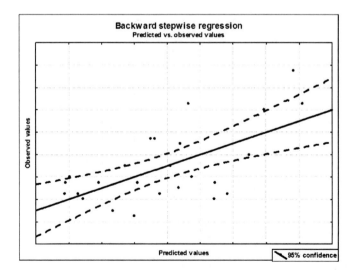

Fig. 3.50 Predicted vs. observed values (backward stepwise regression)

Finally, we proceed with the standard multiple regression method, in which case we will consider all the nine explanatory variables. The result of applying this method is shown in Table 3.26.

Table 3.26 Standard multiple linear regression

	Coefficient b	Standard error of b	t-test (15)	p-level
Intercept	137.229	207.946	0.659	0.5193
Age	-2.475	4.367	-0.566	0.5792
Sex	-1.388	13.597	-0.102	0.9200
Height	-0.308	0.864	-0.356	0.7262
Weight	2.878	1.846	1.559	0.1398
BMP	-1.797	1.115	-1.610	0.1280
Fev_1	1.494	0.970	1.539	0.1445
RV	0.178	0.186	0.953	0.3552
FRC	-0.163	0.475	-0.344	0.7353
TLC	0.114	0.477	0.239	0.8140

One can easily observe that in this case none of the predictors passes the threshold of 5% for the significance p-level.

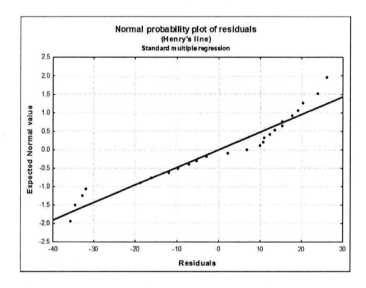

Fig. 3.51 Normal probability plot of residuals (standard multiple regression)

However, given that the p-level is not an absolute criterion for acceptance of the model, we take into consideration the forecast thus obtained, especially since the graphical representation of residuals (illustrated in Fig. 3.51) shows an acceptable degree of accuracy of the model.

Moreover, the proportion of the variation explained by the statistics R^2 equals 66.27%, so 66% of the original variability has been explained, the rest of 34% are left to the residual variability.

Finally, we illustrated in Fig. 3.52 the predicted values vs. the expected values in order to reconfirm the acceptability of the standard multiple regression model, which includes all predictors.

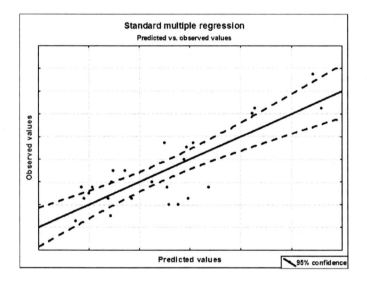

Fig. 3.52 Predicted vs. observed values (standard multiple regression)

The corresponding regression equation for this case is given by:

$$PEmax = 137.23 - 1.79 \times BMP + 2.87 \times Weight + 1.49 \times FEV_1 + 0.17 \times RV -$$

$$- 2.47 \times Age - 0.3 \times Height - 0.16 \times FRC + 0.11 \times TLC - 1.38 \times Sex \quad (3.54)$$

Remark 3.11. We use the multiple regression equation to obtain the values of the dependent variable for any individual values of the explanatory variables. In this way, for a certain object with known predictive attributes, the value of the unknown attribute (considered as an outcome attribute) is thus inferred. In the above case, for a patient with cystic fibrosis, whose predictive values of the nine medical (clinical) parameters are known, one can predict with sufficient accuracy the PEmax value, by entering in the regression equation its individual predictive values. In this case, we say that one can obtain a *prognostic index*, based on known data.

At the end of this short presentation concerning the multiple (linear) regression, we present some standard considerations regarding the regression analysis:

- If we have to choose between several predictive variables, without having any prior knowledge, we must know that there is no definitive rule concerning their choice. In principle, one uses as a threshold value the p-level equaling 0.05, but there are cases when, due to the role gained in practice by a certain predictor, we can consider for it the value of p equaling 0.2 or even greater.
- When the dataset size is small, and we have a large number of variables being considered, we face the problem of choosing a fair number of predictors for the regression model. One suggests avoiding the application of a regression model to small datasets. Alternatively, if we choose to apply regression, it is suggested to decide in advance the maximum size of model that is feasible. Unfortunately, there is no clear rule, but a guideline might be to choose no more than $n/10$ variables, where n is the dataset size.
- An automatic procedure for selecting a model, based on appropriate statistic software, is normal and useful, but we must not ignore the practitioners' common sense concerning the final assessment and validation of the model.
- The strongly correlated explanatory variables will be selected in such a way to avoid their simultaneous inclusion in the model (e.g., weight and height were not included together in the above model, since it is well-known that they are very highly correlated, $r = 0.92$, in Table 3.19), in order to avoid redundancy.
- We have to *a priori* check whether there is indeed a linear relation between the dependent variable and each predictor, since such a relationship is assumed in the regression analysis.
- It is assumed that the effects of each explanatory variable are independent. If we suspect the existence of a relationship between two predictors (this is not determined solely based on the correlation, but on intrinsic knowledge pertaining to the intimate nature of the investigated phenomenon), we need to add an interaction term to the model (i.e., a new variable which is a certain combination of them, e.g., their product).
- For more safety, we need to consider the possibility of checking the capacity of the model on another dataset, if possible.

3.6.2 Logistic Regression

In the preceding section we have presented some classical notions concerning the multiple (linear) regression, showing how to obtain the linear equation that establishes the relationship between several random variables, when the dependent variable is a continuous variable, thus extending the method of linear regression to multiple variables. There are many research areas, however, including health care, economics, physics, meteorology, astronomy, biology, etc., in which the dependent variable is no longer a continuous variable, but a binary (categorical) one. We can mention, as examples, the response to the treatment of a patient, the presence or absence of a myocardial infarction, the categories of patients presenting a particular symptom, the customer loyalty to a supermarket, the classification of stars, etc. In this case, when the dependent variable refers to two values (categories), we cannot use standard multiple regression for such data, but instead we can use an approach somewhat similar, but distinct, known as *multiple linear logistic regression*. Thus, instead of predicting the values of the outcome variable depending on the explanatory variables values, we will predict a transformation of the dependent variable. This transformation is called the *logit* transformation, written *logit* (p), where p is the proportion of objects with a certain characteristic (e.g., p is the probability of an individual having myocardial infarction, or p is the probability for a customer to remain loyal to a supermarket or product, etc.). To understand the rationale of this procedure, recall that, if we quantify the dependent categorical variable using the values 1 and 0, respectively, representing two possible situations A and B, then the mean of these values in a dataset of objects represents exactly the proportion of objects corresponding to the two situations.

Returning to the transformation defined by *logit* (p), we mention its formula:

$$logit(p) = \ln\left(\frac{p}{1-p}\right). \tag{3.55}$$

When using the logistic regression method and ending computations, we obtain the value of *logit* $(p) = \alpha$ as a linear combination of explanatory variables. Under these circumstances, we can calculate the actual value of the probability p, using the following formula:

$$p = e^{\alpha}/(1 + e^{\alpha}). \tag{3.56}$$

Example 3.8. We consider the database regarding 299 individuals with various liver diseases. Both the classical serum enzymes: total bilirubin (TB), direct bilirubin (DB), indirect bilirubin (IB), alkaline phosphatase (AP), leucine-amino-peptidase (LAP), gamma-glutamyl-transferase (GGT), aspartate-amino-transferase (AST), ala-nine-amino-transferase (ALT), lactic dehydrogenase (LDH), prothrombin index (PI), and the clinical parameters: glycemia, cholesterol, albumin, gamma and age are seen as patients attributes. Consider the diagnosis of liver cancer (*hepatocellular carcinoma*, also called *malignant hepatoma*) (HCC) as the dependent variable of the regression model and the 15 attributes above (i.e., serum enzymes and clinical parameters) as explanatory variables. Here, the parameter p is the probability of an individual having liver cancer, based on the risk factors corresponding to the 15 predictive variables. The figure below (Fig. 3.53) illustrates the observed values vs. predicted values, showing an acceptable accuracy concerning the forecasting power of this method.

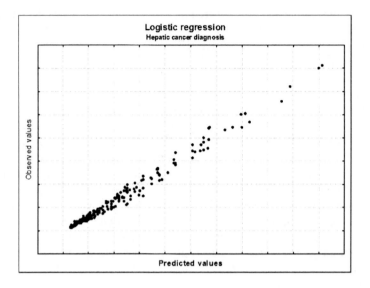

Fig. 3.53 Predicted vs. observed values (logistic regression)

The graph displaying the distribution of residuals (Fig. 3.54) confirms the above statement concerning the model accuracy.

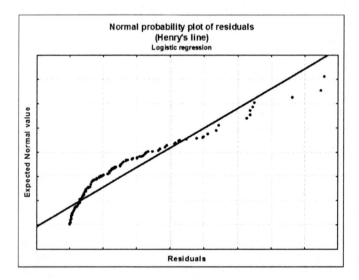

Fig. 3.54 Normal probability plot of residuals (logistic regression)

The logistic regression equation attached to this dataset is given by:

$$logit(p) = 13.97 + 0.69 \times TB + 2 \times DB -$$

$$- 0.007 \times AP - 0.002 \times LDH - 0.05 \times PI + 1.81 \times Gamma - 0.12 \times Age \quad (3.57)$$

Note that, although 15 predictors have been considered in the regression study, only 7 have been included in the logistic regression equation. This is due to the statistical significance criterion, based on the p-level (standard statistical threshold equaling 0.05). In other words, the 7 predictors above have a p-level less than 0.05, when entered in the model.

Example 3.9. The purpose of this study is to establish the influence, if there is one, of smoking, obesity and snoring on hypertension. The idea is to estimate the probability of occurrence of hypertension based on the explanatory variables mentioned above, regarded as risk factors for this heart disease (see (279), (10)). Using logistic regression, we get the equation:

$$logit(p) = -2.378 - 0.068 \times Smoking + 0.695 \times Obesity + 0.872 \times Snoring, \tag{3.58}$$

equation from which we can obtain the probability of a subject to develop hypertension, based on his (her) individual values of the three explanatory variables, considered as risk factors for hypertension, and coded as: 0 = non-smoker, 1 = smoker, 0 = normal weight, 1 = over-weight, 0 = no-snoring, 1 = snoring.

If we wish to perform a comparison between smokers and non-smokers, regarding the risk of having hypertension, we compare the equations:

$$logit(p_{smoker}) = -2.378 - 0.068 + 0.695 \times Obesity + 0.872 \times Snoring, \quad (3.59)$$

$$logit(p_{non-smoker}) = -2.378 + 0.695 \times Obesity + 0.872 \times Snoring, \quad (3.60)$$

As it can be seen without any difficulty, we considered the (coded) variable 'smoking' equaling first 1 and then 0.

It follows that:

$$logit(p_{non-smoker}) - logit(p_{smoker}) = 0.068, \tag{3.61}$$

from where:

$$\ln \left[\frac{p_{non-smoker}(1 - p_{smoker})}{p_{smoker}(1 - p_{non-smoker})} \right] = 0.068, \tag{3.62}$$

or:

$$\left[\frac{p_{non-smoker}(1 - p_{smoker})}{p_{smoker}(1 - p_{non-smoker})} \right] = 1.070, \tag{3.63}$$

value (well-known as *odds ratio*) that can be interpreted as a measure of the risk of hypertension among non-smokers compared with smokers (for much more details, see (10)).

Remark 3.12. 1) The regression logistic model enables us to predict the probability of a certain outcome, based on the values of the predictive variables, allowing thus to distinguish categories of objects depending on that outcome. For example, in a medical model case, we can distinguish (discriminate) individuals who can develop a certain disease compared with the others. Just as with the multiple regression, we can use the logistic regression model as a *prognostic (diagnostic)* index for a certain group of objects. Basically, we define:

$$PI = \ln\left(\frac{p}{1-p}\right) = b_0 + b_1 x_1 + b_2 x_2 + ... + b_k x_k \qquad (3.64)$$

where *PI* (prognostic index) is the logit transformation of the probability p that an object has a certain characteristic, and the model contains k explanatory variables. We will calculate *PI* for all the objects in the study and compare the distributions among those, with and without the characteristic of interest. Thus, we can discover how good the separation (discrimination) is between the two groups, and can identify the best cut-off value to maximize the discrimination. If all the explanatory variables are binary, then *PI* may have few distinct values. If, instead, one or more of the explanatory variables in the model are continuous, the score *PI* will be continuous as well.

2) Another approach to the problem of discriminating between groups of objects, based on the use of several variables, is known in Statistics as the *discriminant analysis*. The discriminant analysis belongs to the methodological field of *multivariate exploratory techniques*, with which we will deal in more detail later in this chapter.

3.6.3   Cox Regression Model

One of the main branches in Statistics, with very interesting applications, especially in the medical studies and mechanical systems, is represented by the *survival analysis*. In medical research it is known by this (original) name - *"survival analysis"*, while in engineering sciences it is called the *"reliability theory"*, and in economical studies or sociology it is known as the *"duration analysis"*. Whatever the context, the 'cornerstone' of this theory is the concept of "death", "failure", "drop", "absence", "out of service", etc., which is regarded as the basic *event* in the survival analysis.

To better understand what it is about, let us consider the medical research field, which was actually the starting point to develop this statistical branch. Thus, the problem of analyzing survival times, as its name indicates, refers, in principle, to the survival of a patient following a serious surgical operation, a treatment of certain diseases with lethal end, e.g., cancer, AIDS, etc. Basically, one records the time since the beginning of the medical process (surgery, treatment, etc.) until death (i.e., time to death), period called '*survival time*', the corresponding study being accordingly called '*survival analysis*'. Let us mention that, in many clinical studies, survival time often refers to the time to death, to development of a particular symptom, or to relapse after the remission of the disease. In this context, by '*probability of surviving*' we mean the proportion of the population of such people (i.e., subject to certain common medical experiment, e.g., heart transplant, chemotherapy, etc.) who would survive a given length of time in the same circumstances. The classical technique for calculating the probability of surviving a given length of time is briefly described below. Thus, we denote by X the random variable representing the survival time. The survival probability is calculated by dividing time in many small intervals $(0, t_1), ..., (t_{k-1}, t_k), ...$, and estimating the probability:

$$P\{X \leq t_n\} = P\{X \leq t_1\} \cdot P\{X \leq t_2 \mid X = t_1\} \cdots P\{X \leq t_n \mid X = t_{n-1}\}.$$

A common problem of both clinical studies (i.e., survival time) and engineering (i.e., safe operation duration of a mechanism) is to determine the effect of continuous variables (independent variables) on the survival time, in particular identifying the existence of correlation between predictors and survival time.

The classical method of survival analysis, based on techniques such as, for instance, *Kaplan-Meier survival curve*, *life table analysis*, *logrank test*, *hazard ratio*, etc., cannot be used to explore the simultaneous effect of several variables on survival. It should be emphasized that the multiple regression method cannot be directly used in this case because of at least two reasons:

- The variable describing the survival time is not often Normally distributed (usually, exponential or Weibull);
- Survival analysis uses the so-called censored data, i.e., situations when some observations are incomplete.

To better understand what these data mean in this context, imagine that a group of patients with cancer are monitored in an experiment for a certain period of time (*follow-up study*). After this period passed, the patients who have survived are no longer monitored and, when analyzing the survival time, no one knows exactly whether they are still alive. On the other hand, some patients may leave the group during the surveillance period, without knowing their situation further. The data concerning such patients are called *censored data*.

Survival analysis uses several specialized regressive methods; we mention here only the following:

- *Cox proportional hazards regression analysis, or, in short, Cox proportional hazard model*, (72);
- *Cox proportional hazard model with time-dependent covariates*;
- *Exponential regression model*;
- *Log-normal linear regression model*;
- *Normal linear regression model*.

Next, we shortly review the main elements of the Cox proportional hazard model only.

First, the *survival function* is defined by the probability:

$$S(t) = P\{T > t\},\tag{3.65}$$

where t represents, generally, the time, and T is the time to death. The *lifetime distribution* is given by:

$$F(t) = 1 - S(t),\tag{3.66}$$

where $f(t) = \dfrac{d}{dt}F(t)$ represents the *mortality (death) rate* (i.e., number of deaths per unit time).

Next, the *hazard function* is given by the formula:

$$\lambda(t) = P\{t < T < t + dt\} = \frac{f(t)dt}{S(t)} = -\frac{S'(t)dt}{S(t)}.\tag{3.67}$$

The hazard function thus represents the risk to die within a very short period of time dt, after a given time T, obviously assuming survival up to that point. The Cox proportional hazard model is very general among the regressive models, since it is not based on any prior assumption on the distribution of survival. Instead, it is based only on the assumption that the hazard is a function solely of the independent variables (predictors, covariance) Z_1, Z_2 ,..., Z_k, i.e.,:

$$h(t; Z_1, Z_2, ..., Z_k) = h_0(t) \cdot \exp(b_1 Z_1 + b_2 Z_2 + ... + b_k Z_k),\tag{3.68}$$

which, by taking logarithms becomes:

$$\ln\left[\frac{h(t; Z_1, Z_2, ..., Z_k)}{h_0(t)}\right] = b_1 Z_1 + b_2 Z_2 + ... + b_k Z_k,\tag{3.69}$$

being thus a semi-parametric model -no particular type of distribution is assumed for the survival times, but a strong assumption is made that the effects of the different variables on survival are constant over time and are additive in a particular scale. The term $h_0(t)$ is called *baseline hazard*, or *underlying hazard function*, representing the hazard for a certain individual when all the independent variables are zero.

Note that, however, we have to take into account two conditions:

- There must be a multiplicative relationship between $h_0(t)$ and the log-linear function of covariates -*hypothesis of proportionality*, through the prism of hazard.
- There must be a log-linear relationship between hazard and the independent variables.

An interesting example of using the Cox proportional hazard model in clinical studies concerns a long randomized trial comparing azathioprine and *placebo* in the treatment of patients with primary biliary cirrhosis (for details, see (60), (10)).

Remark 3.13. 1) The selection of the explanatory variables for inclusion in the model obeys the same rule as for the multiple (linear) regression.

2) The sign of the regression coefficients b_i must be interpreted as follows. A positive sign means that the hazard is higher and, consequently, the prognosis worse, for subjects with higher values of that variable (characteristic). Conversely, a negative sign of a certain coefficient indicates a relatively low hazard concerning that variable.

3) As with multiple linear regression and logistic regression, the combination of regression coefficients and values of variables can be used as a prognostic index (*PI*).

Thus, we can define the prognostic index by:

$$PI = b_1 Z_1 + b_2 Z_2 + ... + b_k Z_k. \tag{3.70}$$

Consequently, the survival function:

$$S(t) = \exp\left[-H_0(t)\right]^{\exp(PI)}, \tag{3.71}$$

can be calculated, where $H_0(t)$, called the *cumulative underlying hazard function*, is a step function over time.

3.6.4 Additive Models

Suppose, just as with multiple regression, that we deal with a dependent variable Y and k explanatory variables (predictors) $X_1, X_2,..., X_k$. Unlike the case of linear models, when dealing with additive models we consider a relationship between the dependent variable and the predictors given by:

$$Y = f_1(X_1) + f_2(X_2) + ... + f_k(X_k) + \varepsilon, \tag{3.72}$$

where f_j, $j = 1, 2,..., k$ generally represent smooth functions (i.e., functions that have derivatives of all orders, or to be of class C^∞), in some cases functions of class C^1, and ε is a standard Normally distributed random variable $N(0,1)$. It is easy to see that an additive model is the generalization of the multiple linear regression model (for $\varepsilon = 0$). In other words, instead of a single coefficient per explanatory variable, for the additive models we have an unspecified function per each predictor, which will be estimated in order to optimally forecast the dependent variable values.

Remark 3.14. 1) The additivity hypothesis $\sum f_i(X_i)$ is a restriction of the general case of a predictive model of type $Y = f(X_1, X_2, ..., X_k)$.
2) The parameter functions f_i (i.e., functions as parameters) of the additive model are estimated up to an additive constant.
3) We can mention in this context the *generalized additive models* ((178), (179)). A *generalized linear model* is represented by the following equation:

$$Y = g(b_0 + b_1 X_1 + b_2 X_2 + ... + b_k X_k), \tag{3.73}$$

where where g is an undefined smooth function. If we denote by g^{-1} the inverse function for g, function called *link function*, then we can write the above equation in a slightly modified form:

$$g^{-1}(E[Y]) = b_0 + b_1 X_1 + b_2 X_2 + \ldots + b_k X_k, \tag{3.74}$$

where $E[Y]$ represents the expectation of the dependent random variable Y. Now, let us combine an additive model with a generalized linear model. We then obtain the equation of this new model, known as *generalized additive model*, given by:

$$g^{-1}(E[Y]) = f_1(X_1) + f_2(X_2) + \ldots + f_k(X_k). \tag{3.75}$$

The basic problem in these models is the way to estimate the parameter functions f_i of the model. The best known method of evaluating the functions f_i is represented by an interpolation based on scatterplot smoother, using cubic spline functions. For example, considering a simple model with only two functions f_1 and f_2, having the form $Y = f_1(X_1) + f_2(X_2) + \varepsilon$, then, using the spline approximation, we obtain the following formulas for the two functions:

$$f_1(X) = \delta_1 + X \cdot \delta_2 + \sum_{j=1}^{q_1-2} R(X, X_j^*) \delta_{j+2}, \tag{3.76}$$

$$f_2(X) = \gamma_1 + X \cdot \gamma_2 + \sum_{j=1}^{q_2-2} R(X, X_j^*) \gamma_{j+2}, \tag{3.77}$$

where δ_j and γ_j are the unknown parameters of f_1 and f_2, q_1, q_2 represent the number of the unknown parameters, and X_j^* are the interpolation knots for the two functions.

An illustrative example for the use of a generalized additive model in practice is given by the problem of estimating the wooden mass of a tree, based on its circumference and height. In this case one can use the generalized additive model instead of the 'classical' method of calculating the volume of a cylinder. Thus, its corresponding equation is given by:

$$ln(E[Wooden\ mass]) = f_1(Circumference) + f_2(Height),$$

assuming that the wooden mass has a Gamma distribution (technical details are to be found in (411)).

3.6.5 Time Series: Forecasting

During the observation of several real-world events or phenomena, it is found that the data that interests us, and are collected for a certain analysis, have a chronology that allows us to analyze their evolution (trend) as time goes by. Such a sequence of measurements that follow non-random orders, denoted by $(\xi_t, t \in T)$, where $T \subset \mathbf{R}$ refers to time, is called *time series*, or *dynamic series*. Unlike the analyses of random samples of observations, that are discussed in the context of most other statistics, the analysis of time series is based on the assumption that successive values in the dataset represent consecutive measurements taken at equally spaced time intervals. We remark that the data ξ_t may relate to observations in discrete time, i.e., days, weeks, quarters, years, etc., or can be considered as data obtained in continuous time (continuous-time observations).

Let us now mention two main goals of the time series analysis:

- *Identifying the nature of the phenomenon*, represented by the sequence of observations;
- *Forecasting possible future values*, starting from the observations already known.

For carrying out the above mentioned tasks, a previous identification of the observed time series pattern is necessary. Once the pattern identified and described, we can integrate it in certain well-defined classes of similar phenomena, and interpret it in order to predict future values of the studied phenomenon.

Regarding the problem of identifying a time series pattern, the underlying assumption is that the data consist of a systematic pattern (i.e., a set of identifiable components) and random noise, seen as a disturbance which makes the identification of the genuine data form difficult. The methods used to solve this problem will appeal, therefore, to different data filtering techniques to remove noise.

Unfortunately, there is no definite science for identifying the hidden pattern in data, after removing noise. Accordingly, the methods are chosen depending on the problem to be solved. Thus, if we meet a monotone trend (ascending or descending, besides some singular values), then the problem of estimating it concretely and, afterwards, the prognosis stage, are not difficult. But if instead, the time series has many errors, then, as a first step, we can try to smooth out the data, i.e., to perform a local approximation of the data in order to remove the non-systematic components (e.g., the *moving average* technique consisting in replacing a value by the simple or weighted average of n neighboring values).

From the point of view of time series applications in the data mining context, we want to mention the following matter. Thus, any sequential combination of time and numbers can be regarded as a time series. For example, the historical oil price fluctuations (e.g., Nymex Crude, IPE Crude, Brent Crude, etc.) can be considered as time series. The business analysts will study these time series to advise the oil companies regarding the oil consumption forecast, taking into account various seasonal and long-term conditions (e.g., hot or cold season, political changes, trends in regional economic growth, etc.).

From the graphical point of view, the time series are most often visualized by tables or graphs, e.g., tables or flow charts of passengers at an airport in a given period, chart of temperatures for a given month of the year, recorded in the last century, chart of the frequency of developing a particular disease over a period of time, graph displaying the values of certain clinical parameters regularly recorded, chart regarding the evolution of the exchange rate, etc. We present, for instance, such a graphic regarding clinical studies. The table below (Table 3.27) shows the values

Table 3.27 Annual values of three serum enzymes (ALT, AST, GGT)

Serum enzyme	Jan	Feb	Mar	Apr	May	Jun	Jul	Aug	Sep	Oct	Nov	Dec
ALT	29	51	56	10	22	12	10	8	12	30	24	15
AST	29	50	39	15	22	12	14	19	11	8	6	12
GGT	67	40	127	55	98	86	74	48	55	44	34	65

for the main serum enzymes, indicating cholestatic and hepatocellular injury (in the liver diseases case). These clinical observations are able to support the diagnosis process of chronic hepatitis C. Specifically, these medical analyses are carried out during 12 consecutive months from patients with chronic hepatitis C, the monthly analyses referring to the following serum enzymes: alanine-amino-transferase (ALT), aspartate-amino-transferase (AST) and gamma-glutamyl-transpepdiase (GGT).

In the figure below (Fig. 3.55) we represented the corresponding graphs of the annual curves, corresponding to the three serum enzymes.

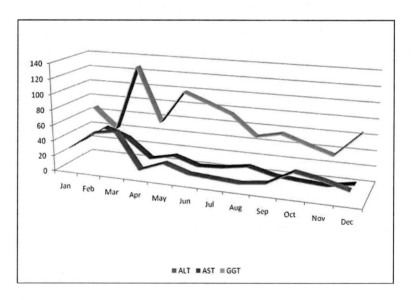

Fig. 3.55 Graphs of three serum enzymes (ALT, AST and GGT) -annual values

Returning to the important issues concerning time series, we list below the following most common problems we are facing:

- *Forecasting*, referring to the estimation of future values ξ_{T+h}, $h > 1$, based on known data $\xi_1, \xi_2, ..., \xi_T$. Often, instead of indicating a certain forecasted value, one indicates *interval forecasts* (or, *interval forecasting*), because sometimes one is more interested in predicting intervals for future values than simply value estimates;

- *Trend analysis*, referring to the situation when several time series relate to the same time. Under these circumstances, it is sometimes necessary to consider cases where two or more such time series take the same direction (are strongly correlated), although at first glance this fact seems inexplicable;
- *Seasonal decomposition*, referring to the decomposition of the original time series, seen as an amalgam of different tendencies, in well-structured components and analyzing them in their global inter-dependence. Without doing such decomposition, it is difficult to identify the basic pattern in data.

In general, a time series can be decomposed into:

- A *seasonal component* S_t,
- A *trend component* T_t,
- A *cyclical component* C_t,
- A *random error*, or *irregular component* ε_t,

where t stands for the particular point in time. We mention here two models of seasonal decomposition:

(a) *Additive model*, given by $\xi_t = T_t \cdot C_t + S_t + \varepsilon_t$;

(b) *Multiplicative model*, given by $\xi_t = T_t \cdot C_t \cdot S_t \cdot \varepsilon_t$.

- *Distinction* between short-time and long-time observations, which refers to the separation of persistent relations over time, observed in collected data, from conjectural relations.
- *Causality relationship*, which can be observed between two or more time series. Furthermore, when determining a causality relationship, one studies the 'dephasation' that occurs between cause and effect, in the involved time series.

Time series are also known as *dynamic series* because they illustrate the kinetics (dynamics) of a phenomenon evolving in real time. Therefore, in connection with time series, we can talk about dynamic models, i.e., models that 'capture the movement' of a particular phenomenon in relation to time.

From the multitude of dynamic models based on time series, we recall here only three types: *adjustment models*, *autopredictive models* and *explanatory models*. Next, we briefly present the principles of each of the three types of models.

- *Adjustment models.* In this case, based on observations obtained by analyzing the actual data, we can design a mathematical model, illustrated by an equation of the form:

$$\xi_t = f(t, u_t), \tag{3.78}$$

where f is a function determined by a finite number of unknown parameters, and u_t represents a random variable with the mean equaling zero, chosen according to the actual situation modeled in this way (e.g., disturbing factor/noise/error). We note that the assumptions on the variable u_t, as well as the estimate of the function f, are made starting with the so-called *global adjustments* in which all observations are equally taken into consideration, thus having the same roles in estimates, or the so-called *local adjustments*, in which each observation is playing its own part in determining the model parameters.

- *Autopredictive models.* In these models it is assumed that the present is influenced by the past, so mathematically speaking, such a model is illustrated by an equation of the form:

$$\xi_t = f(\xi_{t-1}, \xi_{t-2}, ..., u_t), \qquad (3.79)$$

where here u_t stands for the disturbing factor, being represented by a random variable.

- *Explanatory models.* For these models, the underlying equation becomes:

$$\xi_t = f(x_t, u_t), \qquad (3.80)$$

where x_t is an observable variable, called *exogenous variable*, and u_t represents again the disturbing factor. Mainly, these models are divided into two cases: the *static* ones, in which the exogenous variable x_t does not contain information about the ξ_t's past, and u_t are mutually independent, and the *dynamic* ones, in which either x_t contains information about the ξ_t's past or u_t are autocorrelated.

We mention here, for instance, just a classic case of autopredictive model, namely the ARIMA (*Auto-Regressive Integrated Moving Average*) model or, as it is often known, the *Box-Jenkins model*. This model, developed by Box and Jenkins (40), is, at least theoretically, the largest known class of models for forecasting, extremely common today, gaining enormous popularity in many areas and research practice, because of its great power and flexibility. Thus, from the rich class of ARIMA models of type (p, d, q), we apply in a simple medical situation the so-called ARIMA $(1, 1, 1)$ '*mixed*' model, given by the equation:

$$\widehat{Y}(t) = \mu + Y(t-1) + \Phi \cdot (Y(t-1) - Y(t-2)) - \theta \cdot \varepsilon(t-1), \qquad (3.81)$$

where $\widehat{Y}(t)$ represents the forecast for the time series at time t, μ is a constant, Φ denotes the autoregressive coefficient, $\varepsilon(t-1)$ denotes the error at period $(t-1)$, and θ is the coefficient of the lagged forecast error. We illustrate the above model by a practical application regarding a classical treatment of liver diseases (329). Thus, we are interested in forecasting the health evolution of a patient with chronic hepatitis C, which followed a standard 6 months treatment with interferon. The importance of this forecasting consists in the fact that, on the one hand, the treatment with interferon is very expensive and, on the other hand, a non-negligible fraction of patients do not respond positively to this treatment. Accordingly, the continuation of this treatment is not effective in any regard. Such being the case, it is important to provide a predictive model, sufficiently reliable, forecasting the clinical behavior of a patient in the next 3 to 6 months (based on measuring the main serum enzymes), in order to decide if it is about a so-called *'responder'*, i.e., a patient with positive response to treatment, or a *'non-responder'*, i.e., a patient with a negative response to treatment, for which further treatment no longer serves any purpose. For modeling the clinical behavior of patients, the classical ARIMA (1, 1, 1) 'mixed' model has been chosen. Since both the computerized data processing related to this model and the forecasting degree of confidence require a relatively high number of temporal data (at least 20), the *B-spline cubic interpolation* has been used, inducing thus sufficiently smooth curves to the knots, being therefore compatible with the clinical process associated with the treatment. Starting from the data recorded in the first 6 months, and dividing this period in small intervals and, finally, interpolating the corresponding knots by B-cubic spline, a sufficiently large database has been created, in order to properly implement the ARIMA model. In the following figure (Fig. 3.56) the graph corresponding to the B-spline interpolating curve for the clinical score, representing a probable forecasting of the health trend of a patient, is presented.

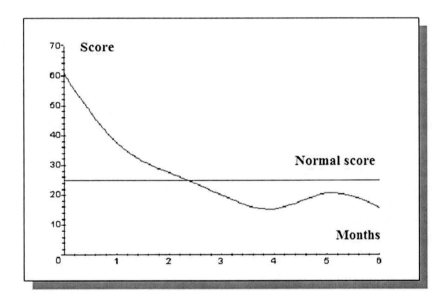

Fig. 3.56 B-cubic spline interpolation of 6 months recorded data

Starting from these data processed by B-spline interpolation, we present below (Fig. 3.57) both the ARIMA forecasted curve and the confidence interval (95%), corresponding to the next 6 months (weeks 25-49), regarding the clinical behavior of a virtual *'responder'* patient, under interferon treatment.

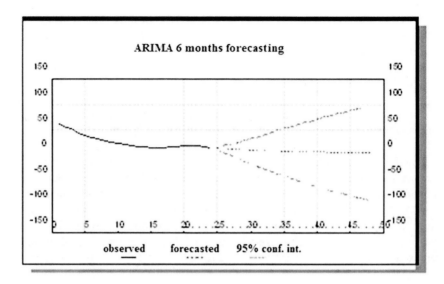

Fig. 3.57 ARIMA forecasting for a virtual '*responder*' patient (6 months)

3.7   Multivariate Exploratory Techniques

Multivariate exploratory techniques are those methods specifically designed to discover hidden patterns in multidimensional data, including, among others: *factor analysis* (FA), *principal components analysis* (PCA), *canonical analysis* (CA) and *discriminant analysis* (DA). Other techniques applicable to multidimensional data, such as time series, linear models, non-linear and additive models, have been already mentioned above, and others, such as cluster analysis and classification trees, will be discussed later in this book. Let us mention here other techniques, which will not be presented in this book, such as *correspondence analysis*, *general* CHAID *models*, *multidimensional scaling*, the *reliability theory*, which, in turn, are often used in data mining.

3.7.1   Factor Analysis

Factor analysis (factor analytic techniques, in a broad sense) is seen from the data mining point of view mainly as a tool for solving the following two problems:

- *Reducing* the number of attributes of objects (variables), in order to increase the data processing speed;
- *Detecting* structure hidden in the relationships between attributes (variables), that is to classify attributes of objects.

Factor analysis (term introduced by Thurstone, 1931, (382)) refers to a variety of statistical techniques used to represent a set of variables depending on a smaller number of hypothetical variables, called *factors*. A simple example to illustrate the role of data reduction and identification of the relationships structure, obtained by using factor analysis, regards the process of designing the standard customer typology (e.g., bank, insurance company, supermarket, mobile network operator, etc.). Thus, in the process of building the customers database, it is possible that certain attributes, such as the annual income and the annual tax appear simultaneously. Since the two attributes are connected by the formula for computing income tax, only one attribute is necessary, the other being redundant, so it can be removed without loss of information. As the two attributes (variables) are correlated, their relationship is very well summarized by the regression line passing through the 'cloud' of points generated by the pairs of data, and can therefore be used to detect the (linear) structure of the relationship between them. In fact, in this way the two variables are reduced to a single factor, which is a linear combination of the two initial variables.

Remark 3.15. For more than two variables, the philosophy of reducing their number to a single factor remains the same. Thus, for three (correlated) variables, for instance, we can consider their regression line (line passing through the 'cloud' of points generated by triplets of data in the three-dimensional space created by them), and thus we reduce them to a single factor, i.e., a linear combination of them.

In conclusion, as we already said, factor analysis is the statistical methodology that summarizes the variability of the attributes of data, regarded as random variables, with the aid of a limited number of other variables -the *factors*. The attributes (variables) considered in the analysis are expressed by linear combinations of factors, adding a term referring to the model error. Let us remember that factor analysis has been and is still extensively used in various fields such as, for instance, psychology (e.g., psychometrics -see C. Spearman (355), (356), (357), also known as a pioneer of factor analysis), social sciences, marketing, production management, operations research, etc.

To better understand the way of working in factor analysis, let us consider the following example. Suppose that the staff of a chain of supermarkets want to measure the satisfaction degree of customers relative to the provided services. Thus, we consider two 'factors' measuring the customers' satisfaction: (*a*) satisfaction regarding the way of serving a client (service quality), and (*b*) satisfaction regarding the quality of marketed products. This is carried out by using a sample involving, say, $N = 1,000$ customers, who have to answer a questionnaire with, say, $M = 10$ 'key' questions which can measure the customers' satisfaction. We will consider the response of each customer as a 'score' regarding the respective matter, score which is seen as an *observed* (or *observable*) variable. Because customers were randomly selected from a large 'population', one can assume that the 10 responses (scores) are random variables. Suppose also that the average score per customer per question may be viewed as a linear combination of the two types of satisfaction (factors of satisfaction-*unobserved* (or *unobservable*) variables). For instance, for question #k,

$k = 1, 2 ,...., 10$, we have: {7 × *satisfaction regarding the service* + 5 × *satisfaction regarding product*}, where the numbers 7 and 5 are called *factor loadings* and are identical for all customers. Let us remark that there may be different loadings for different questions, here 7 and 5 are factor loadings relating to question #k. Two customers with the same degree of satisfaction in both directions may have different scores to the same question included in the questionnaire, because individual opinions differ from the average opinion, this difference representing the error.

We further present the corresponding mathematical model regarding the factor analysis adapted to the above example. Thus, for each customer #i ($i = 1, 2,..., N$) the M scores are given by the equations:

$$x_{1,i} = b_1 + a_{1,1} \cdot s_{1i} + a_{1,2} \cdot s_{2i} + \varepsilon_{1,i}, \ i = 1, 2, ..., N, \tag{3.82}$$

$$x_{2,i} = b_2 + a_{2,1} \cdot s_{1i} + a_{2,2} \cdot s_{2i} + \varepsilon_{2,i}, \ i = 1, 2, ..., N, \tag{3.83}$$

....

$$x_{M,i} = b_M + a_{M,1} \cdot s_{1i} + a_{M,2} \cdot s_{2i} + \varepsilon_{M,i} \ i = 1, 2, ..., N, \tag{3.84}$$

where:

- $x_{k,i}$ represents the score corresponding to question #k for customer #i;
- s_{1i} represents the degree of satisfaction concerning the service quality provided for customer #i;
- s_{2i} represents the degree of satisfaction regarding the product quality assessed by customer #i;
- a_{kj} represents the factor loadings for question #k corresponding to factor j;
- ε_{ki} represents the model error (i.e., the difference between customer #i score and the average opinion for question #k for all customers whose satisfaction regarding services and products quality are the same as for customer #i);
- b_k represents some additive constants (more specifically, the average opinion for question #k).

In matrix language, the above equations are rewritten into the following form:

$$X = b + AS + \varepsilon, \tag{3.85}$$

where:

- X is the matrix of *observed* random variables;
- b is the vector of *unobserved* constants;
- A is the matrix of factor loadings (*unobserved* constants);
- S is a matrix of *unobserved* random variables;
- ε is a matrix of *unobserved* random variables (*error matrix*).

Factor analysis aims to estimate the matrix A of factor loadings, the vector b of averages and the variance of errors ε.

Remark 3.16. We have to emphasize the distinction between the 'philosophy' underlying the factor analysis techniques and the effective application of these techniques on actual data. Practically, factor analysis can only be applied using specialized software, especially designed for this purpose. For technical details regarding the principles of factor analysis, the reader may consult, for instance, (67), (214), (215), (134), (365), (272), while specialized software packages can be found in *Statistica*, *SAS*, *SPSS*, etc.

3.7.2 Principal Components Analysis

In conjunction with factor analysis, we now present some aspects of *principal components analysis* (PCA), since the latter can be regarded as a factor analysis technique when the total variation of data is taken into consideration.

In essence, principal component analysis aims to reduce the number of variables initially used, taking into account a smaller number of 'representative' and uncorrelated variables. As a consequence of this approach, we obtain a classification of variables and cases.

To better understand this topic, let us consider that we want to buy a certain product from a store and, in the beginning, we are only interested in its two characteristics A and B. In this case we can consider the 'spreading cloud' of points generated by the pairs of data corresponding to the two attributes. We further consider the line crossing the center of the 'cloud' of points (in particular, the 'cloud' centroid), that is their regression line, which is representative for the two attributes. Now, suppose that we consider another feature of the product, denoted by C. If in this case we consider only the pairs of regressions between the three attributes, we do not obtain a satisfactory choice because we have no overview of all the three attributes. We instead need something helping us to 'aggregate' all three attributes simultaneously. The problem becomes more complicated if we consider an even higher number of attributes. In this case we need a 'score' to characterize the object, instead of pairs of regressions. From a geometric point of view, this score may be generated by a line, or lines (factor axes), passing through the centroid of the 'cloud' of points generated by the data *tuples*. Thus, starting from the initial data space, one considers a subspace generated by a set of new axes, called *factor axes*, subspace in which the initial space is projected. In principle, the PCA technique searches for the line that best fits the 'cloud' of points in the vector space of objects and attributes. Mathematically speaking, if we consider p attributes and q objects, the PCA technique for identification of the factors relates to the diagonalization of the symmetric matrix which represents the correlation matrix (covariance matrix). We recall that, because covariance is calculated only for pairs of (statistical) variables, in case of three variables X, Y and Z, the covariance matrix is given by:

$$Cov(X,Y,Z) = \begin{pmatrix} cov(X,X) & cov(X,Y) & cov(X,Z) \\ cov(Y,X) & cov(Y,Y) & cov(Y,Z) \\ cov(Z,X) & cov(Z,Y) & cov(Z,Z) \end{pmatrix},$$

for *n* variables doing the same. In our case, if the *standardized* matrix X (i.e., with dimensionless elements about the respective means) represents the data corresponding to the *q* objects and *p* attributes, then $X^T \cdot X$ represents the covariance matrix, and the actual issue concerns its diagonalization. The result will be a new set of variables -*principal components*- which represent linear combinations of the original attributes, and are uncorrelated. We thus obtain a space with smaller dimension, in which the objects and attributes are projected, and which keeps maximum of the data variability. Schematically, the PCA can be summarized in the following two steps:

- identifying the eigenvectors of the covariance matrix;
- building the new space generated by eigenvectors.

The figure below (Fig. 3.58) synthetically illustrates the principle of the PCA methodology.

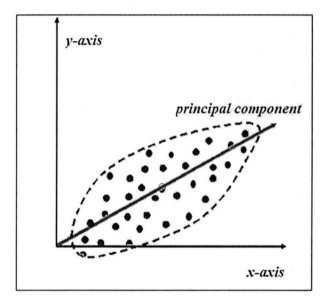

Fig. 3.58 PCA technique illustration

In the following figures, we schematically illustrate the steps of a PCA analysis. Technically speaking, the first principal component represents the combination of variables which 'explains' the largest variance of data. The second principal component 'explains' the next largest variance of data, being independent of the first one, and so forth. In principle, we can consider as many principal components as the number of existing variables. The first figure (Fig. 3.59) shows the dataset using the corresponding 'cloud' of spread data.

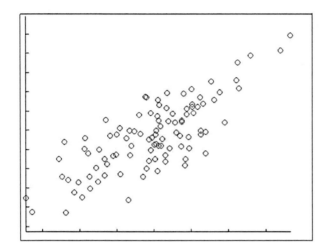

Fig. 3.59 Plot of original data

After computing the covariance matrix and the two principal components, the following figure (Fig. 3.60) illustrates the original data with eigenvectors added.

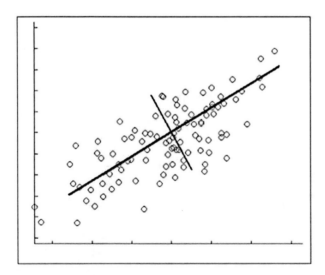

Fig. 3.60 Plot of original data with eigenvectors added

The oblique line (the thick black line), which passes through the center of the 'cloud' of points, explaining the largest variance, is the first principal component. The second principal component is perpendicular to the first (independent of it) and explains the rest of variance. Finally, by multiplying the initial data with the

principal components, the data will be rotated so that the principal components form
the axes of the new space, as seen in Fig. 3.61 below.

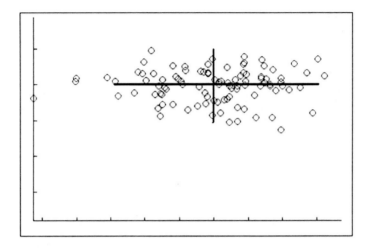

Fig. 3.61 Data transformed with two principal components

Finally, recall that PCA is also known, depending on versions, as *Hotelling transform* (HT), or *Karhunen-Loeve transform* (KLT). For technical details regarding
calculations and the main algorithms used (e.g., covariance method, correlation
method) the reader is advised to consult (194) (91) (209) (202), or the online texts "A
tutorial on Principal Components Analysis", L. Smith, URL: http://www.cs.otago.
ac.nz/cosc453/student_tutorials/principal_components.pdf, "A tutorial on Principal
Components Analysis", Jon Shlens, URL: http://www.cs.princeton.edu/picasso/
mats/PCA-Tutorial-Intuition_jp.pdf (accessed in 2010).

3.7.3   Canonical Analysis

Suppose that the managers of a supermarket are interested to explore the satisfaction degree of customers regarding the service quality provided to them. To achieve
this task, customers are asked to complete a questionnaire with a number of questions related to the satisfaction regarding the service quality. At the same time, they
are also asked to answer another set of questions related to the measurement of
the satisfaction degree regarding other areas than the service (e.g., product quality,
product diversity, etc.). The issue is to identify possible connections between the
satisfaction related to service and the satisfaction regarding other aspects of the supermarket activity. Let us now recall the principles of the simple or multiple (linear)
regression. There we handled a set of one or more predictor (explanatory) variables

and a dependent variable (criterion), which was determined by the former. Just as it can easily be seen, here we are in the situation when the dependent variable would be represented only by the degree of satisfaction regarding the service quality. But in this case, there are also other dependent variables (product quality, product diversity, etc.), so we are in the position to have a set of explanatory variables and a set of dependent variables instead of a single dependent variable. To solve this issue, we generalize the linear multiple regression technique, being interested to correlate a set of dependent variables, each of them weighted, with a set of predictor variables, also weighted. Note that this idea is somewhat similar to that introduced by H. Hotelling (1936), (194), known as *canonical correlation analysis* (CCA). Formally, given a set of explanatory variables $\{X_1, X_2,..., X_q\}$ and a set of dependent variables $\{Y_1, Y_2,..., Y_p\}$, we have to determine the concrete form of the following equation:

$$a_1 \cdot Y_1 + a_2 \cdot Y_2 + ... + a_p \cdot Y_p = b_1 \cdot X_1 + b_2 \cdot X_2 + ... + b_q \cdot X_q, \qquad (3.86)$$

which establishes the relationship between the two sets of variables.

From a computational point of view, there are specialized software for solving this problem (e.g., Statistica-*Multivariate exploratory techniques*, SPSS-*Canonical analysis*, Matlab -*Multivariate analysis*, etc.).

3.7.4 Discriminant Analysis

Suppose we have to classify a hurricane (labeling it with an appropriate category, e.g., from 1 to 5, according to the *Saffir-Simpson Hurricane Scale*). To achieve this task we have available a certain number of measurements related to different meteorological features preceding the onset of hurricane (discriminant predictive variables). The study we have to perform aims to determine which variables are the best predictors of the hurricane category, in other words, which variables effectively distinguish (discriminate) between different categories of hurricanes (e.g., the intensities of their sustained winds). Similarly, in commerce, we can analyze what characteristics (discriminant variables) make a difference in terms of customer's rationality, when choosing between many categories of products a specific one (in a broader sense, related to the *free will and rationality* concept). In the health care field, also, a doctor may be interested in identifying the medical features that possibly drive a patient to partial or complete recovery.

As it can be seen from the examples above, the discriminant analysis is basically a method for classifying objects in certain classes, based on the analysis of a set of predictor variables -the model inputs. The model is based, in principle, on a set of observations for which the corresponding classes are *a priori* known, forming thus the training dataset. Based on training, one builds a set of discriminant functions, of the form:

$$L_i = b_1 \cdot X_1 + b_2 \cdot X_2 + ... + b_n \cdot X_n + c, \; i = 1, 2, ..., k, \qquad (3.87)$$

where $X_1, X_2, ..., X_n$ are the *predictor variables* (discriminating between classes), b_1, $b_2, ..., b_n$ represent the *discriminant factors*, and c is a constant. Each discriminant function L_i corresponds to a class Ω_i, $i = 1, 2, ..., k$, in which we have to partition the observations. A new object (observation) will be classified in that category for which the corresponding discriminant function has a maximum.

In the same context, we also mention the equivalent topics: *Linear Discriminant Analysis* (LDA), as well as *Fisher's linear discriminant*, used with these terminologies in automatic learning. As fields of practical applications for the discriminant analysis one can mention the *face recognition* domain (see, for instance http://www.face-rec.org/), *marketing* (e.g., distinction between customers, product management, etc.), *medical studies* (e.g., medical diagnosis, identifying influential observations, etc.), mentioning that there are specialized software for this purpose (e.g., Statistica, SPSS, Matlab, etc.).

3.8 OLAP

The OLAP (*On-Line Analytical Processing*) technique, or FASMI (*Fast Analysis of Shared Multidimensional Information*), or even OLTP (*On Line Transaction Processing*) was initiated by E.F. Codd, (65), considered the father of relational databases, and refers to the method that allows users of multidimensional databases to generate on-line descriptive or comparative summaries ("illustrations") of data and other analytic queries. OLAP is included in the broader field of *Business Intelligence* (BI), its typical applications being circumscribed in the following domains: business reporting for sales, marketing, management reporting, business performance management (BMP), budgeting and forecasting, financial reporting, etc. Note that, despite its name, it is not always about on-line data processing, excepting the case of dynamically updating a multidimensional database, or the existence of possible queries. Unlike the relational databases case, where data are put into tables, OLAP uses a multidimensional array to represent data, since multidimensional arrays can represent multivariate data. At the core of OLAP is the concept of the OLAP *cube* (also called *multidimensional cube*, or *hypercube*). Before showing how to build such a hypercube, let us see how to convert tabular data into a multidimensional array. Thus, starting from tabular data, we can represent them as a multidimensional array, taking into account two main aspects:

1. First, we have to identify which attributes will be considered as *dimensions* (i.e., data elements that categorize each item in a dataset into non-overlapping regions), and which attributes will be seen as *target attributes*, whose values will appear as entries in the multidimensional array.

   - The 'dimensions' attributes must have discrete values;
   - The 'target' attributes are either a count or continuous variable;
   - It is possible to have no 'target' attribute, excepting the count of objects that have the same set of attribute values.

2. Secondly, we need to find the value of each entry in the multidimensional array, by either summing the values of the 'target' attributes, or counting the number of all objects that have the attribute values corresponding to that entry (for more technical details, see (378)).

As mentioned above, the basic procedure underlying the OLAP technology is the building of the "OLAP hypercube", which "geometrically" structures the data. Thus, a "cube" of data means a multidimensional representation of data in a 'cube' shape (obviously, only in the three-dimensional space we speak of a cube, but the idea of 'cube' can be easily extrapolated to more than three dimensions -multidimensional hypercube), together with all possible aggregates. By all possible aggregates, we mean the *aggregates* that result by selecting a proper subset of the dimensions, and summing over all remaining dimensions, taking into account a particular purpose (*idem*).

Basically, there are three types of OLAP techniques:

- *Multidimensional* OLAP (MOLAP-**M**ultidimensional **O**n-**L**ine **A**nalytical **P**rocessing), which is the 'classical' form of OLAP and uses data features such as time periods (time intervals), location, product codes, etc., representing the common attributes in the area of interest, as dimensions of the cube; the manner in which each dimension is aggregated is predefined by one or more hierarchies.
- *Relational* OLAP (ROLAP-**R**elational **O**n-**L**ine **A**nalytical **P**rocessing) directly uses relational databases, the database and dimension tables being stored as relational tables, and new tables are created to record the aggregated information.
- *Hybrid* OLAP (HOLAP-**H**ybrid **O**n-**L**ine **A**nalytical **P**rocessing) is a combination of ROLAP and MOLAP. Thus, given a database, one splits the data into a relational storage and a specialized storage (e.g., relational tables for certain types of data and, on the other hand, specialized storage for other types of data).

Remark 3.17. 1) MOLAP is used especially for smaller databases (quick calculation of totals and effective responses, reduced storage space); ROLAP is considered more accessible, although a large volume of data cannot be processed efficiently, and performance in responding to queries is not so bright; HOLAP, as a hybridization of the two previous techniques, is naturally positioned between them, having a fast processing speed and good accessibility.
2) The difficulty in implementing OLAP comes from forming the queries, choosing the database and developing the scheme. Thus, most modern OLAP products come with huge libraries of pre-configured queries.

Fig. 3.62 below presents some of the most well-known commercial OLAP products.

• Business Objects	• MicroStrategy
• Celequest	• Oracle Business Intelligence
• IBM Cognos TM1	Discoverer
• DataWarehouse Explorer (CNS International)	• ProClarity (Microsoft Business Intelligence)
• Oracle Hyperion	• SAP BW Business Warehouse
• IBM DB2 Cube Views	• Oracle Essbase
• IBM DB2 Data Warehouse Edition	• Mondrian OLAP server (Pentaho)
• OLAP ModelKit™	• SAS® OLAP Server
• Information Builders	• SunSystems (Systems Union Inc.)
• Microsoft Analysis Services (SQL Server)	• BI2M (B&M Services)
• InstantOLAP	• Decision Technology Business Intelligence software
	• Beyond 20/20

Fig. 3.62 Well-known OLAP products

Let us notice that among companies that sell OLAP products we meet names like: Microsoft, Oracle, SAP, IBM, SAS, etc.

To get an idea about the current situation of the market of OLAP products, we present below the "Top Ten" regarding the shares of 2006, as published in 2007, for commercial OLAP products, according to the *BI Verdict* (former *The OLAP Report*)-Business Application Research Center (http://www.bi-verdict.com/).

Table 3.28 *Top Ten* OLAP vendors (2006)

VENDOR	Market position (#)	Share (%)
Microsoft ecosystem	1	31.6
Hyperion Solutions	2	18.9
Cognos	3	12.9
Business Objects	4	7.3
MicroStrategy	5	7.3
SAP	6	5.8
Cartesis	7	3.7
Applix	8	3.6
Infor	9	3.5
Oracle	10	2.8

(more details at: http://www.bi-verdict.com/fileadmin/FreeAnalyses/market.htm?user_id=).

We illustrate the main OLAP features, with a very suggestive and nice example, (378)), regarding the *Iris* Plants Database (R.A. Fisher -1936, URL: http://archive.ics. uci.edu/ml/datasets/Iris, (113)). Wishing to 'sweeten' this rather technical model, we chose to present in Fig. 3.63 a picture of these beautiful plants, used as an "artistic illustration" of the database.

Fig. 3.63 *Iris* plants

This database contains information regarding three types of *Iris* flowers (150 flowers, totally):

1. *Setosa*
2. *Virginica*
3. *Versicolour*

and four main attributes, fully characterizing the *Iris* flower:

- Petal length;
- Petal width;
- Sepal length;
- Sepal width.

Fig. 3.64 shows a 'sample' of this database, ready to be "data mined".

Fisher (1936) iris data: length & width of sepals and petals, 3 types of Iris				
1	2	3	4	5
Sepal length	Sepal width	Petal length	Petal width	IRIS type
5.0	3.3	1.4	0.2	SETOSA
6.4	2.8	5.6	2.2	VIRGINIC
6.5	2.8	4.6	1.5	VERSICOL
6.7	3.1	5.6	2.4	VIRGINIC
6.3	2.8	5.1	1.5	VIRGINIC
4.6	3.4	1.4	0.3	SETOSA
6.9	3.1	5.1	2.3	VIRGINIC
6.2	2.2	4.5	1.5	VERSICOL
5.9	3.2	4.8	1.8	VERSICOL
4.6	3.6	1.0	0.2	SETOSA
6.1	3.0	4.6	1.4	VERSICOL
6.0	2.7	5.1	1.6	VERSICOL
6.5	3.0	5.2	2.0	VIRGINIC
5.6	2.5	3.9	1.1	VERSICOL
6.5	3.0	5.5	1.8	VIRGINIC
5.8	2.7	5.1	1.9	VIRGINIC
6.8	3.2	5.9	2.3	VIRGINIC
5.1	3.3	1.7	0.5	SETOSA
5.7	2.8	4.5	1.3	VERSICOL
6.2	3.4	5.4	2.3	VIRGINIC
7.7	3.8	6.7	2.2	VIRGINIC
6.3	3.3	4.7	1.6	VERSICOL
6.7	3.3	5.7	2.5	VIRGINIC
7.6	3.0	6.6	2.1	VIRGINIC
4.9	2.5	4.5	1.7	VIRGINIC
5.5	3.5	1.3	0.2	SETOSA
6.7	3.0	5.2	2.3	VIRGINIC
7.0	3.2	4.7	1.4	VERSICOL
6.4	3.2	4.5	1.5	VERSICOL
6.1	2.8	4.0	1.3	VERSICOL

Fig. 3.64 Sample of *Iris* database (30 flowers of different types)

In Fig. 3.65 we presented the distributions and correlations (i.e., data points scatter plots) for all four attributes of the *Iris* plant. Thus, we can easily see both the distribution of each characteristic of the flower, and the scatter plots of pairs of data, grouping the *Iris* flower attributes. Hence, we can intuitively get an idea about the distribution of each feature of the flower, and about the connection between its dimensions.

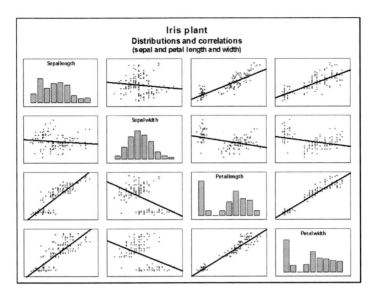

Fig. 3.65 Distributions and correlations of the *Iris* flower dimensions (length and width)

Fig. 3.66 illustrates, using the "Box & Whiskers" visualization technique, some statistical characteristics (mean, mean ± SD, mean ± 1.96 × SD), corresponding to the four attributes.

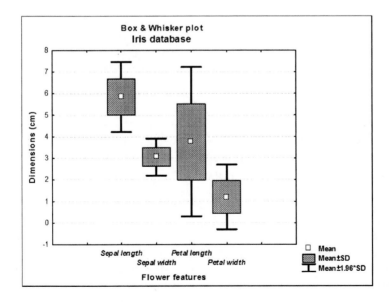

Fig. 3.66 *Iris* flower (Box & Whiskers plot)

Considering only two attributes: length and width of petals, we convert the dataset into a multidimensional array, as follows (summarized in Table 3.29):

- We discretize the attributes regarding the two petal dimensions (width and length), initially represented as continuous data, to obtain categorical values, such as: *low*, *medium*, and *high*;
- We count the plants according to the three size types.

Table 3.29 *Iris* plants -attributes discretization

Petal length	Petal width	Flower types	Count attribute
Low	Low	Setosa	46
Low	Medium	Setosa	2
Medium	Low	Setosa	2
Medium	Medium	Versicolour	43
Medium	High	Versicolour	3
Medium	High	Virginica	3
High	Medium	Versicolour	2
High	Medium	Virginica	3
High	High	Versicolour	2
High	High	Virginica	44

In this way, each triplet (petal length, petal width, flower type) identifies an item in the multidimensional array, counted then in the last column. Fig. 3.67 illustrates this process through the OLAP 'cube'.

The main OLAP operations are:

- *Building the data hypercube*, i.e., the 'geometric' data structure that allows fast analysis of data. The OLAP cube can be thought of as an extension to the multidimensional array of a spreadsheet, hence the name of *hypercube*. Technically, the data cube is a multidimensional representation of data, together with all possible aggregates, i.e., the aggregates that result by selecting a proper subset of the dimensions and summing over all remaining dimensions (e.g., in the *Iris* plant case, if the type species represents the chosen dimension, then, by summing over all other dimensions, the result will be a one-dimensional entry with three entries, each of which giving the number of flowers of each type. In marketing studies, an OLAP cube corresponding to the financial analyses of a company may have as dimensions: product code, time-period, geographical regions, type of revenue, costs, etc. When choosing, for instance, the time-period as the main dimension, then we can sum over all other remaining dimensions, representing thus all possible aggregates).

- *Slicing*, that is the operation of selecting a group of cells from the hypercube, by specifying a particular value for one or more dimensions. Thus, a *slice* is a subset of the hypercube corresponding to a single value for one or more dimensions not in the subset.
- *Dicing* involves selecting a subset of cells by specifying a range of attribute values, in other words, the dicing procedure is a slicing on more than two dimensions.
- *Roll-up* involves computing all of the data relationships for one or more dimensions; a computational relationship or formula might be defined (e.g., for sales data, we *roll-up* (aggregate) the sales across all the time-period in a year).
- *Drill-down/up* means a specific analytical technique whereby the user navigates among levels of data ranging from the most summarized (*up*) to the most detailed (*down*) (e.g., given a view of the data where the time dimension is broken into months, we could split the monthly sales totals (drill down) into daily sales totals).
- *Pivot*, that is changing the dimensional orientation of a report or page display.

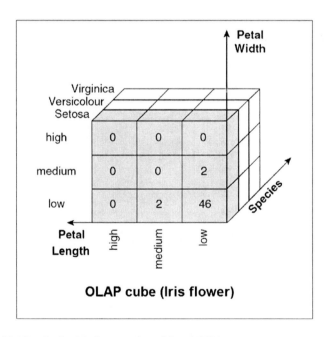

Fig. 3.67 OLAP cube for *Iris* flower (adapted from (378))

Remark 3.18. 1) The specific terminology concerning the OLAP technique may be seen at "OLAP and OLAP Server Definitions", Copyright January 1995, The OLAP Council (http://www.olapcouncil.org/research/glossaryly.htm); "Glossary of Data Mining Terms", University of Alberta, Copyright August 5th, 1999, Osmar R. Zaane (http://webdocs.cs.ualberta.ca/~zaiane/courses/cmput690/glossary.html).

2) Slicing and dicing can also be accompanied by aggregation over some dimensions.

3) Data from the hypercube can be updated at any time (possibly by different people), can be dynamically linked with other hypercubes, and 'alerts' may be sent when certain totals become 'outdated' because of subsequent updates.

For example, after the building of the data 'cube' for the *Iris* plant, using the slicing operation, we obtain the following 'tables-slices' (Fig. 3.68, Fig. 3.69, and Fig. 3.70).

		Width		
		low	medium	high
Length	low	46	2	0
	medium	2	0	0
	high	0	0	0

Setosa

Fig. 3.68 Setosa cube slice

		Width		
		low	medium	high
Length	low	0	0	0
	medium	0	43	3
	high	0	2	2

Versicolour

Fig. 3.69 Versicolour cube slice

		Width		
		low	medium	high
Length	low	0	0	0
	medium	0	0	3
	high	0	3	44

Virginica

Fig. 3.70 Virginica cube slice

The figures below present sample interfaces of three BI commercial products, which include the OLAP technology:

- The commercial product IBM Cognos TM1 software (http://www-01.ibm.com/so ftware/data/cognos/products/tm1/features-and-benefits.html).
- The commercial product Oracle Essbase software (http://www.oracle.com/appser ver/business-intelligence/essbase.html).
- The commercial product SAP Business Information Warehouse (SAP BW) (http://en.sap.info/putting-group-operations-on-track/1768).

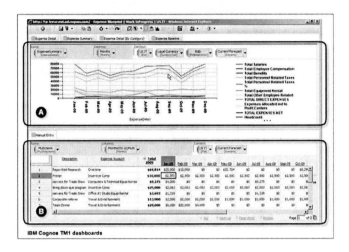

Fig. 3.71 IBM Cognos TM1

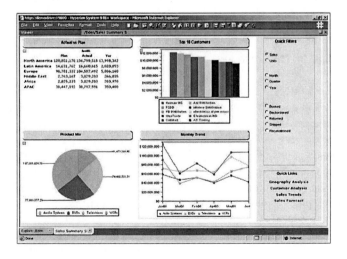

Fig. 3.72 ORACLE Essbase

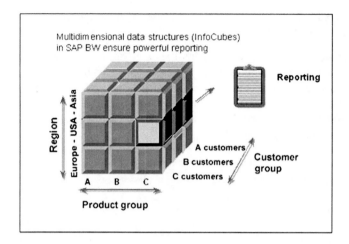

Fig. 3.73 SAP Business Information Warehouse

We list below some of the most popular OLAP links:

- Alta Plana-OLAP (http://altaplana.com/olap/);
- Pentaho BI and Reporting (www.pentaho.org);
- OLAP Info.de (http://www.olapinfo.de/);
- BI Verdict (http://www.bi-verdict.com/);
- Business Intelligence: OLAP Tool Selection (http://www.1keydata.com/datawa-rehousing/toololap.html):
    - OLAP Business Solutions (http://www.obs3.com/);
    - OLAP Business Intelligence (http://www.bipminstitute.com/olap/olap-in-bus-iness-intelligence.php).

3.9 Anomaly Detection

The last issue addressed in this chapter refers to the process of detecting anomalies or outliers in data. The simplest definition for the somewhat equivalent terms: *anomaly*, *extreme value*, or *outlier*, is given in Statistics and refers to that value which is found "very far" from the rest of data, representing actually a 'singularity' (an isolated point) of the dataset. In other words, they are atypical, infrequent observations, that is data points which do not appear to follow the characteristic distribution of the rest of the data. Their existence may reflect genuine properties of the underlying phenomenon, or be due to measurement errors or other facts which should not be modeled. Typically, we consider that *outliers* represent a random error that we would like to be able to control. The occurrence of these values is conceptually normal, even the Gaussian (Normal) distribution assumes their existence (the extremities of the Gauss (bell) curve). Needless to say, outliers are capable of considerably changing the data pattern. We displayed below (Fig. 3.74) the existence of anomalies/outliers in data, visualizing them by the black spots.

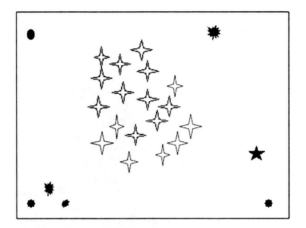

Fig. 3.74 Anomalies/outliers in data

A very simple example in this sense may refer to data representing the height of a population. Although most individuals have heights more or less close to the average height, there are also rare cases of very small or very great heights. Fig. 3.75 illustrates such an existence of anomalies/outliers in a database concerning individuals' heights, together with the Box & Whisker representation of data.

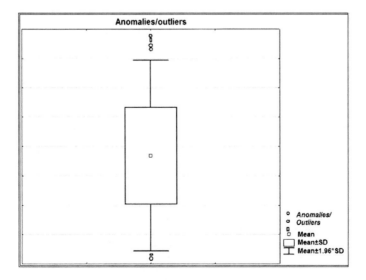

Fig. 3.75 Anomalies/outliers in a database (Box & Whiskers plot)

It should be noted that the non-Gaussian distributions, i.e., asymmetric distributions (see Section 3.5.), may have such values (placed in the curve's 'tail'). By visually examining the data distribution, and corroborating it with the nature of the problem that we have to solve, we can decide to keep or remove them from data.

The histogram below suggestively shows the effective way provided by the graphical representation of data to identify anomalies (extreme left and right columns).

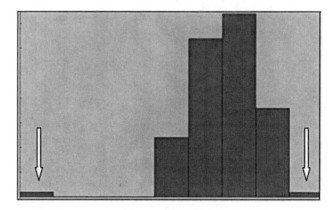

Fig. 3.76 Histogram illustrating the anomalies existence

Fig. 3.77 shows a linear regression model with an extreme value (the black spot outside the 95% confidence interval).

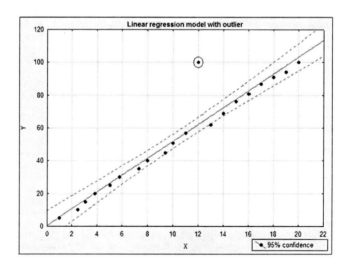

Fig. 3.77 Linear regression model with one outlier

Concluding the discussion on the existence of these atypical values, it is useful to remember that their occurrence can significantly influence the estimation of errors (e.g., Minkowski-R error, the method of least squares ($R = 2$), etc.). Estimates and models that are not significantly influenced by the existence of anomalies/outliers are called *robust* (e.g., robust statistics -(197); robust control models in economics -(171); robust regression -(13)).

We recall in this context that, for instance, the median is more robust than the mean in case of outliers' presence (see example regarding individuals' height, Subsection 3.2.1.); when using Minkowski error, $R = 1$ corresponds to the median, resulting thus a smaller error than for $R = 2$, which corresponds to the mean, (34). Thus, starting from the median, one can consider different types of anomalies (e.g., if the anomaly value is larger than three times the median and less than five times, then the anomaly is of type I, and so on). For multidimensional data, such comparisons can be made on each component. For more details about the statistical approach regarding this issue, see also (269).

Among the applications of anomaly detection, viewed in the light of data mining, we can mention:

- Credit card fraud detection;
- Cryptography and network intrusion detection (intrusion detection systems - IDS);
- Failures detection of different systems (fault detection);
- System health monitoring;
- Video surveillance;
- Prevention of virus and worm attacks;
- Detection of abnormal phenomena with major impacts on the local or global ecological balance, (ozone layer depletion, pollution, etc.).

When we conduct an analysis to identify anomalies in data (using an unsupervised technique), one starts from the assumption that there will be many more 'normal' values in data (their vast majority) than atypical values.

Starting from this assumption, there are two stages to detect anomalies:

1. Construction of the pattern profile of 'normal' data;
2. The use of this profile to identify anomalies, based on measuring the difference between the genuine data pattern and the 'normal' pattern.

As techniques used in this process we may include:

- Graphical methods;
- Statistical methods;
- Measuring distance-based methods;
- Models-based methods.

A. *Graphical methods.* From the graphical methods used in detecting anomalies, the most used are:

- *Box & Whiskers plot* -see Fig. 3.78

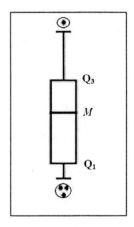

Fig. 3.78 Box & Whiskers plot (outliers are circled)

- *Two-dimensional scatterplots* -see Fig. 3.79

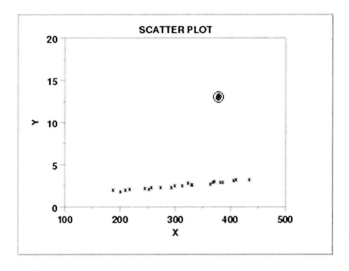

Fig. 3.79 Scatterplot (outlier is circled)

- *Convex hull* -see Fig. 3.80
  In this last case, the only notable problem is that there are anomalies within the convex cover (the two black spots inside the convex hull).

  The graphical method has the advantage that it is very suggestive, but, at the same time, it is subjective, leading thus to errors in detecting anomalies.

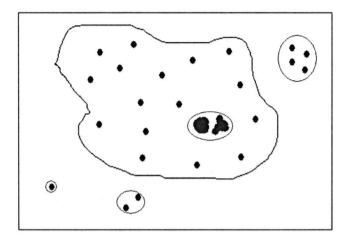

Fig. 3.80 Convex hull (outliers are circled)

B. *Statistical methods.* Regarding the use of statistical methods, in principle one first assumes a certain pattern of data distribution (the pattern of typical data), and then one uses different statistical tests to identify abnormal values in relation to this model. The statistical tests concern:

- The distribution type;
- The distribution parameters;
- The confidence interval.

Depending on the result of the data pattern analysis, corresponding to the studied phenomenon, one can identify those possible values, atypical for the given situation (i.e., anomalies), based on statistical testing. In the following lines we present two of the most popular statistical methods for anomaly detection.

- *Grubbs' test for outliers* (*maximum normed residual test*, (158), (363)) applies to univariate data, relying on the assumption that data are Normally distributed, and detecting an anomaly at each step. Concretely, the Grubbs' test compares the following statistical hypotheses:

  - H_0: There are no anomalies in data (*null hypothesis*),
  - H_1: There is at least one anomaly in data (*alternative hypothesis*),

  using the statistics $G = \dfrac{\max |X - \overline{X}|}{SD}$, where $\overline{X}$ represents the sample mean and SD denotes the standard deviation (*two-sided test*). Thus, one rejects the null hypothesis if $G > \dfrac{N-1}{N} \cdot \sqrt{\dfrac{t^2_{(\alpha/2N,N-2)}}{N-2+t^2_{(\alpha/2N,N-2)}}}$, where t represents the *t-Student*'s distribution and $t^2_{(\alpha/2N,N-2)}$ denotes the critical value of the t-distribution with $(N-2)$ degrees of freedom, and a significance level of $\alpha/2N$ (for further details, see (118)). At each step, using the test above, one possible anomaly is identified, it is afterwards removed from data, and the procedure is repeated (modifying the parameters) until there is none anymore.

• *Likelihood test* applies if the available dataset contains elements that belong to two distinct distributions, one denoted M -distribution law of most data, and the other denoted A -distribution law of anomalies, which are supposed to be in the minority. Since the dataset contains elements belonging to both categories, it results that the overall distribution of data, denoted D, is a mixed one, and might be a combination such as $D = (1 - \alpha) \cdot M + \alpha \cdot A$.

Specifically, the M distribution is estimated from the available data using classical statistical methods, while the A distribution is usually assumed to be the uniform distribution (starting from the assumption that anomalies are uniformly distributed, which is not always consistent with reality). Obviously, if we are in the situation to have a deeper understanding of the phenomenon, one can consider a specific distribution of the corresponding anomalies. The next step consists in the computation of the likelihood corresponding to the D distribution of data at time t. We recall that the term 'likelihood' was introduced in Statistics by Sir Ronald A. Fisher in 1922, (112), in the context of the "maximum likelihood" method, technique used by Carl Friedrich Gauss in the development of the '*least squares*' procedure (around 1794); it is worth mentioning that this method represents one of the oldest and most effective estimation techniques. Briefly, in case of discrete distributions, if we consider the sample $\{x_1, x_2,, x_n\}$ from a discrete random variable X, with the probability mass function $p(x, \theta)$, depending on the parameter θ, then the corresponding *likelihood* is given by:

$$L(\theta) = L(x_1, x_2, ..., x_n; \theta) = P\{X = x_1, X = x_2, ..., X = x_n; \theta\} =$$
$$p(x_1, \theta) p(x_2, \theta)...p(x_n, \theta).$$

On the other hand, for continuous distributions, with the probability density function $f(x, \theta)$, which depends on the parameter θ, the corresponding likelihood is given by:

$$L(\theta) = L(x_1, x_2, ..., x_n; \theta) = P\{x_1 < X_1 < x_1 + h, ..., x_n < X_n < x_n + h; \theta\} =$$
$$= h^n f(x_1, \theta) f(x_2, \theta)...f(x_n, \theta).$$

Returning now to the likelihood test applied to anomalies, the likelihood corresponding to the mixed D distribution of data at time t is given by the formula:

$$L_t(D) = \prod_{i=1}^{N} p_D(x_i) = \left[(1 - \alpha)^{|M_i|} \cdot \prod_{x_i \in M_i} p_{M_i}(x_i) \right] \cdot \left(\alpha^{|A_i|} \cdot \prod_{x_i \in A_i} p_{A_i}(x_i) \right).$$

We remark that, usually, in order to simplify formulas, the likelihood function is not used, but instead its logarithm (*log-likelihood*) is employed. Coming back again, after this short parenthesis, to the likelihood test, it is assumed that, initially, all data are 'typical' data (obeying the distribution law M), so there is no anomaly in data. Accordingly, one calculates the likelihood function $L_t(D)$ at time t. Then, each element x_t (corresponding to the M distribution) is temporary transferred to the A distribution, and the likelihood function $L_{t+1}(D)$, corresponding to time $(t + 1)$, is recalculated. Now, the difference $\Delta = L_t(D) - L_{t+1}(D)$ is calculated, and the following rule: "If $\Delta > c$ (c is a constant threshold), then x_t is considered (definitely) an anomaly, otherwise it is considered as typical value" is applied (for details, see (378)).

Remark 3.19. The weaknesses of the statistical testing methods can be summarized, in principle, in the following two aspects:

1. The tests are usually applied in case of a single attribute and not to the entire sequence of attributes;
2. The data distribution is generally unknown, but only estimated. In the case of large size datasets, this estimate is very difficult to be made (the legendary *'curse of dimensionality'*).

C. *Measuring distance-based methods.* For the methods based on distance measuring, data are represented as vectors, belonging to a certain metric space. We recall two classical methods for anomaly detection:

- *Nearest-neighbor method.* In the pattern recognition field, the "k-nearest neighbor" (k-NN) method represents the technique to classify an object based on the closest (k) objects in its neighborhood Thus, we lay stress on grouping objects based on their (closest) neighborhoods -for technical details about this methodology, see Chapter 5, Section 5.6. In case of anomaly detection, one starts from the initial step of computing the distances between each two elements of the dataset (aiming thus to identify the *'neighbors'*). Then, one moves onto the next step, focused on the way of defining that 'thing' that might be called an anomaly in data. Thus, an anomaly can be defined, for example, as the value for which there is a significant smaller number of 'neighbors' in the dataset (i.e., below a predefined threshold). There are other methods of identifying anomalies, on a case-by-case basis. It is to be noted that, when using this method, its effectiveness is disappointing when considering spaces whose dimensions are large, in which case the notion of proximity (neighborhood) is difficult to be properly defined (see the *curse of dimensionality*). In this case, one can use the dimensionality reduction technique by projecting the initial space on a smaller dimensional space, in which one can operate more efficiently with the projections of the presumed anomalies -see also (378).
- *Clustering method.* This well-known technique aims to divide the dataset into data clusters, based on the similarity between them, and to identify anomalies (atypical values) by highlighting their singular position towards the clusters consisting of 'normal' (typical) values (see Fig. 3.81). Schematically, the available

data are grouped into more or less homogeneous clusters, based on measuring the similarity between them (considering in this regard a similarity 'threshold' -for technical details, see Chapter 5, Section 5.8.). Then, one considers as anomalies those points belonging to clusters with a sufficiently small number of members (starting from clusters with one element, if any). Comparing the distance between these apparent (presumed) anomalies and the 'typical' clusters, we will decide whether or not those values truly represent anomalies.

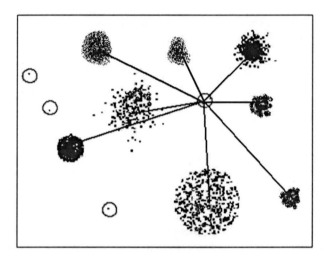

Fig. 3.81 Clustering technique for anomaly detection

D. *Models-based methods.* The methods based on models use data mining (modeling) techniques to identify anomalies in huge datasets, mostly consisting of 'normal' data (typical for the phenomenon under analysis). Here we summarize some of these techniques and show how to apply them in this matter.

- *Classification* uses a large enough number of data, both 'typical' and 'atypical', to build a classification model with two decision categories: (*a*) 'normal' data, and (*b*) 'anomalies'. Then, the model is applied to new data, circumscribed to the same context, aiming to detect any possible anomaly (for details, see Chapter 4 and Chapter 5, Section 5.2.).
- *Machine learning*. In this case, different kinds of learning machines (e.g., artificial neural networks, support vector machines, etc.) are trained in anomaly recognition and then, in detecting them in new datasets.
- *Autoregressive models*, concerning the detection of changes that might intervene in the phenomenon corresponding to some used time series, for instance, in forecasting problems to detect anomalies (see also Subsection 3.6.5.).

Finally, we mention some websites associated with anomaly detection:

- Cisco Anomaly Detection and Mitigation (https://www.cisco.com/en/US/ products/ps5879/Products_Sub_Category_Home.html).
- Oracle Data Mining Concepts Anomaly detection (http://download.oracle.com/ docs/cd/B28359_01/datamine.111/b28129/anomalies.htm).
- IBM -Proventia Network Anomaly Detection System (ADS) (http://www-935. ibm.com/services/uk/index.wss/offering/iss/y1026942).
- SAS -Social Network Analysis (http://www.sas.com/solutions/fraud/ social-network/index.html).

Chapter 4
Classification and Decision Trees

Abstract. One of the most popular classification techniques used in the data mining process is represented by the *classification and decision trees*. Because after accomplishing a classification process, a decision is naturally made, both labels are correctly inserted in its name, though they are usually used separately (i.e., classification trees or decision trees). From now on, we will call them just decision trees, since it represents the final goal of this model. The greatest benefit to use decision trees is given by both their flexibility and understandability. This chapter will present a short overview concerning the main steps in building and applying a decision tree in real-life problems.

4.1 What Is a Decision Tree?

Of the classification methods presented in Subsection 1.4.1 we will investigate here the decision trees model. In principle, decision trees are used to predict the membership of objects to different categories (classes), taking into account the values that correspond to their attributes (predictor variables). As we have mentioned above, the decision tree method is one of the main data mining techniques. The flexibility of this technique makes it particularly attractive, especially because it presents the advantage of a very suggestive visualization (a 'tree' which synthetically summarizes the classification). However, it should be stressed that this technique must necessarily be corroborated with other traditional techniques, especially when their working assumptions (e.g., assumptions about data distribution) are checked. Nevertheless, as an experimental exploratory technique, especially when traditional methods cannot be available, decision trees may successfully be used, being preferred to other classification models. Although decision trees are not so widespread in the pattern recognition field from a probabilistic statistical point of view, they are widely used in other domains such as, for instance, medicine (diagnosis), computer science (data structures), botany (classification), psychology (behavioral decision theory), etc. We illustrate the use of decision trees with a classic example concerning the diagnosis of myocardial infarction, given in a standard reference, (43). Thus, when a heart

F. Gorunescu: Data Mining: Concepts, Models and Techniques, ISRL 12, pp. 159–183.
springerlink.com   © Springer-Verlag Berlin Heidelberg 2011

attack patient is admitted to a hospital, dozens of tests are performed to assess its medical state. Among them, one can mention: heart rate, blood pressure, electrocardiogram (ECG or EKG), etc. Other features such as patient's age, sex, medical history, etc. are also recorded. In addition, patients can be subsequently tracked to see if they survive the heart attack, say, at least 30 days. This complex medical process aims to identify both the risk factors and the profile of the patient with high risk of myocardial infarction. Initially (1984), a relatively simple decision tree has been built, focused on three issues, synthesized in the following assertion: "**If** the patient's minimum systolic blood pressure over the initial 24-hour period is greater than 91, **then if** the patient's age is over 62.5 years, **then** if the patient displays sinus tachycardia, **then and only then** the patient is predicted not to survive for at least 30 days".

Let us see how we can formulate the problem and, especially, how we can "grow a tree from a seed" in this case. The figure below (Fig. 4.1) shows the schema behind the growing process of a decision tree (i.e., training dataset and the corresponding decision tree). The dataset consists of both continuous attributes (age), and categorical attributes (car type and risk of accident).

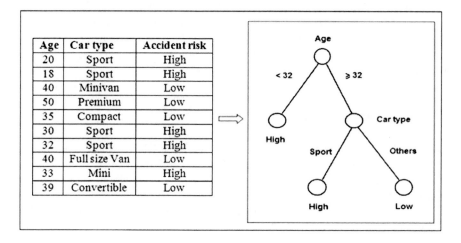

Fig. 4.1 Decision tree schema (training dataset and tree diagram)

Recall in this context that the procedure to build (to "grow") a decision tree represents an inductive process and, therefore, the established term is "*tree induction*". As it is easy to see from the figure above, the classification obtained by the decision tree induction can be characterized by the following:

- Each tree's (internal) node (i.e., non-terminal node) expresses the testing based on a certain attribute;
- Each tree's 'branch' expresses the test's result;
- The 'leaf' nodes (i.e., terminal nodes) represent the (decision) classes.

Remark 4.1. 1) The decision trees have three classical approaches:

1. *Classification trees*, term used when the prediction result is the class membership of data;
2. *Regression trees*, when the predicted result can be considered as a real number (e.g., oil price, value of a house, stock price, etc.);
3. *CART* (or *C&RT*), i.e., *Classification And Regression Tree*, (43), when we take into consideration both cases above.

2) Starting from a given dataset, we can build more than one decision tree; each tree structure depends on the order in which we choose the splitting attributes. Thus, in the figure above, one decision tree starts with attribute "Age" , followed by attribute "Car type", whilst another decision tree could be made following the sequence "Car type" and "Age".

Regarding the problem of decision tree induction, we present below some of the well-known algorithms (software), used over time (Fig. 4.2).

Hunt	SPRINT
CART (C&RT)	QUEST
ID3	DTREG
C4.5 and C5.0	THAID
SLIQ	CHAID

Fig. 4.2 Well-known decision trees algorithms

4.2 Decision Tree Induction

Since there currently are specialized software to build decision trees, we will not enter into technical details concerning this issue. What we wish to remember here is the general way of inducing a decision tree, and, therefore, we further present Hunt's algorithm <*Hunt's Concept Learning System*, 1966>, (199), one of the first algorithms to build a decision tree; thus, we illustrate the clear and simple philosophy behind this process.

Conceptually, Hunt's algorithm lies in the following steps:

Hunt's algorithm

1. Denote by D_t the set of training objects (data) that reach node t;
2. If D_t is an empty set, then t is a terminal node (a leaf node), labeled by the class Φ_t;

3. If D_t contains objects that belong to the same class C_t, then t is also a leaf node, labeled as C_t;
4. If D_t contains objects that belong to more than one class, then we use an attribute test to split the objects into smaller subsets.

We recursively apply this procedure to each subset of objects (non-terminal node).

Fig. 4.3 synthetically illustrates this classification process.

Age	Car type	Accident risk
20	Sport	High
18	Sport	High
40	Minivan	Low
50	Premium	Low
35	Compact	Low
30	Sport	High
32	Sport	High
40	Full size Van	Low
33	Mini	High
39	Convertible	Low

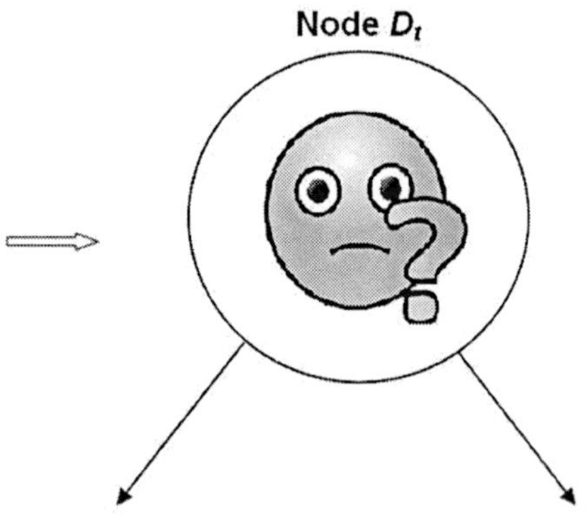

Fig. 4.3 How to split a node?

Example 4.1. An example of applying the above algorithm, aiming to identify the customer profile regarding the place where he/she goes shopping (store (shop) or su-permarket), is shown in the Table 4.1 below (adapted from (378)). Thus, a database with past records concerning individuals who went shopping either to shop or su-permarket has been considered.

The predictive attributes consist of taxable income (continuous variable), car ownership and marital status (categorical variable). The two decision classes are: buy from shop (YES), or does not buy from shop (NO), in other words he/she is a supermarket customer.

Building a decision tree based on these already recorded data (training data), the supermarket managers can decide if a certain person, unknown yet from the point of view regarding the way of shopping, is likely to buy from their supermarket.

The graphical scheme regarding the decision tree induction process, trained on the above dataset, is shown in Fig. 4.4.

Table 4.1 Training dataset for building a decision tree

Taxable income	Car ownership	Marital status	Buy from shop
125,000	Yes	Single	NO
100,000	No	Married	NO
70,000	No	Single	NO
120,000	Yes	Married	NO
95,000	No	Divorced	YES
60,000	No	Married	NO
220,000	Yes	Divorced	NO
85,000	No	Single	YES
75,000	No	Married	NO
90,000	No	Single	YES

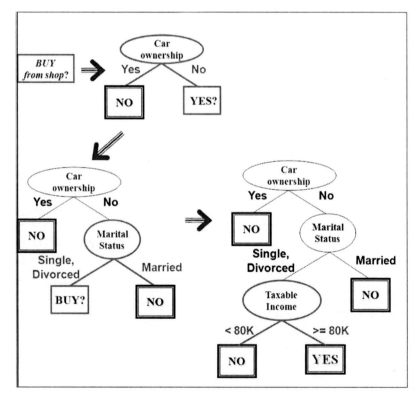

Fig. 4.4 Scheme of a decision tree induction

Starting from the example shown above, we now try to synthesize the methodology concerning the induction of a decision tree. As it is easily observed, a decision tree represents a tool to discriminate between classes, that recursively divides the training dataset until getting the (final) 'leaves', i.e., the terminal nodes that consist of either objects of the same category or objects belonging to a dominant (majority) category. In this respect, any node of the tree which is not a 'leaf' is a split (partition) point based on a test attribute that determines the way to divide that node.

The idea underlying the optimal node splitting (partitioning) is similar to the 'greedy algorithm', i.e., a 'top-down' recursive construction, using the "divide and conquer" strategy. We recall that the greedy algorithms are algorithms that use meta-heuristics to solve problems by identifying local optima, and trying to find the global optimum, based on this approach. A classic example of such an approach is represented by the famous "traveling salesman problem" (TSP), in which, at each step, the nearest city that has not been traveled yet is visited. Regarding the concept "divide and conquer" (D&C), recall that it is based on the famous Latin syntagma "divide et impera", i.e., "divide and rule", and consists in the recursive division of a problem in two or more similar sub-problems, until they reach the degree of simplicity that allows us to obtain their solutions; afterwards, starting from the solutions of these sub-problems, one tries to solve the original problem.

The above algorithm has many classification variants, from which we mention here:

- ID3, C4.5 and C5.0 -Machine learning;
- CART (C&RT) -Statistics;
- CHAID -Pattern recognition.

In principle, the methodology concerning the decision tree induction consists of two phases:

- Building the initial tree, using the available training dataset until each 'leaf' becomes 'pure' (homogeneous) or almost 'pure';
- 'Pruning' the already 'grown tree' in order to improve the accuracy obtained on the testing dataset.

We do not enter here into technical details regarding these two fundamental aspects of building a decision tree, since the literature is rich enough in information about this subject. We will synthetically present only a scheme of the classical algorithm underlying this procedure. This algorithm refers to binary decision trees, (340).

Tree building algorithm
Make Tree (Training Data T)
{

       Partition(T)

}

SPRINT algorithm (node S)
Partition(Data S)
{

    if (all points in S are in the same class) **then**

      return

    for each attribute A **do**

      evaluate splits on attribute A;

    use the best found split to partition S into S_1 and S_2

    Partition(S_1)

    Partition(S_2)

}

Remark 4.2. 1) To measure the effectiveness of the node splitting, several splitting indices (criteria) have been proposed.

2) During the tree building, the task at each node consists in the determination of the splitting point (i.e., the corresponding attribute's value) which divides in the best way the objects at that node, aiming to obtain an optimal purity (homogeneity) of the subsequent nodes.

3) Mathematically speaking, each object (record) from the training dataset is represented by a vector of type $(\mathbf{x}; y_j) = (x_1, x_2, ..., x_k; y_j)$, where there are m different classes (categories) of objects y_j, $j = 1, 2,..., m$, and k attributes on which the classification (decision) process is based; in this context we may note that even the variables y_j are attributes as well, but they are class 'label' attributes. Based on the (predictive) attributes $x_1, x_2, ..., x_k$, a unique class (label) y_j of the m available classes is assigned to the object represented by the vector $(\mathbf{x}; y_j)$.

Regarding the split indices (criteria), seen as measures of node 'impurity' or 'goodness-of-fit measures', we mention here the most commonly used:

- The *GINI (impurity) index*, used mainly in CART (C&RT) and SPRINT algorithms, represents a measure of how often a randomly chosen object from the training dataset could be incorrectly labeled if it were randomly labeled according to the distribution of labels in the dataset. As an impurity measure, it reaches a value of zero when only one class is present at a node. Conversely, it reaches its maximum value when class sizes at the node are equal, (89);

- The *entropy* (*impurity*), used mainly in ID3, C4.5 and C5.0, is based on selecting the splitting point that maximizes the information gain (i.e., maximum entropy reduction) (*idem*). Similarly, the minimum value is zero, when all records belong to one class, implying most information;
- The *misclassification measure*, based on classification error as its name suggests, is sometimes used to measure the node 'impurity' (*idem*). Once again, the minimum value is zero, when all records belong to one class, implying most information;
- *Chi-square measure*, which is similar to the standard *Chi-square* value computed for the expected and observed classifications, (184);
- *G-square measure*, which is similar to the *maximum-likelihood Chi-square* (*idem*).

All these measures are defined in terms of the class distributions of objects, considered both before and after splitting. We will now present in brief the first three above criteria for node splitting.

4.2.1   GINI Index

Let us denote by $f(i, j)$ the frequency of occurrence of class j at node i or, in other words, the proportion of objects belonging to class j that are distributed to node i (for m distinct classes of objects). Then, the *GINI index* is given by:

$$I_G(i) = 1 - \sum_{j=1}^{m} f^2(i, j). \tag{4.1}$$

When a "parent" node is split into p partitions ("children"), the quality of split is given by the *GINI splitting index*:

$$GINI_{split} = \sum_{i=1}^{p} \frac{n_i}{n} GINI(i). \tag{4.2}$$

The optimal split of a node is that ensuring the lowest GINI splitting index (ideally, zero).

Example 4.2. For a better understanding of the manner to compute the GINI index, on the one hand, and the way of using it to optimally split the tree nodes, on the other hand, we present below a very simple example, adapted from (340). Thus, we consider a sample of a dataset regarding the way of shopping, namely from shop or supermarket; the training dataset is given in Table 4.2. We consider the node splitting based on continuous attributes only, given here by "Taxable income".

From the training dataset we consider as possible splitting points (values) all the values taken by the continuous attribute, that is: 64,000; 21,000; 34,000; 55,000; 46,000; 15,000.

Table 4.2 Training dataset for node splitting (GINI index for continuous attributes)

Taxable income	Marital status	Buy from shop
64,000	Married	YES
21,000	Divorced	NO
34,000	Single	NO
55,000	Divorced	NO
46,000	Single	YES
15,000	Married	NO

Remark 4.3. This choice represents the so-called *"brute-force"* method, since it is the simplest but, at the same time, the most computationally expensive, the overall complexity being $O(N^2)$, where N represents the number of candidate splitting points. The alternative to this approach is represented by the reduction of the number of candidate splitting points. We can do this in two different ways. The first one is based on sorting all the possible splitting points in an ascendant order and, next, considering as candidate splitting values the midpoints between two adjacent sorted values; in this way, the overall computation complexity is $O(N * logN)$ -see (378). The second option uses a division of the set of possible splitting points into a certain number of intervals (not necessarily equal), and the candidate splitting values are identified by taking the endpoints of each interval.

What is the optimal splitting point, however, when using the 'brute-force' approach, for instance? To answer this question, we will compute both the GINI index and the GINI splitting index for each point separately. The corresponding calculations are shown below. Thus, from the training dataset it is easy to observe that the relative frequency for each class at node i is the following (see Table 4.3).

Based on the results from Table 4.3 and applying the appropriate formulas, we get:

$$I_{GINI}(Taxable\ income \leq 15,000) = 1 - [(1/1)^2 + (0/1)^2] = 0$$

$$I_{GINI}(Taxable\ income > 15,000) = 1 - [(3/5)^2 + (2/5)^2] = 12/25$$

$$GINI_{split}(Taxable\ income = 15,000) = (1/6) \times 0 + (5/6) \times (12/25) = 0.4$$

$$I_{GINI}(Taxable\ income \leq 21,000) = 1 - [(2/2)^2 + 0^2] = 0$$

$$I_{GINI}(Taxable\ income > 21,000) = 1 - [(2/4)^2 + (2/4)^2] = 1/2$$

$$GINI_{split}(Taxable\ income = 21,000) = (2/6) \times 0 + (4/6) \times 1/2 = 0.33$$

$$I_{GINI}(Taxable\ income \leq 34,000) = 1 - [(3/3)^2 + (0/3)^2] = 0$$

Table 4.3 Frequency of occurrence of class j at node i

Taxable income	Buy from shop = NO	Buy from shop = YES
$\leq 15,000$	1	0
$> 15,000$	3	2
$\leq 21,000$	2	0
$> 21,000$	2	2
$\leq 34,000$	3	0
$> 34,000$	1	2
$\leq 46,000$	3	1
$> 46,000$	1	1
$\leq 55,000$	4	1
$> 55,000$	0	1
$\leq 64,000$	4	2
$> 64,000$	0	0

$$I_{GINI}(Taxable\ income > 34,000) = 1 - [(1/3)^2 + (2/3)^2] = 4/9$$
$$GINI_{split}(Taxable\ income = \textbf{34,000}) = (3/6) \times 0 + (3/6) \times (4/9) = 0.22$$

$$I_{GINI}(Taxable\ income \leq 46,000) = 1 - [(3/4)^2 + (1/4)^2] = 3/8$$
$$I_{GINI}(Taxable\ income > 46,000) = 1 - [(1/2)^2 + (1/2)^2] = 1/2$$
$$GINI_{split}(Taxable\ income = 46,000) = (4/6) \times (3/8) + (2/6) \times (1/2) = 0.42$$

$$I_{GINI}(Taxable\ income \leq 55,000) = 1 - [(4/5)^2 + (1/5)^2] = 8/25$$
$$I_{GINI}(Taxable\ income > 55,000) = 1 - [(0/1)^2 + (1/1)^2] = 0$$
$$GINI_{split}(Taxable\ income = 55,000) = (5/6) \times (8/25) + (1/6) \times 0 = 0.26$$

$$I_{GINI}(Taxable\ income \leq 64,000) = 1 - [(4/6)^2 + (2/6)^2] = 4/9$$
$$I_{GINI}(Taxable\ income > 64,000) = 1 - [0^2 + 0^2] = 1$$
$$GINI_{split}(Taxable\ income = 64,000) = (6/6) \times (4/9) + 0 \times 1 = 0.44$$

As we can see, the lowest GINI splitting index is obtained for the taxable income equaling 34,000. Accordingly, we consider as optimal splitting point the value of 34,000, and the tree split in this case is illustrated in Fig. 4.5.

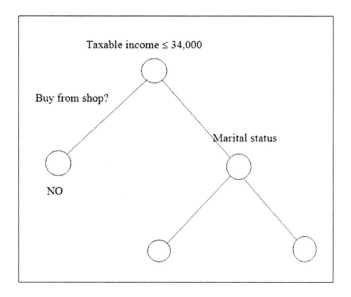

Fig. 4.5 Splitting node using continuous attributes

Similarly, we then proceed to recursively divide the tree's node, using the other attributes of objects in the training dataset, in our case the "marital status".

4.2.2 Entropy

Entropy index is used to select the optimal value for node splitting based on the information maximization or, equivalently, the maximum entropy reduction at that node. Thus, the splitting point chosen based on this method should maximize the information necessary to classify objects in the resulting partition. Therefore, if all objects have the same class label, then the node entropy (impurity) is zero, otherwise it is a positive value that increases up to a maximum when all classes are equally distributed. It is worth mentioning that the term 'entropy' from decision trees is related to the term "information gain" from Information theory and Machine learning, name often synonymous with "*Kullback-Leibler divergence*". Under these circumstances, *information gain* can be used to select an optimal sequence of attributes in order to most rapidly build a decision tree, i.e., to most rapidly reach the tree's leaves.

The entropy formula is given by;

$$Entropy(i) = I_E(i) = -\sum_{j=1}^{m} f(i,j) \cdot \log_2 [f(i,j)], \tag{4.3}$$

where, similarly to the GINI index, $f(i,j)$ is the frequency of occurrence of class j at node i (i.e., the proportion of objects of class j belonging to node i).

When a "parent" node is split into p partitions, the quality of split is given by the *entropy splitting index*:

$$Entropy_{split} = \sum_{i=1}^{p} \frac{n_i}{n} I_E(i) \tag{4.4}$$

Again, the optimal split of a node is the one that insures the lowest entropy splitting index (ideally, zero).

Example 4.3. Let us consider an example regarding continuous attributes, given here by the "Taxable income", to compute the entropy index and the subsequent splitting index (data from Table 4.3).

$I_E(Taxable\ income \leq 15,000) = -(1/1) \times log_2(1/1) = 0$

$I_E(Taxable\ income > 15,000) = -(3/5) \times log_2(3/5) - (2/5) \times log_2(2/5) =$
$= 0.97$

$I_E(Taxable\ income = 15,000) = (5/6) \times 0.97 = 0.81$

$I_E(Taxable\ income \leq 21,000) = -(2/2) \times log_2(2/2) = 0$

$I_E(Taxable\ income > 21,000) = -(2/4) \times log_2(2/4) - (2/4) \times log_2(2/4) = 1$

$I_E(Taxable\ income = 21,000) = (4/6) \times 1 = 0.67$

$I_E(Taxable\ income \leq 34,000) = 0$

$I_E(Taxable\ income > 34,000) = -(1/3) \times log_2(1/3) - (2/3) \times log_2(2/3) =$
$= 0.91$

$I_E(Taxable\ income = \mathbf{34,000}) = (3/6) \times 0.91 = \mathbf{0.46}$

$I_E(Taxable\ income \leq 46,000) = -(3/4) \times log_2(3/4) - (1/4) \times log_2(1/4) =$
$= 0.81$

$I_E(Taxable\ income > 46,000) = -(1/2) \times log_2(1/2) - (1/2) \times log_2(1/2) = 1$

$I_E(Taxable\ income = 46,000) = (4/6) \times 0.81 + (2/6) \times 1 = 0.87$

$I_E(Taxable\ income \leq 55,000) = -(4/5) \times log_2(4/5) - (1/5) \times log_2(1/5) =$
$= 0.72$

$I_E(Taxable\ income > 55,000) = 0$

$I_E(Taxable\ income = 55,000) = (5/6) \times 0.72 = 0.60$

$I_E(Taxable\ income \leq 64,000) = -(4/6) \times log_2(4/6) - (2/6) \times log_2(2/6) =$
$= 0.92$

$I_E(Taxable\ income > 64,000) = 0$

$I_E(Taxable\ income = 64,000) = (6/6) \times 0.92 = 0.92$

As we can see, the lowest entropy splitting index is obtained for the taxable income equaling 34,000. Accordingly, we consider as optimal splitting point the value of 34,000, which is the same as for the GINI index.

4.2.3 Misclassification Measure

Another impurity index that is sometimes used for splitting nodes is based on the misclassification measure. This index measures the classification error that can be made at a node using a certain splitting point, and is given by:

$$Error(i) = I_M(i) = 1 - \max_j f(i, j), \tag{4.5}$$

recalling that $f(i, j)$ is the proportion of objects of class j that are assigned to node i.

When a "parent" node is split into p partitions, the quality of split is given by the *error splitting index*:

$$Error_{split} = \sum_{i=1}^{p} \frac{n_i}{n} I_M(i). \tag{4.6}$$

Once again, we can see that the maximum error is obtained when the objects of different categories are equally distributed in that node, thereby providing the poorest information, whilst the minimum error is obtained if all objects at the node belong to the same category, thereby providing the richest information. To conclude, we choose the split that minimizes error.

Example 4.4. Consider again the example regarding the "Taxable income" (data from Table 4.3).

$I_M(Taxable\ income \leq 15,000) = 1 - max\{1,0\} = 1 - 1 = 0$

$I_M(Taxable\ income > 15,000) = 1 - max\{3/5, 2/5\} = 1 - 3/5 = 2/5$

$I_M(Taxable\ income = 15,000) = (1/6) \times 0 + (5/6) \times (2/5) = 1/3$

$I_M(Taxable\ income \leq 21,000) = 1 - max\{1,0\} = 1 - 1 = 0$

$I_M(Taxable\ income > 21,000) = 1 - max\{1/2, 1/2\} = 1 - 1/2 = 1/2$

$I_M(Taxable\ income = 21,000) = (2/6) \times 0 + (4/6) \times (1/2) = 1/3$

$I_M(Taxable\ income \leq 34,000) = 1 - max\{1,0\} = 1 - 1 = 0$

$I_M(Taxable\ income > 34,000) = 1 - max\{1/3, 2/3\} = 1 - 2/3 = 1/3$

$I_M(Taxable\ income = \mathbf{34,000}) = (3/6) \times 0 + (3/6) \times (1/3) = \mathbf{1/6}$

$I_M(Taxable\ income \leq 46,000) = 1 - max\{3/4, 1/4\} = 1 - 3/4 = 1/4$

$$I_M(Taxable\ income > 46,000) = 1 - max\{1/2, 1/2\} = 1 - 1/2 = 1/2$$
$$I_M(Taxable\ income = 46,000) = (4/6) \times (1/4) + (2/6) \times (1/2) = 1/3$$

$$I_M(Taxable\ income \le 55,000) = 1 - max\{4/5, 1/5\} = 1 - 4/5 = 1/5$$
$$I_M(Taxable\ income > 55,000) = 1 - max\{0, 1\} = 1 - 1 = 0$$
$$I_M(Taxable\ income = \mathbf{55,000}) = (5/6) \times (1/5) + (1/6) \times 0 = \mathbf{1/6}$$

$$I_M(Taxable\ income \le 64,000) = 1 - max\{4/6, 2/6\} = 1 - 4/6 = 2/6$$
$$I_M(Taxable\ income > 64,000) = 1 - max\{0, 0\} = 1 - 0 = 1$$
$$I_M(Taxable\ income = 64,000) = (6/6) \times (2/6) + 0 \times 1 = 1/3$$

Note that there are two values of the "Taxable income" providing a minimum value of error: 34,000 and 55,000 respectively.

At the end of this section we will show how to concretely build a decision tree, in two different cases: (*a*) for credit risk estimation, and (*b*) for *Iris* plant. In the first case, we used the entropy measure and we presented in details the corresponding computations. In the second case, we used the GINI index and we presented the corresponding decision tree only.

Example 4.5. Credit risk estimation -mixed binary/ternary decision tree; entropy measure. In the banking sector, the credit risk estimation when granting a loan to a certain individual is a particularly important issue. Starting from a fictitious database containing 20 individuals (see Table 4.4), we build a decision tree to assist the staff of a bank to safely grant loan, based on customer profile (three risk categories), and resulted from the decision tree. In this case, we consider four predictive attributes used to classify customers:

1. *credit history* -past behavior of an individual when his/her loan has been granted,
2. *debts* -past debts of the possible debtor,
3. *secured loan* -existence of some asset (e.g., a car or property) as collateral for the loan,
4. *taxable income*,

and a decision attribute regarding the risk estimation with three values (*low, moderate* and *high*).

For this example we use entropy as a measure of impurity.

First, we use the entropy splitting index to choose the attribute with which we begin to split the tree root.

Technically, we compute the entropy splitting index for all the four predictive attributes.

Table 4.4 Training dataset for credit risk

#	Credit history	Debts	Secured loan	Taxable income	Risk
1	bad	many	No	9,600-18,000	high
2	unknown	many	No	18,000-30,000	high
3	unknown	few	No	18,000-30,000	moderate
4	unknown	few	No	9,600-18,000	high
5	unknown	few	No	over 30,000	low
6	unknown	few	Yes	over 30,000	low
7	bad	few	No	9,600-18,000	high
8	bad	few	No	over 30,000	moderate
9	good	few	No	over 30,000	low
10	good	many	Yes	over 30,000	low
11	good	many	No	18,000-30,000	high
12	good	many	No	18,000-30,000	moderate
13	good	many	No	over 30,000	low
14	bad	many	No	18,000-30,000	high
15	unknown	many	No	18,000-30,000	high
16	unknown	few	No	18,000-30,000	moderate
17	bad	few	Yes	18,000-30,000	moderate
18	unknown	few	Yes	over 30,000	low
19	good	few	Yes	9,600-18,000	low
20	bad	many	No	9,600-18,000	high

A. Credit history

$I_E(Credit\ history = bad) = -(4/6) \times log_2(4/6) - (2/6) \times log_2(2/6) = 0.92$

$I_E(Credit\ history = unknown) = -(3/8) \times log_2(3/8) - (2/8) \times log_2(2/8) - (3/8) \times log_2(3/8) = 1.56$

$I_E(Credit\ history = good) = -(1/6) \times log_2(1/6) - (1/6) \times log_2(1/6) - (4/6) \times log_2(4/6) = 1.25$

$GAIN_{split}(Credit\ history) = (6/20) \times 0.92 + (8/20) \times 1.56 + (6/20) \times 1.25 = 1.29$

B. Debts

$I_E(Debts = many) = -(6/9) \times log_2(6/9) - (1/9) \times log_2(1/9) - (2/9) \times log_2(2/9) =$
$= 1.22$

$I_E(Debts = few) = -(2/11) \times log_2(2/11) - (4/11) \times log_2(4/11) - (5/11) \times$
$\times log_2(5/11) = 1.49$

$GAIN_{split}(Debts) = -(9/20) \times 1.22 - (11/20) \times 1.49 = 1.32$

C. Secured loan

$I_E(Secured\ loan = Yes) = 0.72$

$I_E(Secured\ loan = No) = 1.46$

$GAIN_{split}(Secured\ loan) = 1.28$

D. Taxable income

$GAIN_{split}(Taxable\ income) = 0.79$

Because the lowest index corresponds to taxable income, this will be the attribute based on which we divide the tree root. There are three categories regarding the "taxable income" that are taken into account: {9,600-18,000}, {18,000-30,000}, and {over 30,000}, so there are three 'children' nodes corresponding to these cases. Table 4.5 displays this situation in the case of "taxable income" {9,600-18,000}.

Table 4.5 Training dataset corresponding to "taxable income" {9,600-18,000}

#	Credit history	Debts	Secured loan	Taxable income	Risk
1	bad	many	No	9,600-18,000	high
4	unknown	few	No	9,600-18,000	high
7	bad	few	No	9,600-18,000	high
19	good	few	Yes	9,600-18,000	low
20	bad	many	No	9,600-18,000	high

The entropy splitting index corresponding to the remaining three attributes are given by:

$I_E(Credit\ history = bad) = 0$

$I_E(Credit\ history = unknown) = 0$

$I_E(Credit\ history = good) = 0$

$GAIN_{split}(Credit\ history) = 0$

$GAIN_{split}(Debts) = 0.55$

$GAIN_{split}(Secured\ loan) = 0$

Since both $GAIN_{split}(Credit\ history) = 0$ and $GAIN_{split}(Secured\ loan) = 0$, we have to choose which attribute will be used to further split the node. Let us select "Secured loan" as splitting attribute. Thus, this branch of the decision tree is split into two leaves: Yes $\rightarrow$ {Risk = low} and No $\rightarrow$ {Risk = $high$}.

Next, we consider the case of "taxable income" {18,000-30,000}. The corresponding dataset is given by Table 4.6 below.

Table 4.6 Training dataset corresponding to "taxable income " {18,000-30,000}

#	Credit history	Debts	Secured loan	Taxable income	Risk
2	unknown	many	No	18,000-30,000	high
3	unknown	few	No	18,000-30,000	moderate
11	good	many	No	18,000-30,000	high
12	good	many	No	18,000-30,000	moderate
14	bad	many	No	18,000-30,000	high
15	unknown	many	No	18,000-30,000	high
16	unknown	few	No	18,000-30,000	moderate
17	bad	few	Yes	18,000-30,000	moderate

Computing the related entropy splitting index, we have:

$I_E(Credit\ history = bad) = 1$

$I_E(Credit\ history = unknown) = 1$

$I_E(Credit\ history = good) = 1$

$GAIN_{split}(Credit\ history) = 1$

$GAIN_{split}(Debts) = 0.45$

$GAIN_{split}(Secured\ loan) = 0.87$

In this case, "Debts" is selected as splitting attribute, since its entropy splitting index is the lowest one. Accordingly, if {Debts = few} then we obtain a tree leaf {Risk = low}, otherwise we have to further split the node. Thus, for those with "taxable income" {18,000-30,000} and many debts, we have:

Table 4.7 "Taxable income" {18,000-30,000} and Debts = *many*

#	Credit history	Debts	Secured loan	Taxable income	Risk
2	unknown	many	No	18,000-30,000	high
11	good	many	No	18,000-30,000	high
12	good	many	No	18,000-30,000	moderate
14	bad	many	No	18,000-30,000	high
15	unknown	many	No	18,000-30,000	high

It is easy to observe that in this case we obtain the following node split: {Credit history = *bad*} → {Risk = *high*}, {Credit history = *unknown*} → {Risk = *high*}, {Credit history = *good*} → {Risk = *high* 50% / Risk = *moderate* 50%}.

Next, let us select the splitting node for a "taxable income" {over 30,000}. The corresponding data is presented below.

Table 4.8 Training dataset corresponding to "taxable income " {over 30,000}

#	Credit history	Debts	Secured loan	Taxable income	Risk
5	unknown	few	No	over 30,000	low
6	unknown	few	Yes	over 30,000	low
8	bad	few	No	over 30,000	moderate
9	good	few	No	over 30,000	low
10	good	many	Yes	over 30,000	low
13	good	many	No	over 30,000	low
18	unknown	few	Yes	over 30,000	low

The corresponding entropy splitting indices for the predictive attributes are the following:

$$I_E(Credit\ history = bad) = 0$$

$$I_E(Credit\ history = unknown) = 0$$

$$I_E(Credit\ history = good) = 0$$

$$GAIN_{split}(Credit\ history) = 0$$

$$I_E(Debts = few) = 0.72$$

$$I_E(Debts = many) = 0$$

$$GAIN_{split}(Debts) = 0.51$$

$$I_E(Secured\ loan = Yes) = 0$$

$$I_E(Secured\ loan = No) = 0.72$$

$$GAIN_{split}(Secured\ loan) = 0.41$$

Taking into account the values above, we choose as splitting attribute "Credit history", with the smallest value. Consequently, the node is split into three leaves: {Credit history = *bad*} → {Risk = *moderate*}, {Credit history = *unknown*} → {Risk = *low*}, and {Credit history = *good*} → {Risk = *low*}.

Finally, after passing through all the predictive attributes and corresponding splitting points, we have obtained the decision tree illustrated in Fig. 4.6

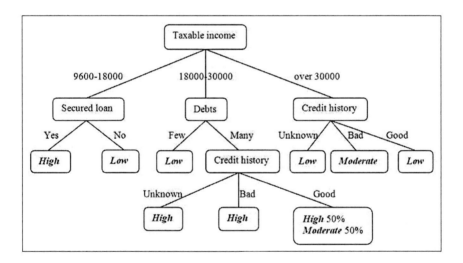

Fig. 4.6 Decision tree corresponding to *credit risk*

Example 4.6. Iris plant -binary tree; GINI index. We consider now the dataset related to the *Iris* plant, aiming to classify flowers depending on the four predictive attributes: petal length and width, and sepal length and width. To accomplish this task, we have chosen this time the GINI index as impurity measure. For the sake of simplicity, we present the decision tree only, together with the histogram illustrating the distribution of the three flower types in each node. The role of histograms is to show the leaves impurity, thus illustrating the classification accuracy. Table 4.9 summarizes in short the tree structure (i.e., child nodes, observed class *n*'s, predicted classes, and splitting conditions for each node).

Table 4.9 *Iris* plant -tree structure

Left branch	Right branch	Class SET (n)	Class VER (n)	Class VIR (n)	Predicted class	Split value	Split variable
2	3	50	50	50	SET	2.45	Petal length
-	-	50	0	0	SET	-	-
4	5	0	50	50	VER	1.75	Petal width
-	-	0	49	5	VER	-	-
-	-	0	1	45	VIR	-	-

Fig. 4.7 illustrates the tree structure presented above, showing in addition the corresponding histogram regarding the frequency of each flower type per node. As we mentioned above, we can in this way see the 'purity' of each node, especially the terminal nodes (the tree leaves).

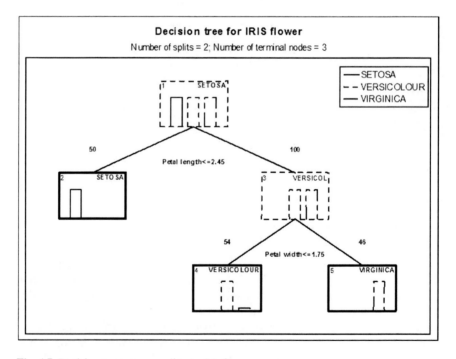

Fig. 4.7 Decision tree corresponding to *Iris* flower

4.3 Practical Issues Regarding Decision Trees

4.3.1 Predictive Accuracy

Obviously, the purpose of building a decision tree is to obtain a precise, as good as possible, prediction for a new entry data. Although it is extremely difficult, if not impossible, to define in an 'absolute' manner what the valence of the predictive accuracy is, we can however consider, in practical terms, some indicators of the prediction process, known as prediction 'costs'. Thus, an optimal prediction will therefore imply minimum erroneous classification costs. Essentially, the idea of prediction costs generalizes the fact that a better prediction would lead to a more reduced rate of wrong classification. From practice it resulted that, in most cases, it is not the classification rate only that is important, but also the consequences of an incorrect classification. Think only about a wrong medical diagnosis in the case of an incurable disease (e.g., cancer). What is the worst situation for a doctor: to say that a patient suffers from that disease when, in fact, he/she does not, or to decide that the patient is not suffering from that disease and the patient actually has the disease? This is not just a case of a wrongly diagnosed patient, but it is about the serious consequences of such a fact.

We present below the main costs involved in the classification process.

- *Prior probabilities* (or *a priori probabilities*) are those parameters that specify how likely it is, without using any prior knowledge, that an object will fall into a particular class. Thus, the general idea is that the relative size of the prior probabilities assigned to each class can be used to "adjust" the importance of misclassification for each class. Usually, when we do not have prior knowledge about the studied phenomenon, allowing us to make a clear choice, prior probabilities are chosen proportional to the number of objects in each class.
- *Misclassification costs* refer to the fact that the classification process usually requires a more accurate sorting of objects for some classes than for others, for reasons unrelated to relative class sizes. Referring to the example regarding the diagnosis of an incurable disease, liver cancer, for instance, naturally needs to be more accurately predicted than chronic hepatitis. Misclassification costs are chosen to reflect the importance of each class. When there is no preference for a certain class, we will take them equal.

It is worth mentioning that the relationships between prior probabilities and misclassification costs represent a quite complex issue in all but the simplest situations -for details, see (43), (317).

4.3.2 STOP Condition for Split

After establishing the criterion for splitting the nodes, we are facing the problem of selecting an appropriate way for stopping the process. A feature of the decision trees induction refers to the fact that the splitting process runs until all terminal nodes (namely 'leaves') are homogeneous ('pure') in terms of components, as long as there is no stop condition, halting the tree induction. In this case, each node will contain objects of the same class, fact that is most often counterproductive, especially taking into account that the process of tree induction represents the training phase of the classification model, and it is not the main decision tree objective, that of classifying new, unknown objects (real-life problems). In other words, it is not at all interesting that the classifier provides 100% training accuracy (pure 'leaves'), but on the testing dataset the performance is poor; the significant performance is thus the testing performance.

Usually, there are two *stopping rules* for controlling when splitting stops:

- *Minimum n.* One way for controlling when splitting stops is to allow splitting to continue until all terminal nodes ('leaves') are pure or contain no more than a specified minimum number of objects (hence the name minimum n). Thus, the splitting process will stop when all 'leaves' containing more than one class have no more than the specified number of objects.
- *Proportion of objects.* Another way for controlling when splitting stops is to allow splitting to continue until all terminal nodes are pure or contain no more cases than a specified minimum proportion of the sizes of one or more classes. Thus, the splitting process will stop when all 'leaves' containing more than one class have no more objects than the specified proportion of the class sizes for one or more classes.

4.3.3   Pruning Decision Trees

Once the decision tree built, based on the objects belonging to the training dataset, it is natural to think that it will, more or less, reflect the features of these objects. Many of its branches will be strongly influenced by anomalies that may be located in the training dataset, due to 'noise' or some outliers escaping the initial data-filtering process, if it has been previously performed. We note that, in principle, one can build a tree starting from the raw data directly, without prior processing, so that the decision tree built using them will fully reflect the particularities of this training dataset. Since a decision tree is built with the aim of being applied in various situations, implying different datasets, it is necessary to avoid a too good 'fit' (*overfitting* phenomenon) with the dataset used to train it. On the other hand, when the decision tree (the classifier) is too simple compared with the data used in the training process and, consequently, both the training and the testing errors are inadmissibly large, we are dealing with the reverse situation, that is a bad 'fit' (*underfitting* phenomenon) of the tree with data. However, we most often meet the first situation, the overfitting case. In this case, one uses the well-known method of 'trimming' (the *'pruning'* process) the tree. In principle, statistical methods are used to remove insignificant branches, providing redundant information, or which do not follow the general pattern of data, in this way obtaining a tree not too 'leafy', but with greater scalability and classification speed.

There are two types of pruning a decision tree:

- *Pre-pruning* (early stopping rule), which means that we practically stop the tree 'growth' during the tree induction process, deciding to halt splitting the node, so that it becomes a 'leaf', labeled with the name of the class with most elements. The main conditions for stopping the tree growth are either that all objects belong to the same class, or all the attribute values are the same. We note that there are also more restrictive stopping conditions to halt the tree growth (see, for instance, the stopping rules above).
- *Post-pruning*, which is used after the tree growth is completed, being a "bottom-up" approach, based on the error classification value. Thus, a node will be 'pruned' by removing its branches, consequently becoming 'leaf', labeled in the same manner as above, if the classification error is reduced by this operation.

Remark 4.4. Usually, we use the post-pruning approach, subsequent to the complete tree induction. The pre-pruning method follows its own 'philosophy' regarding the tree splitting, and can be controlled during this process, since the beginning of the tree induction.

As one can easily observe, in the pre-pruning case we need to quantify the classification error at each step of 'cutting' unneeded branches, and we also need criteria for its use to determine whether or not the classifier performance increases. In this regard, we mention the following:

- *Generalization errors* (testing errors) are based on estimating the testing performance and comparing it with the one achieved in practice. In some cases we use, in addition, a validation error (obtained on an extra validation dataset).
- *Occam's Razor* (principle of parsimony), originating in the work of the English logician, William of Ockham (c. 1285 - 1347/1349), a Franciscan philosopher and theologian, method with many applications in biology, medicine, religion, philosophy, statistics, etc. (see (349), (35), (127), (82)). It is based, in the decision trees case, on the following principles:

  - Given two decision trees of similar generalization errors, one should prefer the simpler tree over the more complex tree.
  - A more complex tree has a greater chance to be fitted accidentally by errors in data (i.e., to be errors dependent).
  - Accordingly, we should include the classifier complexity when evaluating a decision tree.

- *Minimum description length* -MDL, (318), is a criterion for the selection of models, regardless of their complexity, and represents a formalization of the above Occam's Razor in which the best hypothesis for a given dataset is the one that leads to the largest compression of the data. If assessing the optimal classifier, one uses the formula: Cost(Model, Data) = Cost(Data|Model) + Cost(Model), where Cost(Data|Model) encodes the misclassification errors and Cost(Model) uses node encoding (number of children) plus splitting condition encoding (more technical details in (319), (159); (378), (309)).

4.3.4 Extracting Classification Rules from Decision Trees

Once a decision tree built, the model is then used in optimal decision making. Knowledge involved in this 'arboreal' structure may be easily 'read' by climbing down along the tree's branches to its leaves (do not forget that, usually, a decision tree has its root on top), thereby extracting classification rules of IF-THEN form. A rule is extracted starting the 'path' from the top (where the tree's root is situated) and ending it at each leaf. Any pair of values of an attribute along this path will form a conjunction in the rule antecedent (condition), and the leaf containing the predictive class (which provides its label) will form the rule consequent.

We illustrate the process of extracting classification rules from decision trees using the example concerning the credit risk. For the sake of simplicity, we will only take into consideration a 'sub-tree', displayed in the figure below (Fig. 4.8).

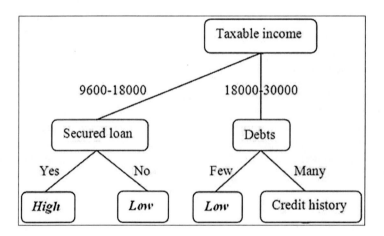

Fig. 4.8 'Sub-tree' used for classification rules extraction

The classification rules corresponding to the three leaves above are the following:

- IF 'Taxable income' = "9,600-18,000" and 'Secured loan' = "Yes" THEN 'Risk' = "*High*";
- IF 'Taxable income' = "9,600-18,000" and 'Secured loan' = "No" THEN 'Risk' = "*Low*";
- IF 'Taxable income' = "18,000-30,000" and 'Debts' = "Few" THEN 'Risk' = "*Low*";
- IF 'Taxable income' = "18,000-30,000" and 'Debts' = "Many" THEN 'Risk' involves "*Credit History*", and the splitting process will continue.

4.4   Advantages of Decision Trees

In the next chapter we will present other well-known models of classifiers. However, by comparing the classification models used in data mining, we remark some advantages of decision trees, which we list below.

- They are easy to understand and interpret, their graphical form representing a 'trump' in this respect;
- Inexpensive to be built. They require a small amount of training data compared with other classification techniques;
- They allow the use of both numerical and categorical data without any restriction;
- They represent models of the "white-box" type, in which the logic underlying the decision process can be followed easily, the classification rules being understood 'at first sight'. Unlike decision trees, other used classification techniques, such as artificial neural networks, act as "black-box" models, not directly providing the user with the classification rules;
- They use classical statistical techniques to make the model validation possible;
- They are robust, fast and process well large datasets;
- Finally, their accuracy is comparable to other classification techniques for many simple datasets.

Fig. 4.9 Tree of "Knowledge" (by decision)

Chapter 5
Data Mining Techniques and Models

Abstract. Data mining can also be viewed as a process of model building, and thus the data used to build the model can be understood in ways that we may not have previously taken into consideration. This chapter summarizes some well-known data mining techniques and models, such as: Bayesian classifier, association rule mining and rule-based classifier, artificial neural networks, k-nearest neighbors, rough sets, clustering algorithms, and genetic algorithms. Thus, the reader will have a more complete view on the tools that data mining borrowed from different neighboring fields and used in a smart and efficient manner for digging in data for hidden knowledge.

5.1 Data Mining Methods

In the previous chapter we presented the method concerning decision trees, regarded as a special classification technique, the reason why we dedicated it the entire chapter. As we showed in the first chapter, there are other advanced techniques that are used in data mining, both in classification and other areas of automatic data exploration, well-known methods, such as:

- Bayesian classifier/Naive Bayes;
- Neural networks;
- Support vector machines;
- Association rule mining;
- Rule-based classification;
- k-nearest neighbor;
- Rough sets;
- Clustering algorithms;
- Genetic algorithms.

A big part of this chapter will be devoted to the conceptual outline of these techniques, aiming to familiarize the reader with their main characteristics, using for this purpose simple and easy to understand examples, illustrating the methodology.

F. Gorunescu: Data Mining: Concepts, Models and Techniques, ISRL 12, pp. 185–317.
springerlink.com   © Springer-Verlag Berlin Heidelberg 2011

Note that for these techniques there is an entire library of specialized software, the goal of this presentation is to introduce the reader into the principles underlying those software and not with the actual design of the corresponding algorithms.

5.2 Bayesian Classifier

Let (Ω, Σ, P) be a probability space and $A_1, A_2,..., A_n$ a partition of the sample space Ω. We remind now a particularly important result of the Probability theory, based on which we can calculate the probability of any event belonging to the σ-field Σ.

Theorem 5.1. *(Total probability formula) Let B be an arbitrary event and $\{A_1, A_2,..., A_n\}$ a partition of the sample space Ω. Then:*

$$P\{B\} = \sum_{i=1}^{n} P\{B|A_i\}P\{A_i\}, \; P\{A_i\} > 0. \tag{5.1}$$

Recall that the English Presbyterian minister and mathematician Thomas Bayes discovered the following famous result, particularly important through its applications, which practically 'reverses' the total probability formula, (25) - published posthumously.

Theorem 5.2. *(Bayes' formula -1763) Let B be an arbitrary event and $\{A_1, A_2,..., A_n\}$ a partition of the sample space Ω. Then:*

$$P\{A_i|B\} = \frac{P\{B|A_i\}P\{A_i\}}{\sum_{i=1}^{n} P\{B|A_i\}P\{A_i\}}, \; P\{B\} > 0, P\{A_i\} > 0, \; i = 1, 2, ..., n. \tag{5.2}$$

Usually, $P\{A_i|B\}$ is known as *posterior probability (posterior)* because it is derived from, or depends upon, the specified value of B, $P\{A_i\}$ as *a priori probability (prior probability)* because it does not take into account any information about B, $P\{B|A_i\}$ as *likelihood*, and $P\{B\}$ as *evidence*. In this context, the Bayes' formula may be written as:

$$posterior = \frac{likelihood \times prior\ probability}{evidence}.$$

Remark 5.1. One can weaken the conditions of the two previous results if, instead of a partition of the sample space Ω we consider $\{A_1, A_2,..., A_n\}$ as a family of events, such as:

$$A_i \cap A_j = \emptyset, \forall i \neq j \text{ and } A_1 \cup A_2 \cup ... \cup A_n \supseteq B.$$

Example 5.1. ((287)) A factory produces the same items using three machines B_1, B_2 and B_3, whose production capacities are 60%, 30% and 10%, respectively (these percentages are virtually the probabilities that a certain item comes from one of three machines). Each machine has a rate of failure equaling 6%, 3% and 5%, respectively. What are the probabilities that randomly selected defective items were produced by machines B_1, B_2 and B_3, respectively? Thus, applying Bayes' formula, we have:

$$P\{B_1|A\} = \frac{P\{A|B_1\}P\{B_1\}}{P\{A\}} = \frac{0.06 \times 0.6}{0.05} = \frac{36}{50} = 72\%,$$

$$P\{B_2|A\} = \frac{9}{50} = 18\%,$$

$$P\{B_3|A\} = \frac{5}{50} = 10\%.$$

Example 5.2. The staff of a company consists of 45% men and 55% women. A survey showed that 4% of men and 6% of women frequently use the company phones for personal purposes. Accordingly, the company managers can evaluate, based on the above method, both the proportion of people who frequently use the company phones for personal business (i.e., 5.1%) and the proportion of men, for instance, who do so (i.e., 35.3%).

Remark 5.2. The Bayes' formula is sometimes presented in a simplified form, in conjunction with the conditional probability formula. Thus, based on the conditional probability formula, linking two events A and B, $(P\{B\} \neq 0)$, given by:

$$P\{A|B\} = \frac{P\{A \cap B\}}{P\{B\}}, \tag{5.3}$$

where $P\{A|B\}$ is the probability of the event A given the event B, one defines the simplified Bayes' formula by:

$$P\{B|A\} = \frac{P\{A|B\} \cdot P\{B\}}{P\{A\}}. \tag{5.4}$$

Example 5.3. ((378)) Suppose that from given statistics, it is known that meningitis causes stiff neck (torticollis) 50% of the time, that the proportion of persons having meningitis is 1/50,000, and that the proportion of people having stiff neck is 1/20. Then, the percentage of people who had meningitis and complain about neck immobility will equal 0.02%. The computation is simple, being based on the simplified Bayes' formula. Thus, if we note:

- $P\{M|S\}$ = probability that a person had meningitis, conditioned by the existence of stiff neck;
- $P\{S|M\}$ = probability that a person complains about stiff neck, conditioned by the existence of meningitis;
- $P\{S\}$ = proportion of people who complain about stiff neck;
- $P\{M\}$ = proportion of people who had meningitis.

Then:

$$P\{M|S\} = \frac{P\{S|M\} \cdot P\{M\}}{P\{S\}} = \frac{\dfrac{1}{2} \cdot \dfrac{1}{50,000}}{\dfrac{1}{20}} = 0.0002.$$

Bayesian decision theory is a fundamental statistical method in the pattern classification field. In the decision theory, the traditional goal is to minimize the probability to make a wrong decision, or the *expected risk*.

Basically, the Bayesian decision rule can be summarized by the following algorithm:

- Let D_k be the decision rule regarding the 'natural' state A_k.
- Given a measurement x, the error related to the state A_k is defined by: $P\{error|x\} = 1 - P\{A_k|x\}$.
- The probability to fail is *minimized*.
- The Bayesian decision rule is given by the assertion: "*Choose D_k if $P\{A_k|x\} > P\{A_j|x\}$, $\forall j \neq k$*", or, equivalently "*Choose D_k if $P\{x|A_k\}P\{A_k\} > P\{x|A_j\}P\{A_j\}$, $\forall j \neq k$*".

Let us consider a set of data to be classified using a Bayesian classifier, and let us assume in this respect that each attribute (including the attribute corresponding to the class label) is a random variable. Given an object with attributes $\{A_1, A_2,..., A_n\}$, we wish to classify it in class C. The classification is correct when the conditional probability:

$$P\{C|A_1, A_2, ..., A_n\},$$

reaches its maximum. The concrete problem which arises in the classification process is to estimate this probability directly from the data, with the goal to maximize it. With this aim in mind, we apply Bayes' formula as follows:

- Compute the posterior probabilities $P\{C_j|A_1, A_2, ..., A_n\}$ for all classes C_j, using Bayes' formula:

$$P\{C_j|A_1, A_2, ..., A_n\} = \frac{P\{A_1, A_2, ..., A_n|C_j\} \cdot P\{C_j|\}}{P\{A_1, A_2, ..., A_n\}}. \tag{5.5}$$

- Choose the class C_k that maximizes $P\{C_j|A_1, A_2, ..., A_n\}$ (equivalently, the class C_k that maximizes $P\{A_1, A_2, ..., A_n|C_j\} \cdot P\{C_j\}$).

From the above facts we have to compute the probability $P\{A_1, A_2, ..., A_n|C_j\}$. A possible approach in this regard uses the so-called *naive Bayes* classification (also known as *Idiot's Bayes*, (169)), which assumes, very often without any basis, the independence of the events, hence the term 'naive'. In this case we assume the mutual independence of attributes (obviously, a false assumption most of the time) for a given class C, namely:

$$P\{A_1, A_2, ..., A_n | C\} = P\{A_1 | C\} \cdot P\{A_2 | C\} \cdots P\{A_n | C\}. \qquad (5.6)$$

Finally, we estimate the probabilities $P\{A_i | C_j\}$ for all attributes A_i and classes C_j, so that a new and unknown object will be classified to class C_k if the probability corresponding to that class:

$$P\{C_k\} \cdot \prod P\{A_i | C_k\}, \qquad (5.7)$$

is maximal among the others.

Example 5.4. Let us take again the example presented in the previous chapter, related to the identification of the customer profile regarding the place where he/she goes shopping (store or supermarket), for which the training dataset is given below.

Table 5.1 Training dataset for shopping choice

Taxable income	Car ownership	Marital status	Buy from shop
125,000	Yes	Single	NO
100,000	No	Married	NO
70,000	No	Single	NO
120,000	Yes	Married	NO
95,000	No	Divorced	YES
60,000	No	Married	NO
220,000	Yes	Divorced	NO
85,000	No	Single	YES
75,000	No	Married	NO
90,000	No	Single	YES

For the sake of simplicity, we considered in what follows $1,000 \sim 1K$. As shown in the table, there are two distinct classes: buy from shop [YES], and does not buy from shop [NO]. The probabilities of the two classes are $P\{NO\} = 7/10$ and $P\{YES\} = 3/10$.

Regarding the conditional probabilities of the form $P\{A_i | C_j\}$, in the case of discrete attributes, they will be naturally computed as:

$$P\{A_i | C_j\} = \frac{|A_{ij}|}{N_{C_j}}, \qquad (5.8)$$

where $|A_{ij}|$ represents the number of objects having attribute A_i and belonging to class C_j. Thus, using this formula, we get, for instance:

$P\{Marital\ status = "Married"|NO\} = 4/7$

$P\{Car\ ownership = "Yes"|YES\} = 0$

In the continuous attributes case, in order to estimate the conditional probabilities $P\{A_i|C_j\}$ we need to identify the type of attribute distribution, viewed as a continuous random variable. Usually, excepting the cases where we have an *a priori* knowledge of them, it is assumed that all continuous attributes are Normally distributed, the only issue in this case being the estimation of its parameters (i.e., mean and variance). Once the probability density function estimated, we can evaluate the conditional probability $P\{A_i|C_j\}$ for each class separately. In our case, the attribute *"Taxable income"* is considered as continuous random variable, with the density given by:

$$P\{A_i|C_j\} = \frac{1}{\sqrt{2\pi}\sigma_{ij}} \exp\left(-\frac{(A_i - \mu_{ij})^2}{2\sigma_{ij}^2}\right). \tag{5.9}$$

Thus, the mean of the variable *"Taxable income"*, conditioned by class [NO], equals $110K$, and its variance equals $2,975K^2$. Consequently, we compute the conditional probability:

$$P\{Taxable\ income = 120K|NO\}=$$

$$= \frac{1}{\sqrt{2\pi}(54.54K)} \exp\left(-\frac{(120K - 110K)^2}{2 \cdot 2,975K^2}\right), \tag{5.10}$$

that is 0.0072%. Similarly, we can compute the conditional probability corresponding to class [YES].

Finally, all the situations met in this case are displayed below.

$P\{Car\ ownership = "Yes"|NO\} = 3/7$

$P\{Car\ ownership = "No"|NO\} = 4/7$

$P\{Car\ ownership = "Yes"|YES\} = 0$

$P\{Car\ ownership = "No"|YES\} = 1$

$P\{Marital\ status = "Single"|NO\} = 2/7$

$P\{Marital\ status = "Divorced"|NO\} = 1/7$

$P\{Marital\ status = "Married"|NO\} = 4/7$

$P\{Marital\ status = "Single"|YES\} == 2/7$

$P\{Marital\ status = "Divorced"|YES\} = 1/7$

$P\{Marital\ status = "Married"|YES\} = 0$

Alternatively, *"Taxable income"* conditioned by class [YES] has the mean $90K$ and the variance $25K^2$.

Let us now analyze the way in which this classifier deals with a new case. Suppose we have to classify an individual who has the following attributes:

- Car ownership = "No";
- Marital status = "Married ";
- Taxable income = $120K$.

We have:

$$P\{No, Married, 120K|NO\} = P\{No|N\} \times P\{Married|NO\} \times P\{120K|NO\} =$$
$$= 4/7 \times 4/7 \times 0.000072 = 0.0024\%$$

$$P\{No, Married, 120K|YES\} = P\{No|YES\} \times P\{Married|YES\} \times P\{120K| |YES\} = 0$$

To conclude, since:

$$P\{No, Married, 120K|NO\} \times P\{NO\} > P\{No, Married, 120K|YES\} \times P\{YES\},$$

we have:

$$P\{NO| No, Married, 120K\} > P\{YES| No, Married, 120K\},$$

and hence, we classify an unknown individual with these attributes in the category [NO], in other words he/she probably does not buy from shop.

Finally, let us review the main advantages of the (naive) Bayesian classification:

- Robust to isolated noise in data;
- In case of missing values, ignores the corresponding objects during the process of computing probabilities;
- Robust to irrelevant attributes.

Remark 5.3. The Naive Bayesian classifier has been used as an effective classifier for many years. Unlike many other classifiers, we saw above that it is easy to construct, as the structure is given *a priori*, and hence no structure learning procedure is required. To solve their main problem given by the fact that all attributes are independent of each other, one can consider other options, such as the *Bayesian (belief) networks*. For technical details, see for instance, (221), (75).

5.3   Artificial Neural Networks

The domain of Artificial Neural Networks (ANNs) or Neural Networks (NNs), is still, at about 70 years after its 'attestation', a research field not yet "classicized". In recent decades, NNs appear as a practical technology, designed to successfully solve many problems in various fields: neural science, mathematics, statistics, physics, computer science, engineering, biology, etc. NNs are applied in modeling, time series analysis, pattern recognition, signal processing, control theory, etc., due to their fundamental characteristic, the ability to *learn* from training data "with or without a teacher". Although they process information based on the '*black-box*' principle and, unlike other 'transparent' techniques, such as decision trees, they do not directly 'unveil' the way they '*think*', their effectiveness in the above mentioned domains is undeniable.

Synthetically speaking, NNs represent non-programmed (non-algorithmic) adaptive information processing systems. NNs learn from examples and behave like 'black boxes', the way they process information being inexplicit. We can consider NNs as a massively parallel distributed computing structure, an information processing paradigm with ancestors such as: mathematics, statistics, computer science, neurosciences, etc., inspired by the way the human brain processes information. In principle, the similarity between NNs and the way of action of the human brain may be condensed in the following two aspects:

- Knowledge is acquired by the network through the *learning* (*training*) process;
- The intensities of the inter-neuron connections, known as (*synaptic*) *weights*, are used to store acquired knowledge.

This subsection will be devoted to a synthetic presentation of the principles underlying NNs, and to some applications illustrating their undeniable practical valency. For more details concerning the neural networks, see, for instance, (106), (15), (34), (317), (161), (181), (419).

5.3.1 Perceptron

We could set the title of this subsection, in a suggestive way, as: "*From McCulloch and Pitts's artificial neuron to Rosenblatt's perceptron*", because we intend to present both the history of this very interesting field of research and the concept underlying it.

Let us first consider the biological concept underlying NN. Thus, NNs process information in a similar way the human brain does, manner that was the basis of their construction. Basically, each *neuron* is a specialized cell which can propagate an electrochemical signal. In the human brain, a typical neuron collects signals from others through a host of fine structures called *dendrites*. The neuron sends out spikes of electrical activity through a long, thin stand known as an *axon*, which splits into thousands of branches. At the end of each branch, a structure called a *synapse* converts the activity from the axon into electrical effects that inhibit or excite activity from the axon in the connected neurons. When a neuron receives excitatory input that is sufficiently large compared with its inhibitory input, it sends ('*fires*') a spike of electrical activity (an electrochemical signal) down its axon. "Learning" occurs by changing the effectiveness of the synapses so that the influence of one neuron on another changes. Summarizing the above considerations, the neuron receives information from other neurons through dendrites, processes it, then sends response-signals through the axon, moment when the synapses, by altering some inhibition/excitation 'thresholds', control the action upon the connected neuron. By fine-tuning at the synapses level, based on learning from past experience, one obtains optimal outputs as answers to the inputs received. Thus, a neuron is either inhibited or excited, depending on the signal received from another neuron, and, according to it, it will answer or not, influencing the action of neurons connected into the network. Fig. 5.1 synthetically illustrates the basic architecture of a natural (biological) neuron.

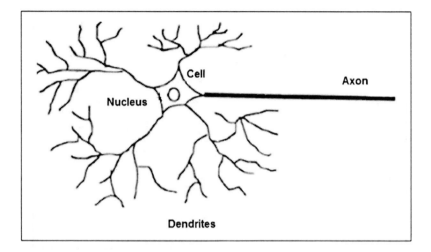

Fig. 5.1 Structure of the biological neuron

Next figure (Fig. 5.2) illustrates the synaptic liaisons that connect neurons into a neural network.

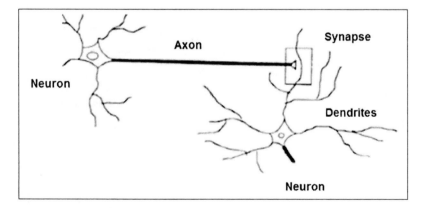

Fig. 5.2 Synaptic connections

According to Kohonen, (217), the first artificial neuron was produced in 1943 by the neurophysiologist Warren McCulloch and the logician Walter Pitts. Although they derived theorems and assumptions related to models of neural computing, few of them were implemented since the technology available at that time did not allow them to do too much. We could see this approach as a simple mathematical model for the behavior of a single natural neuron in a biological nervous system. The artificial neuron model, which seeks to 'imitate' the concept of the natural (biological) neuron is presented in the following two figures (for both the case of internal 'bias' and external 'bias').

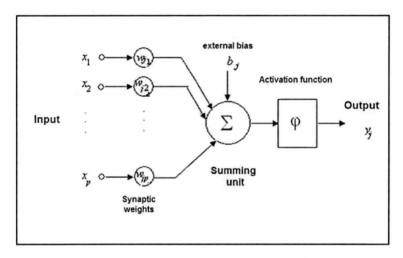

Fig. 5.3 Non-linear neuron model (with external bias)

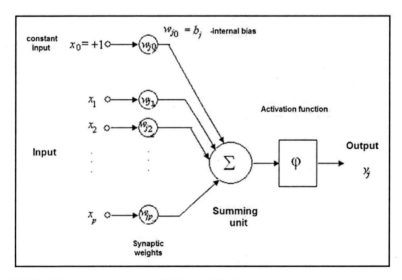

Fig. 5.4 Non-linear neuron model (with internal bias)

An artificial neuron, seen as a single device, has a certain number p of real inputs x_i, weighted by respective weights w_i, summed and then passed on to an activation function φ to produce a certain output, depending on a predetermined 'threshold' T.

As the natural nervous system, a NN is composed of a large number of highly interconnected processing elements -artificial neurons- working in parallel to solve a specific problem. Accordingly, we will see a neuron as the basic component of a neural network rather than a singular device working independently. Thus, in a NN structure, a real input x_i reaching the input of synapse i, connected to neuron j, is multiplied by the synaptic weight w_{ji}. In this mathematical neural model, the (scalar) inputs x_i represent the levels of activity of other neurons which connect to the neuron being modeled and the weights w_{ji} represent the strengths of the interconnections (synapses) between neurons. Let us remark that, unlike a synapse in the brain, the synaptic weight of an artificial neuron may lie in a range that includes negative as well as positive values. Further, a specific predetermined threshold T_j corresponds to each neuron j. The artificial neuron also includes a (fictitious) constant external input b_j, called *bias*, having the effect of increasing or decreasing the net input of the activation, depending on whether it is positive or negative, respectively. In other words, the bias b_j represents the 'threshold' for the neuron to "*fire*".

If the scalar sum of the inputs multiplied by the synaptic weights, which is the dot product, exceeds the threshold value T_j, then it is processed by the activation function φ to produce the output, else a zero value is considered as the output.

Mathematically speaking, we may describe the j-th neuron in a neural network by the following equations:

$$u_j = \sum_{i=1}^{p} w_{ji} \cdot x_i = \mathbf{w}_j \cdot \mathbf{x}^T, \qquad (5.11)$$

where $\mathbf{x} = (x_1, x_2, ..., x_p)$ represents the *input vector*, $\mathbf{w}_j = (w_{j1}, w_{j2}, ..., w_{jp})$ is the *synaptic weight vector* and u_j is the *linear combiner output* due to the input $\mathbf{x}$.

The 'firing' activity of the neuron is given by:

$$y_j = \varphi(u_j + b_j) = \begin{cases} h_j & , u_j + b_j \geq T_j \\ 0 & , u_j + b_j < T_j, \end{cases} \qquad (5.12)$$

where b_j is the bias, φ is the *activation function* (generally chosen to be monotonic) and y_j is the *output signal* of the neuron. Let us note that, usually, T_j equals zero.

We have previously considered the bias b_j as an external parameter (see Fig. 5.3). Alternatively, we may account for its presence as an internal parameter (see Fig. 5.4) regarding the bias $b_j = w_{j0}$ as being weight from an extra input $x_0 = 1$, so that:

$$u_j = \sum_{i=0}^{p} w_{ji} \cdot x_i, \qquad (5.13)$$

and:

$$y_j = \varphi(u_j). \tag{5.14}$$

As shown above, the activation function φ defines the output of an artificial neuron in term of the linear combiner output u. Seven of the more common types of such functions that can be used in most concrete real-world problems are mentioned below.

1. *Heaviside* activation function ('*threshold*' function), given by:

$$\varphi(u) = \begin{cases} 0 & ,u < 0 \\ 1 & ,u \geq 0. \end{cases} \tag{5.15}$$

In this case, also known as the *original McCulloch-Pitts* (MCP) model, we have a binary output describing the *all-or-none property* of the (MCP). An alternative to the Heaviside function is given by:

$$\varphi(u) = \begin{cases} -1 & ,u < 0 \\ +1 & ,u \geq 0. \end{cases} \tag{5.16}$$

A generalization of the threshold function is represented by the *step activation function (hard limiter)*, given by:

$$\varphi(u) = \begin{cases} a & ,u < h \\ b & ,u \geq h. \end{cases} \tag{5.17}$$

2. *Piecewise-linear* activation function (*ramp function*), given by (variant):

$$\varphi(u) = \begin{cases} -a & ,u \leq -c \\ u & ,|u| < c \\ a & ,u \geq c. \end{cases} \tag{5.18}$$

3. *Linear* activation function, given by:

$$\varphi(u) = a \cdot u. \tag{5.19}$$

4. *Gaussian* activation function, given by:

$$\varphi(u) = \exp(-u^2/a). \tag{5.20}$$

5. *Sigmoid* activation function with an S-shape graph (hence the name) is by far the most common form of activation function used in the design of artificial neural networks. An example of such a function is represented by the *logistic function* mapping the interval $(-\infty, \infty)$ onto $(0, 1)$, and given by:

$$\varphi(u) = \frac{1}{1 + \exp(-a \cdot u)}, \tag{5.21}$$

where a is the slope parameter of the sigmoid. Another example of a sigmoid function, used as activation function for NN, is the *hyperbolic-tangent* function, given by:

$$\varphi(u) = \tanh(u) = \frac{e^u - e^{-u}}{e^u + e^{-u}}, \tag{5.22}$$

which differs from the logistic function through a linear transformation.

Remark 5.4. The above neural model is deterministic, that is its input/output behavior is precisely determined for all inputs. A stochastic alternative model is given by the probabilistic decision for the neuron to '*fire*'. Let X denote the binary state of a neuron and u denote the linear combiner output. Then:

$$X : \begin{cases} fire & ,with\ probability\ P\{u\} \\ not\ fire & ,with\ probability\ 1 - P\{u\}. \end{cases} \tag{5.23}$$

Rosenblatt's perceptron model

The basic neuron model, also known as the *McCulloch-Pitts model* in recognition of the pioneering work done by the two scientists at the "dawn" of the NNs era, has fixed weights and binary output with no learning or adaptation. The next step is represented by *Rosenblatt's perceptron model* -invented in 1957 at the Cornell Aeronautical Laboratory, (324), (325). Starting from the previous model, the perceptron possessed a simple learning mechanism, based on feedback of the error difference of the desired and actual outputs. The name *perceptron* comes from Rosenblatt's original purpose: "to distinguish (perception/percept -Latin *perceptum/percipere*) black-and-white images of geometric patterns using binary photosensor inputs and the step activation function (hard limiter)".

The goal of a perceptron is to correctly classify a set of external stimuli into one of the two classes (decision categories) C_1 and C_2. The decision rule for the classification problem is to assign the point $\mathbf{x}$ represented by the input vector $\mathbf{x}=(x_1, x_2, ..., x_p)$ to class C_1 if the perceptron output is +1 and to class C_2 if it is -1 (or, equivalently, 0 and 1). In what follows we will use the internal bias model, that is with the input vector $\mathbf{x}=(x_0, x_1, x_2, ..., x_p)$ and the synaptic weight vector $\mathbf{w}=(w_0, w_1, w_2, ..., w_p)$.

In the simplest form of the perceptron there are two decision regions in the $(p+1)$ -input space, separated by a hyperplane given by:

$$u = \sum_{i=0}^{p} w_i \cdot x_i = 0, \tag{5.24}$$

that is the two regions are linearly separable. Concretely, there exists a weight vector $\mathbf{w}$ such that:

- $\mathbf{w} \cdot \mathbf{x}^T > 0$, for every input vector $\mathbf{x}$ belonging to class C_1.
- $\mathbf{w} \cdot \mathbf{x}^T < 0$, for every input vector $\mathbf{x}$ belonging to class C_2.

The synaptic weights $w_0, w_1, w_2, ..., w_p$ can be adapted using an *iteration-by-iteration* procedure.

The original perceptron learning algorithm, (187), with two decision classes $C_1 \sim P^+$ and $C_2 \sim P^-$, is given below:

Input: A set of positive and negative vector samples, denoted by P^+ and P^-.

1. Initialize the weight vector $\mathbf{w}$ to zero.
2. Choose a vector sample $\mathbf{x}$ from the sample set $P^+ \cup P^-$.
3. If $\{\mathbf{x} \in P^+$ and $\mathbf{w} \cdot \mathbf{x}^T > 0\}$ or $\{\mathbf{x} \in P^-$ and $\mathbf{w} \cdot \mathbf{x}^T < 0\}$ Then GOTO step 2.
4. If $\{\mathbf{x} \in P^+$ and $\mathbf{w} \cdot \mathbf{x}^T < 0\}$ Then $\mathbf{w} = \mathbf{w} + \mathbf{x}$; GOTO step 2.
5. If $\{\mathbf{x} \in P^-$ and $\mathbf{w} \cdot \mathbf{x}^T > 0\}$ Then $\mathbf{w} = \mathbf{w} - \mathbf{x}$; GOTO step 2.

Output: A single weight vector $\mathbf{w}$ for the linear threshold function that classifies the input samples in P^+ and P^-, if such a vector exists.

Remark 5.5. For the perceptron to function properly it is necessary that the two classes C_1 and C_2 are linearly separable enough, otherwise the corresponding decision is beyond the computing capabilities of the perceptron. In Fig. 5.5 (*a* and *b*) we present both pairs of linearly and non-linearly separable patterns.

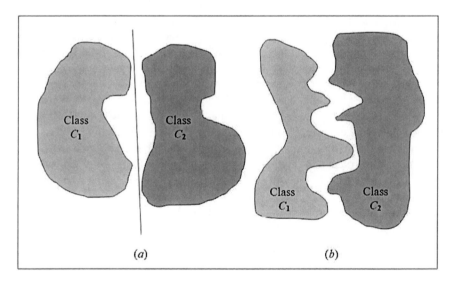

Fig. 5.5 (*a*) Pair of linearly separable patterns, (*b*) pair of non-linearly separable patterns.

To establish the correctness of the *error-correction learning* algorithm, we will present an interesting result stating that, for any dataset which is linearly separable, the learning rule is guaranteed to find a solution in a finite number of steps. Suppose that the input variables of the perceptron originate from a training dataset C which is linearly separable, that is the patterns to be classified must be sufficiently separated from each other to ensure that the decision surface consists of a hyperplane. Accordingly, there is a partition $\{C_1, C_2\}$ of C, with C_1 and C_2 the two corresponding subsets, such that $C = C_1 \cup C_2$ and C_1, C_2 are disjoint sets, and there is a hyperplane between them, representing the decision boundary (see Fig. 5.5 a). Mathematically speaking, if C_1 and C_2 are linearly separable, there exists a weight vector $\mathbf{w}$, such that:

- $\mathbf{w} \cdot \mathbf{x}^T > 0$, for every training pattern $\mathbf{x} \in C_1$,
- $\mathbf{w} \cdot \mathbf{x}^T < 0$, for every training pattern $\mathbf{x} \in C_2$.

The learning rule in the above perceptron learning algorithm can be generalized as follows:

- If the vector $\mathbf{x}_n$ of C is correctly classified by the weight vector $\mathbf{w}(n)$, computed for the nth iteration of the algorithm, then:

  1. $\mathbf{w}(n+1) = \mathbf{w}(n)$ if $\mathbf{w}(n) \cdot \mathbf{x}_n^T > 0$ and $\mathbf{x}_n \in C_1$;
  2. $\mathbf{w}(n+1) = \mathbf{w}(n)$ if $\mathbf{w}(n) \cdot \mathbf{x}_n^T < 0$ and $\mathbf{x}_n \in C_2$.

- Otherwise:

  1. $\mathbf{w}(n+1) = \mathbf{w}(n) - \eta \cdot \mathbf{x}_n$ if $\mathbf{w}(n) \cdot \mathbf{x}_n^T > 0$ and $\mathbf{x}_n \in C_2$,
  2. $\mathbf{w}(n+1) = \mathbf{w}(n) + \eta \cdot \mathbf{x}_n$ if $\mathbf{w}(n) \cdot \mathbf{x}_n^T < 0$ and $\mathbf{x}_n \in C_1$,

  where the *learning-rate* parameter η is constant and independent of the iteration.

We can summarize the above algorithm as follows: "cycle through all the training patterns and test each pattern using the current weight values. If the pattern is correctly classified keep the current weight values, otherwise add the pattern vector multiplied by η to the weight vector if the pattern belongs to C_1, or subtract the pattern vector multiplied by η if the pattern belongs to C_2".
  Since $\|\mathbf{x}_n\|^2 > 0$ and $\eta > 0$, then:

$$\mathbf{w}(n+1) \cdot \mathbf{x}_n^T = \mathbf{w}(n) \cdot \mathbf{x}_n^T - \eta \cdot \mathbf{x}_n \cdot \mathbf{x}_n^T < \mathbf{w}(n) \cdot \mathbf{x}_n^T \quad \text{-first case,}$$

$$\mathbf{w}(n+1) \cdot \mathbf{x}_n^T = \mathbf{w}(n) \cdot \mathbf{x}_n^T + \eta \cdot \mathbf{x}_n \cdot \mathbf{x}_n^T > \mathbf{w}(n) \cdot \mathbf{x}_n^T \quad \text{-second case,}$$

and, therefore, the above procedure tends to reduce the error.

Remark 5.6. Clearly, the value of η is in fact unimportant as long as it is positive, since it rescales the weights. This leaves the location of the decision boundary, given by $\mathbf{w} \cdot \mathbf{x}^T = 0$, unchanged. Thus, when minimizing the perceptron criterion, we can take $\eta = 1$.

Now, let us state the *convergence theorem* for the perceptron.

Theorem 5.3. *(Rosenblatt 1962, (325)). Let the subsets of training vectors C_1 and C_2 be linearly separable. Then, the perceptron converges after a finite number n_0 of iterations.*

Proof. Here we give the proof, in a simple form, based on (182).
  Let us first prove the convergence of a learning rule with fixed increment rate η, choosing $\eta = 1$.

Since we are considering the training dataset linearly separable, then there exists at least one weight vector $\mathbf{w}_0$ for which all training vectors are correctly classified, so that $\mathbf{w}_0 \cdot \mathbf{x}_n^T > 0$ for all vectors $\mathbf{x}_1,..., \mathbf{x}_n$ belonging to the class C_1. Let us consider a positive quantity $\alpha = \min\{\mathbf{w}_0 \cdot \mathbf{x}_n^T; \mathbf{x}_n \in C_1\}$. The learning process starts with some arbitrary weight vector which we can assume to be $\mathbf{w}(0) = \mathbf{0}$. Assume that the perceptron will incorrectly classify the vectors $\mathbf{x}_1, \mathbf{x}_2,...$, i.e., $\mathbf{w}(n) \cdot \mathbf{x}_n^T < 0$ for all $n = 1, 2,...$, and the input vectors $\mathbf{x}_n$ belong to the class C_1. Then, at each step of the algorithm the weight vector is updated using the formula:

$$\mathbf{w}(n+1) = \mathbf{w}(n) + \mathbf{x}_n, \tag{5.25}$$

for $\mathbf{x}_n \in C_1$, representing a vector misclassified by the perceptron. Suppose that, after running the algorithm for some time, each vector $\mathbf{x}_n$ has been presented to the network and misclassified. Then, given the initial condition $\mathbf{w}(0) = \mathbf{0}$, the weight vector at this point will be given by:

$$\mathbf{w}(n+1) = \sum_{j=1}^{n} \mathbf{x}_j. \tag{5.26}$$

We now multiply (scalar product) both sides of this equation by $\mathbf{w}_0$ and obtain:

$$\mathbf{w}_0 \cdot \mathbf{w}^T(n+1) = \sum_{j=1}^{n} \mathbf{w}_0 \cdot \mathbf{x}_j^T \geq n\alpha. \tag{5.27}$$

Now, using the Cauchy-Schwarz inequality, we get:

$$\|\mathbf{w}_0\|^2 \cdot \|\mathbf{w}(n+1)\|^2 \geq \left[\mathbf{w}_0 \cdot \mathbf{w}^T(n+1)\right]^2. \tag{5.28}$$

From (5.27) and (5.28) we have:

$$\|\mathbf{w}_0\|^2 \cdot \|\mathbf{w}(n+1)\|^2 \geq n^2\alpha^2, \tag{5.29}$$

or, equivalently:

$$\|\mathbf{w}(n+1)\|^2 \geq \frac{n^2\alpha^2}{\|\mathbf{w}_0\|^2}. \tag{5.30}$$

It follows that, since $\mathbf{w}_0$ is fixed, the value of $\|\mathbf{w}(n+1)\|^2$ is bounded below by a function which grows linearly with the squared number of iterations n.

Next, keeping this result in mind, we will turn to another consideration of the magnitude of the weight vector $\mathbf{w}(k+1)$.

Thus, from the updating formula (5.25) of the algorithm, we have:

$$\|\mathbf{w}(j+1)\|^2 = \|\mathbf{w}(j)\|^2 + \|\mathbf{x}_j\|^2 + 2\mathbf{w}(j) \cdot \mathbf{x}_j^T \leq$$

$$\leq \|\mathbf{w}(j)\|^2 + \|\mathbf{x}_j\|^2, j = 1, 2,...,n, \tag{5.31}$$

where $\mathbf{w}(j) \cdot \mathbf{x}_j^T < 0$, since the object $\mathbf{x}_j \in C_1$ must have been misclassified by the perceptron.

Thus, the change in value of the norm of the weight vector $\mathbf{w}$ satisfies the inequality:

$$\|\mathbf{w}(j+1)\|^2 - \|\mathbf{w}(j)\|^2 \leq \|\mathbf{x}_j\|^2, j = 1, 2, ..., n. \tag{5.32}$$

Let us denote by $\beta = \max\{\|\mathbf{x}_j\|^2; \mathbf{x}_j \in C_1\}$ the (squared) length of the longest input vector in C_1.

Then:

$$\|\mathbf{w}(j+1)\|^2 - \|\mathbf{w}(j)\|^2 \leq \beta. \tag{5.33}$$

Adding inequalities (5.33) for $j = 1, 2, ..., n$, we have:

$$\|\mathbf{w}(n+1)\|^2 \leq n\beta, \tag{5.34}$$

and so the value of $\|\mathbf{w}(n+1)\|^2$ increases at most linearly with the number of iterations n.

Finally, from (5.30) and (5.34) we get:

$$n\beta \geq \frac{n^2\alpha^2}{\|\mathbf{w}_0\|^2},$$

or, equivalently:

$$n \leq \frac{\beta \|\mathbf{w}_0\|^2}{\alpha^2}.$$

Thus, the number of iterations n cannot grow indefinitely, and so the algorithm must converge in a finite number of steps. From the above inequality we can state that n cannot be larger than:

$$n_{\max} = \frac{\beta \|\mathbf{w}_0\|^2}{\alpha^2}. \tag{5.35}$$

To complete the proof, let us consider the general case of a variable learning-rate parameter $\eta(n)$, depending on the iteration number n. For the sake of simplicity, consider $\eta(n)$ the smallest integer, such that:

$$|\mathbf{w}(n) \cdot \mathbf{x}_n^T| < \eta(n) \cdot \mathbf{x}_n \cdot \mathbf{x}_n^T. \tag{5.36}$$

Suppose that at iteration n we have a misclassification, that is:

$$\mathbf{w}(n) \cdot \mathbf{x}_n^T > 0 \text{ and } \mathbf{x}_n \in C_2,$$

or:

$$\mathbf{w}(n) \cdot \mathbf{x}_n^T < 0 \text{ and } \mathbf{x}_n \in C_1,$$

Then, if we modify, without any loss of generality, the training sequence at iteration $(n+1)$ by setting $\mathbf{x}_{n+1} = \mathbf{x}_n$, we get:

$$\mathbf{w}(n+1) \cdot \mathbf{x}_n^T = \mathbf{w}(n) \cdot \mathbf{x}_n^T - \eta(n) \cdot \mathbf{x}_n \cdot \mathbf{x}_n^T < 0 \text{ and } \mathbf{x}_n \in C_2, \tag{5.37}$$

or:

$$\mathbf{w}(n+1) \cdot \mathbf{x}_n^T = \mathbf{w}(n) \cdot \mathbf{x}_n^T + \eta(n) \cdot \mathbf{x}_n \cdot \mathbf{x}_n^T > 0 \text{ and } \mathbf{x}_n \in C_1, \tag{5.38}$$

that is a correct classification of the pattern $\mathbf{x}_n$.

To conclude, each object is repeatedly presented to the perceptron until it is classified correctly.

Q.E.D.

We have thus proved that, given that a weight vector $\mathbf{w}_0$ (not necessarily unique) exists, for which all training patterns are correctly classified, the rule for adapting the synaptic weights of the perceptron must terminate after at most n_{max} iterations.

Note. Theorem 5.3 (for constant learning-rate parameter η) is known as *fixed-increment convergence theorem* for perceptron.

Below, the *perceptron convergence algorithm* (Lippmann 1987, (239)) with two decision classes C_1 and C_2 is briefly presented.

Input:

$\mathbf{x}(n) = (1, x_1(n), x_2(n), ..., x_p(n))$ -input training vector

$\mathbf{w}(n) = (b(n), w_1(n), w_2(n), ..., w_p(n))$ -weight vector

$y(n)$ -actual response

$d(n)$ -desired response

η -constant learning-rate parameter (positive and less than unity)

1. *Initialization.* Set $\mathbf{w}(0) = \mathbf{0}$. Perform the following computations for time step $n = 1, 2, ...$

2. *Activation.* At time step n, activate the perceptron by applying continuous-valued input training vector $\mathbf{x}_n$ and desired response $d(n)$.

3. *Computation of actual response.* Compute the actual response of the perceptron, given by:

$$y(n) = sign\left[\mathbf{w} \cdot \mathbf{x}(n)^T\right],$$

where $sign[\cdot]$ represents the *signum function*.

4. *Adaptation of the weight vector.* Update the weight vector of the perceptron:

$$\mathbf{w}(n+1) = \mathbf{w}(n) + \eta \left[d(n) - y(n)\right] \cdot \mathbf{x}(n),$$

where:

$$d(n) = \begin{cases} +1, \ \mathbf{x}_n \in C_1 \\ -1, \ \mathbf{x}_n \in C_2 \end{cases}$$

5. *Continuation.* Increment time step n by one and GOTO step 2.

Output: After a finite number of time steps n, the rule for adapting the synaptic weights of the perceptron must terminate.

For more details, see (181).

We have already seen that the perceptron described above represents a network with a single layer of synaptic weights, using only raw input data, and therefore having very limited "thinking" capabilities.

To improve the performance of the original perceptron, Rosenblatt used an extra layer of fixed processing elements $\psi = (\psi_0, \psi_1, ..., \psi_p)$ aiming to transform the raw input data $\mathbf{x} = (x_0, x_1, ..., x_p)$, as shown in Fig. 5.6 (we used again the convention $\psi_0 = 1$ with the corresponding internal bias w_0).

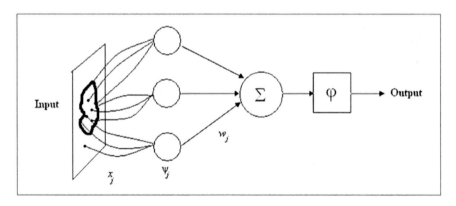

Fig. 5.6 The generalized perceptron scheme

These processing elements typically take the form of fixed weights, connected to a random subset of inputs, and with a binary activation function.

The output of the perceptron is therefore given by:

$$y = \varphi \left(\sum_{i=0}^{p} w_i \psi_i(\mathbf{x}) \right) = \varphi \left(\mathbf{w} \cdot \psi^T \right). \tag{5.39}$$

Since the goal of using the perceptron learning algorithm is to produce an effective classification system, it would be natural to define an *error function* in terms of *'the total number of misclassifications over the training dataset'*. Accordingly, we will introduce a continuous, piecewise-linear error function, called the *perceptron criterion*. We will associate to each training vector $\mathbf{x}_n \in C$ a corresponding target value t_n, defined by:

$$t_n = \begin{cases} +1, \mathbf{x}_n \in C_1 \\ -1, \mathbf{x}_n \in C_2. \end{cases}$$

Since from the perceptron training algorithm we have $\mathbf{w} \cdot \mathbf{x}_n^T > 0$ for vectors belonging to C_1, and $\mathbf{w} \cdot \mathbf{x}_n^T < 0$ for vectors belonging to C_2, it follows that, for all the training patterns $\mathbf{x}_n \in C$ we have $\mathbf{w} \cdot (t_n \cdot \mathbf{x}_n)^T > 0$. If M represents the set of vectors $\mathbf{x}_n$ which are misclassified by the current weight vector $\mathbf{w}$, then we can define the error function (perceptron criterion) by:

$$E(\mathbf{w}) = -\sum_{x_n \in M} \mathbf{w} \cdot (t_n \cdot \mathbf{x}_n)^T,$$

which is usually positive (and equals zero if all the training patterns are correctly classified). To conclude, during the perceptron learning process we try to minimise $E(\mathbf{w})$.

For more details concerning the perceptron convergence theorem see (34) and (181).

Notes and comments. The introduction of the first artificial neuron by McCulloch and Pitts in 1943 could be considered as the neural networks "dawn". Let us mention here that this new notion influenced J. von Neumann to use *idealized switch-delay elements* derived from it in the construction of the EDVAC (*Electronic Discrete Variable Automatic Computer*) that developed out of the celebrated ENIAC (*Electronic Numerical Integrator and Computer*) considered as the first real computer, (17). Let us remember that ENIAC, with its 30 tons and 18,000 vacuum tubes/electron tube (built at the University of Pennsylvania) is considered as the first 'true' computer (electronic "brain") ever, in other words it is seen as the 'father' of modern computers based on silicon chips/integrated circuits and tens of thousands of electronic microelements on square millimeter. As a matter of fact, we should note in this context that the fear of 'bugs' still remaining vivid in the minds of people who do work in computer science, although now referring to something else, has solid causes. Try to imagine the real tragedy involved by the penetration of a real bug among those thousands of tubes and the short circuits caused by 'frying' it, and the shutting down of this colossus eventually.

The perceptron represents the simplest form of a neural network used for the classification of linearly separable patterns. Basically, it consists of a single artificial neuron with adjustable synaptic weights and bias. The perceptron built around a single artificial neuron is limited to performing pattern classification with only two classes. By expanding the output layer to include more than one neuron, we may consider classifications with more than two classes. Almost at the same time when Rosenblatt was developing the perceptron, Widrow and his collaborators were working along similar lines using a system known as *ADALINE* (*ADAptive LINear Element*), which had essentially the same form as the perceptron, but which used a much better learning algorithm (*least mean square*-LMS *algorithm*), the extension of which is used in the multi-layer perceptron, (406), (407). When the perceptron was being studied experimentally in the 1960s, it was found that it could solve many problems very readily. Following an initial period of enthusiasm, as it usually happens, the field went through a period of frustration and 'disrepute', totally undeserved. It was found that other problems, which superficially appeared to be no more difficult, proved impossible to solve using it. During this "black" period when funding and professional support was minimal, important advances were made by relatively few researchers, working solely driven by the philosophy "*art for art's sake*". A real 'kick' was received by the perceptron when Minsky and Papert (1969), (265), published a book, "*Perceptrons*", in which they summed up a general feeling of frustration (against neural networks) among researchers, and was thus accepted by most without further critical analysis. From a formal mathematical point of view, they showed that there are many types of problems for which a perceptron cannot, in any practical sense, be used to solve the task (e.g., learn an XOR function). The real difficulty with the perceptron arises from the fact that the processing elements ψ_j are fixed in advance and cannot be adapted to the particular problem which is being considered. Minsky and Papert discussed a range of different forms of perceptron (corresponding to the form of the functions ψ_j) and, for each of them, provided examples of problems which cannot be solved. This unfortunate situation has a long time overshadowed the neural networks development, but it may have had a good side in overcoming the initial disadvantages of neural networks. The practical solution to the difficulties connected to the use of the perceptron is to allow the processing elements to be adaptive, so that they are chosen as part of the learning process and, thus, leading to the consideration of *multi-layer adaptive networks*.

5.3.2 Types of Artificial Neural Networks

In the previous subsection we presented the structure and usage of a single neuron, constituting the "brick" of a neural network construction. The next step deals with the way in which more neurons can be interconnected to create a complex and functional structure, in other words a real neural network. The basic elements of this construction are: the *input* (units), fed with information from the environment, the "shadow" units within the network (*hidden neurons*), controlling the actions in the network, and the *output* (units), which synthesize(s) the network response. All these neurons must be interconnected in order that the network becomes fully functional.

In seeking to classify the NN models, we can rely on their specific architecture, operating mode and learning manner, (187). Thus, the NN architecture refers to the topological organization of neurons (their number, number of layers of neurons, layer structure, the signal direction and reciprocity). The operating mode refers to the nature of activities during the information processing (dynamic or static for each new input). Finally, the learning paradigm refers to the way NN acquires knowledge from the training dataset.

In this context, a basic issue is the so-called *feedback* (reverse connection) present or not in such systems. Briefly, we say that there is a feedback in a dynamical system when the output of an item belonging to the system has a certain influence on the input of that element, via the so-called feedback loop, see figure below.

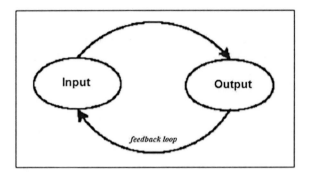

Fig. 5.7 Feedback loop

We say that NN has a *feedforward* type structure when the signal moves from input to output, passing through all the network's hidden units, so the outputs of neurons are connected to the next layer and not to previous ones. These networks have the property that the outputs can be expressed as a deterministic function of inputs. A set of values entering the network is transmitted through the activation functions over the network to its output, by the so-called *forward propagation*. Such a network has a stable operating mode. Fig. 5.8 schematically illustrates this network structure.

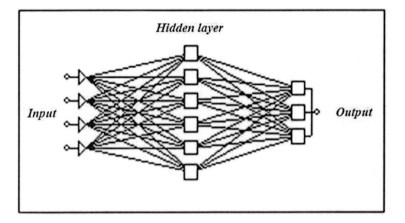

Fig. 5.8 Feedforward network structure

Different from feedforward neural networks, there are networks with feedback loops, the so-called *recurrent neural networks* (RNN). Common examples of such RNNs are *Elman and Jordan networks*, also known as *simple recurrent networks* (SRN). The interested reader may wish to find out more in (165), (247), (263).

Regarding the NN architecture, we will mention three fundamental categories of such networks. But first, recall that in a *layered* NN, the neurons are organized into one or more *layers*.

- *Single-layer feedforward networks.* In this case, the simplest one, there is an *input layer* of the source nodes, followed by the *output layer* of computing nodes. Note that the term *single layer* refers only to the output layer, because it is involved in the computation. Fig. 5.9 illustrates such a network.

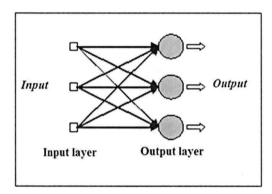

Fig. 5.9 Single-layer feedforward network

- *Multilayer feedforward networks.* Unlike the previous network, in this case there are one or more hidden layers, whose (computing) elements are called hidden neurons, their role being to act between the input layer and the output layer, so that the network performance is improved. Schematically, by the input layer the information from the environment enters the network, representing the inputs of neurons in the second layer (i.e., the first hidden layer), then, being processed by them it will become the input of the next layer (i.e., the second hidden layer), and so on. Fig. 5.10 illustrates such a network with a single hidden layer.

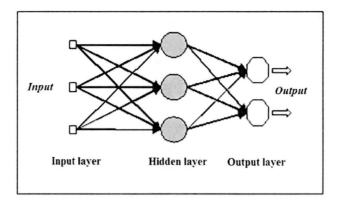

Fig. 5.10 Multilayer feedforward network

- *Recurrent networks.* As we already said, this network differs from those of the feedforward type by the existence of at least one feedback loop. The presence of such a feedback loop is very important both concerning the learning method and its performance. Fig. 5.11 illustrates three cases of such networks.

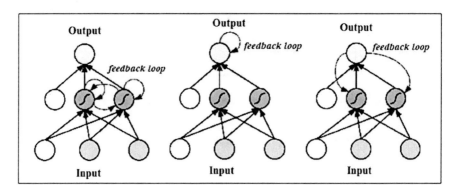

Fig. 5.11 Types of recurrent NNs

The learning paradigm, viewed in the NN context, is the process of the network adaptation to external environment (i.e., the adaptation/tuning/adjustment of its parameters) by a process of stimulation due to the environment. Schematically, the environment stimulates the network (NN receives inputs from the environment), the system parameters receive certain values as reaction to these stimuli, and then NN responds to its external environment with its new configuration. Since there are several ways of setting the network parameters, there will be several types of learning rules. We briefly mention here some of the best known such learning rules:

- *Error-correction learning* is based on a control 'mechanism' of the difference between the actual response and the network response. Technically, the network weights are adapted according to the error of the neurons output;
- *Memory-based learning* (*instance-based learning*) uses the explicit memorization of the training data;
- *Hebbian learning* is based on neurobiological considerations, named in honor of the neuropsychologist D.O. Hebb (*Hebb's postulate of learning*);
- *Competitive learning* is based on competition between neurons, i.e., only one neuron (winner neuron) from a given iteration in a given layer will fire at a time;
- *Boltzmann learning* is based on ideas borrowed from statistical mechanics, and is named in honor of the physicist L. Boltzmann.

(technical details concerning these learning methods can be found in (181)).

Next, we mention two fundamental characteristics of the learning process:

- *Learning with a 'teacher'*. In this case, like in school, complete examples (i.e., with both input and output) are presented to the network for learning, the parameters optimization being performed based on the error measurement, given by the difference between the network output (the student's response) and the expected response (teacher's response);
- *Learning without a 'teacher'*. In this case, only inputs are used (without corresponding outputs), and thus the network adjustment (i.e., the parameters optimization) does not benefit from a teacher's tutoring. There are two categories of such learning, depending on the method by which the parameter adaptation is performed:
  - *Reinforcement learning*, that is learning to map situations to actions, maximizing thus a numerical reward (reinforcement) signal. Basically, the learning of an input-output mapping is performed by repeated interactions with the environment, in order to maximize the performance;
  - *Self-organized* (or *unsupervised*) learning with no external teacher, or referee, to monitor the learning process.

Without entering into details, we briefly introduce the characteristics of neural networks, depending on their learning manners.

- NN with *supervised learning*, in other words, learning with a teacher. In this case, NN is trained to repeatedly perform a task, monitored by a 'teacher', by presenting it examples of pairs of input/output samples. Thus, the decision error, seen as the difference between the expected and the actual response, provided by NN, is computed during the training (learning) iteration. Once this error estimated, it is then used to adjust the synaptic weights according to some learning algorithms. As NN 'learns', a diminishing of error is expected. The learning process continues until a certain acceptable accuracy 'threshold' is reached. Fig. 5.12 schematically shows NN with supervised learning, (181), (419).

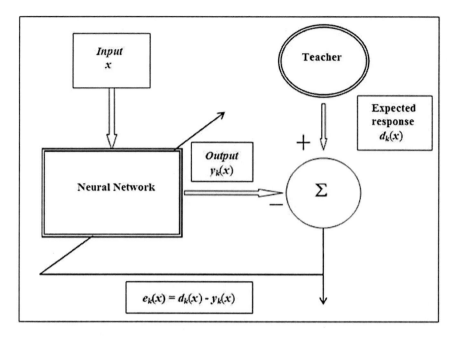

Fig. 5.12 Supervised learning

- NN with *reinforcement learning*, in other words, learning without a teacher. In this case, instead of computing the decision error (as difference between the expected and the actual response) and using it to optimize the network output, the network is warned about its performance, through a "critic"(or "agent"), i.e., signal of type *pass/fail* for each training sample, (373). If a *fail* is assigned, then the network will continue to adjust its parameters until it achieves a 'green light' (i.e., a *pass*), or continues for a predetermined number of tries before accepting another training sample, whichever comes first. Sometimes regarded as a special case of supervised learning (the teacher's role being played in another way, by the "critic"), this procedure has proved to be slow and ineffective in many applications, due either to the lack of a 'teacher' providing the desired response, or the reinforcement that occurs with some delay (*delayed reinforcement*), (181). Despite the inherent difficulties in the way of training it, this learning system is however effective because it relies on its interaction with the environment, being used in complex NN systems, (372), among its applications one could quote the robot control, telecommunications, chess game. In Fig. 5.13 such a type of reinforcement learning is schematically shown, (23), (181).

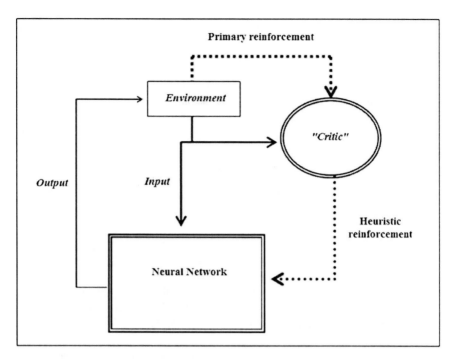

Fig. 5.13 Reinforcement learning

- NN with *self-organizing* (*unsupervised*) learning. In this case, the learning goal is either to model the (input) data distributions or to (automatically) discover certain clusters or structures in the (input) set, based on certain similarities and a competitive learning rule. Thus, it is possible that certain categories are assigned to these clusters/structures, according to their nature and the problem that needs to be solved. Once the groups/clusters structured (i.e., the network adapted (tuned) to the statistical model of the input data), NN can be used for new patterns classification, just as in the case of supervised learning, (34), (181), (186), (419). We schematically present in Fig. 5.14 a diagram illustrating the unsupervised learning model.

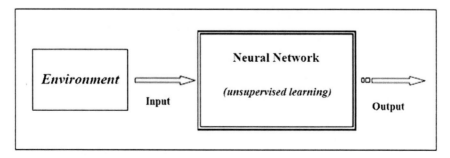

Fig. 5.14 Unsupervised learning

It is now time to briefly present some of the most popular types of neural networks:

- *Single-layer perceptron.* In subsection 5.3.1 we presented the basic concept of a single perceptron, which represents the simplest form of a neural network. Recall that, in principle, it consists of a single neuron with adjustable (synaptic) weights and (internal/external) bias, being used in linearly separable pattern classification problems. With a more complex architecture, the single-layer perceptron (SLP) consists of one or more neurons working in parallel (see Fig. 5.15)

  Technically speaking, a SLP has a single layer of output nodes, the inputs are directly applied through the weights, representing, therefore, the simplest type of a real feedforward network. The weighted sum of inputs (i.e., the dot product of the weight vector and input vector), computed at each node, is compared with a (activation) threshold; according to the result of this comparison, the network will take a certain action ("fire" or not "fire"). The SLP learning rule is the so-called *"delta rule"*, consisting in computing the difference (delta) between the calculated output of the network and the actual output, and using it, regarded as error, to adjust the network parameters (see also the *error-correction learning*, (181)). Because of the significant limitations of such a neural architecture, it was necessary to develop networks with more (multiple) layers of adjustable weights, thereby obtaining "true" NNs.

- *Multi-layer perceptron* -MLP. This type of network consists of several layers of computing units (neurons), connected together in a hierarchical feedforward framework. In principle, the basic features of a MLP can be summarized as follows.

  - Each neuron belonging to the network has a non-linear activation function of class C^1;

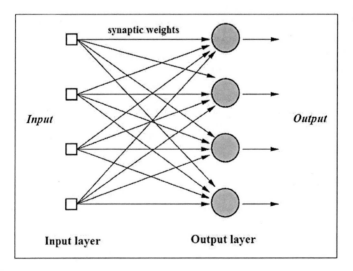

Fig. 5.15 Single-layer perceptron

- MLP contains one or more hidden layers;
- The signal propagates through the network, from input to output, in a forward direction, layer by layer;
- The network provides a high (synaptic) connectivity.

MLP uses several learning techniques, the most popular being the well-known *back-propagation algorithm* (BP), abbreviation for '*backwards propagation of errors*'. In principle, the output values are compared with the actual values and the error is computed based on a predefined error function E. Then, according to the result thus obtained, one acts backward through the network to adjust the weights in order to minimize the error. Basically, in the BP algorithm, the gradient vector of the error function $E = E(w_1, w_2, ..., w_p)$, seen as a function of weights, is computed. This vector points in the direction of steepest descent from the current point, and so, we know that, if we move along it, the error will diminish. A sequence of such moves (slowing as we get near the bottom) will eventually find a minimum of some sort. Technically, when the input $\mathbf{x}_i$ from the training dataset is presented to the network, it produces an output y_i, different in general from the desired response d_i (known by the 'teacher'). Under these circumstances, we want to minimize the error function of the network, usually defined as:

$$E = \frac{1}{2} \sum_i |y_i - d_i|^2. \tag{5.40}$$

Since E is computed exclusively through composition of the node functions, it is a continuous and differentiable function of the (synaptic weights) w_i of the network. We can thus minimize E by using an iterative process of *gradient descent*, for which we need to calculate the *gradient*:

$$\nabla E = \left(\frac{\partial E}{\partial w_1}, ..., \frac{\partial E}{\partial w_p} \right). \tag{5.41}$$

Next, each weight is updated using the increment $\triangle w_i = -\eta \frac{\partial E}{\partial w_i}$. To conclude, the whole learning problem reduces to the question of calculating the gradient of a network error function E with respect to its weights. Once we have computed the gradient, we can adjust the network weights iteratively. In this way we expect to find a minimum of the error function E.

A key observation in the practical use of MLP is that MLP with only two hidden layers is theoretically sufficient to model almost any real-life problem, according to one Kolmogorov's theorem, see (34). Therefore, such a MLP is adopted in several software packages concerned with NNs. Fig. 5.16 illustrates the framework of a MLP with two hidden layers.

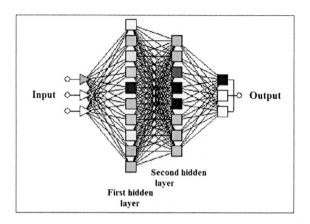

Fig. 5.16 Multi-layer perceptron with two hidden layers

- ADALINE represents a simple network with two layers (i.e., with only an input and output layer, having a single output node), the activation function at the output node being a stair step non-linear function (with binary threshold). There is also an extension of this network, consisting in the connection in series of two or more such networks, known as MADALINE (Multiple-ADALINE, or Many ADAptive LInears NEurons).
- *Radial basis function neural network* -RBFNN. Unlike the networks described above, where the computing units use a non-linear activation function, based on the dot product between the input vector and the weight vector, in this case the activation of a (hidden) unit is based on the distance between the input vector and a prototype vector (center). The basic structure of this network type involves three layers: (*a*) the input layer consisting of the source nodes which connect the network to the environment, (*b*) the hidden layer (the only one) applying a non-linear transformation from the input space on the hidden space, and (*c*) the (linear) output layer, that produces the system output. Thus, instead of using hyperplanes to divide up the problem space as in the MLP case, the RBF divides up the space by using hyperspheres characterised by centres and radii. Technically, considering a number M of *basis functions*, the RBF mapping is given by:

$$y_k(\mathbf{x}) = \sum_{j=1}^{M} w_{kj} \phi_j(\mathbf{x}), \qquad (5.42)$$

with the classical case of *Gaussian basis function* given by:

$$\phi_j(\mathbf{x}) = \exp\left(-\frac{\|\mathbf{x} - \mu_j\|^2}{2\sigma_j^2}\right), \qquad (5.43)$$

where $\mathbf{x}$ is the input vector, μ_j is the vector determining the center of the basis function ϕ_j, and σ_j is the width parameter. For training RBF, a two-stage learning procedure is used. Thus, in the first stage the input dataset is used to determine the parameters of the basis function ϕ_j (i.e., μ_j and σ_j), while in the second stage the weights w_{kj} are found. As in the MLP case, it is convenient to use the *sum-of-squares error* to be minimized in the training process (for more details, see (34), (181)). Without discussing the matter in detail, we present a brief scheme of this type of network (Fig. 5.17).

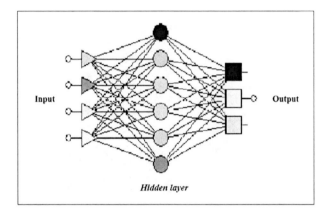

Fig. 5.17 Radial Basis Function neural network

- *Kohonen self-organizing (feature) map* -SO(F)M represents an unsupervised NN trained to obtain a transformation of an incoming signal pattern of arbitrary dimension (generally, large) into a one- or two -dimensional discrete map, performing this transformation adaptively in a topologically order way. This NN has a feedforward structure, with one layer devoted to computation, consisting of neurons arranged in rows and columns (lattice of neurons). Each neuron is fully connected to all the source nodes in the input layer, and each (synaptic) weight vector of each neuron has the same dimension as the input space. The weights are first initialized by assigning them small random values. Once the network has been initialized, the process continues with the following three steps:
  - *Competition*: for each input pattern, the neurons compute their values for a discriminant function, which provides the framework for competition between neurons. Thus, the particular neuron with the largest value will be declared as *winning* neuron;
  - *Cooperation*: the winning neuron will determine the spatial location of a topological neighborhood of excited neurons, providing the basis for cooperation among the neighboring neurons;
  - *Synaptic adaptation*: the excited neurons are enabled to increase their individual values of the discriminant function in relation to the input patterns through suitable adjustments applied to their synaptic weights. In this way, the response of the winning neuron to similar input pattern will be enhanced. More details about SOM can be seen in (218), (181). To conclude, SOM is a computational method for the visualization and analysis of high-dimensional data,

especially experimentally acquired information. Concretely, we can mention: visualization of statistical data and document collections, process analysis, diagnostics, monitoring and control, medical diagnosis, data analysis in different fields, etc. Fig. 5.18 illustrates the framework of a SOM with topological map dimension 6×7.

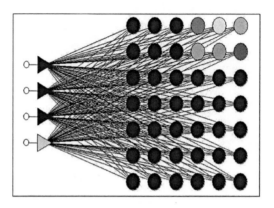

Fig. 5.18 Self-organizing map network

- *Recurrent* NNs, i.e., networks with feedback loops. Regardless of their architecture, these networks have two common features:

  – They incorporate a static MLP in their framework;
  – As a consequence, they take advantage of the non-linear mapping potential provided by MLP.

  Without going into details (see (181)), we mention four types of such networks:

  – *Input-output recurrent model*;
  – *State-space model*;
  – *Recurrent multi-layer perceptron*;
  – *Second-order network*.

- *Hopfield network/model*. Invented by the physicist J. Hopfield (1982), (191), it is a neural network where all connections are symmetrical. Its structure consists of a set of neurons and a corresponding set of unit delays, forming a *multiple-loop feedback system*, the number of feedback loops equaling the number of neurons. Technically, the output of each neuron is fed back through a *unit delay element*, to each of the other neurons. Such a network is globally asymptotically stable, however with limitations concerning its storage capacity to the size of the network (for technical details see, for instance, (181)).

Finally, we will review some classical NN models (along with their inventors' names and year of discovery), thus making a small "escapade" into the amazing history of the *Artificial Intelligence* field; see also, for instance (181), (419):

- *Perceptron* (Rosenblatt, 1957);
- *Adaline, Madaline*, (Widrow, Hoff, 1960-1962);
- *Committee machines/modular networks* (Nilsson's idea 1965/Osherson, 1990);
- *Avalanche* (Grossberg, 1967);
- *Cerebellation* (Marr, Albus, Pellionez, 1969);
- *Wisard* (Wilkie, Stonham, Aleksander, 1974-1980);
- *Backpropagation* (BPN), also known as *Multi-Layer Perceptron* (MLP) (Werbos, Parker, Rumelhart, 1974-1985);
- *Brain State in a Box* (BSB) (Anderson, 1977);
- *Cognitron* (Fukushima, 1975);
- *Neocognitron* (Fukushima, 1978-1984);
- *Adaptive Resonance Theory* (ART) (Carpenter, Grossberg, 1976-1986);
- *Self-Organizing Map* (SOM) (Kohonen, 1982);
- *Hopfield* (Hopfield, 1982);
- *Bi-directional Associative Memory* (Kosko, 1985);
- *Boltzmann/Cauchy machine* (Hinton, Sejnowsky, Szu, 1985-1986);
- *Counterpropagation* (Hecht-Nielsen, 1986);
- *Radial Basis Function Network* (RBFN) (Broomhead, Lowe, 1988);
- *Probabilistic Neural Network* (PNN) (Specht, 1988);
- *Elman network* (Elman, 1990);
- *General Regression Neural Network* (GRNN) (Specht, 1991)/*Modified Probabilistic Neural Network* (MPNN) (Zaknich et al. 1991);
- *Support Vector Machine* (SVM) (Vapnik, 1995).
- *Helmholtz machine* (Dayan, Hinton, Neal, Zemel, 1995)

Remark 5.7. 1) A recent class of (statistical) learning procedures, also included in the general field of neural networks, is represented by the *Support Vector Machines* (SVM), defined as linear feedforward learning machines, incorporating a learning procedure that is independent of the problem dimension. We will talk more broadly about this type of universal feedforward networks in subsection 5.3.5.

2) NN can be implemented both as usual software simulations, related to our main interest -the data mining field, and in the hardware area (software engineering), known as neurocomputers, which are of two types:

- The *fully implemented* type, in which there is a dedicated processor for each neuron;
- The *virtual* type, in which a single controlling microcomputer is used, together with a framework consisting of virtual (fictitious) neurons that are implemented as a series of *look-up tables*, (126), (419).

5.3.3 Probabilistic Neural Networks

In the context of the intensive use of NNs as classifiers, an interesting interpretation of the network outputs is to estimate the probability of class membership. In this case, NN learns, in fact, to estimate a probability density function. Technically, if we are able to estimate the probability density functions of all the (decision) classes, then we can compare their probabilities, and select the most-probable class. Thus, the network attempts to learn an approximation of the probability density function. Such a special case of NN, introduced by Specht (1988), (358), and known as *probabilistic neural network* (PNN), replaces the activation function (usually, sigmoid) with an exponential function.

This particular NN type provides a general solution to pattern classification problems, using a probabilistic approach based on the Bayesian decision theory. As noted in section 5.2., this theory takes into account the relative likelihood of events, and uses *a priori* information to improve the prediction. Thus, the network paradigm uses the Parzen's estimates (kernel density estimation, or *Parzen window*), (290), to obtain the probability density functions corresponding to the decision classes. In his classic paper, (290), Parzen (1962) showed that a special class of estimates asymptotically converge, under certain conditions, to the expected density. Cacoulos (1966), (51), extended Parzen's method for multivariate distributions.

PNN, as a supervised NN, uses a complete training dataset (i.e., with both input and output) in order to estimate probability density functions corresponding to the decision classes. The PNN training methodology is much simpler than for other NNs. The main advantage of PNN is that training requires a single pass over all the training patterns (training a PNN actually consists mostly of copying training cases into the network, and so it is as close to instantaneous as can be expected). Moreover, the decision hyper-surfaces thus obtained certainly tend to the Bayes optimal decision boundaries, when the number of training objects increases. In this context, we also note the fact that the PNN output is probabilistic, making thus the interpretation of the network output very easy. On the other hand, we have to highlight its main disadvantage consisting in the network size. Thus, PNN actually contains the entire set of training cases (i.e., all training samples must be stored and used to classify new objects), and is therefore space-consuming and slow to execute. In other words, PNN requires the 'knowledge baggage' always kept by the 'bearer', being not able to learn once and for all like the other networks.

We now show the manner in which PNN uses the Bayes decision rule. For this, let us consider the general classification problem consisting in determining the category membership of a multivariate sample data, i.e., a p-dimensional vector $\mathbf{x}$, based on a set of measurements. Concretely, we have to classify a certain object, represented by a p-dimensional vector $\mathbf{x} = (x_1, x_2, ..., x_p)$, into one of q possible decision categories (classes/groups), denoted by $\Omega_1, \Omega_2, ..., \Omega_q$, that is we have to decide $D(x) := \Omega_i, i = 1, 2, ..., q$. In this case, if we know:

- the probability density functions $f_1(x)$, $f_2(x)$, ..., $f_q(x)$, corresponding to the categories $\Omega_1, \Omega_2, ..., \Omega_q$,
- the *a priori* probabilities $h_i = P(\Omega_i)$ of occurrence of objects from category Ω_i (*membership probability*),
- the parameters l_i associated with all incorrect decisions given $\Omega = \Omega_i$ (*loss* or *cost* parameters),

then, using the Bayes decision rule, we classify the object $\mathbf{x}$ into the category Ω_i, if the following inequality holds true:

$$l_i \cdot h_i \cdot f_i(\mathbf{x}) > l_j \cdot h_j \cdot f_j(\mathbf{x}), i \neq j.$$

Thus, the decision boundaries between any two classes Ω_i and Ω_j, $i \neq j$, are given by the hyper-surfaces:

$$l_i \cdot h_i \cdot f_i(\mathbf{x}) = l_j \cdot h_j \cdot f_j(\mathbf{x}), \ i \neq j, \text{ and the decision}$$

accuracy will depend on the estimation accuracy of the probability density functions corresponding to the decision classes.

As we have seen above, the key element in using the Bayes decision rule in the PNN case is represented by the technique chosen for estimating the probability density function $f_i(x)$ for each category Ω_i, estimation based on the training dataset. The classical approach in this regard is the use of a sum of small multivariate Gaussian distributions (also known as *Parzen-Cacoulos* or *Parzen like window* classifiers) centered at each training sample (see also RBF), given by:

$$f_{\Omega_i}(\mathbf{x}) = \frac{1}{(2\pi)^{p/2}\sigma^p} \cdot \frac{1}{m_i} \cdot \sum_{j=1}^{m_i} \exp\left(-\frac{d(\mathbf{x}, \mathbf{x}_j)^2}{2\sigma^2}\right), \tag{5.44}$$

$i = 1, 2, ..., q$, where m_i is the total number of training patterns in Ω_i, $\mathbf{x}_j$ is the jth training sample from category Ω_i, p is the input space dimension, and σ is the only adjustable parameter of the network, known as '*smoothing*' (or *scaling*) parameter, obtained by the training procedure. The smoothing parameter σ (seen as standard deviation) defines the width of the area of influence of each decision, and should decrease as the training dataset size increases. We can suggestively see these densities as some 'bells' -see the *Gauss bell* (*bell curve*)- under which we can find various types of patterns (each type of patterns under its own 'umbrella' given by the corresponding bell coverage) and, the more they are, the smaller the bells width is, in order to enter all under the same 'umbrella'. The key factor in PNN is therefore the way to determine the value of σ, since this parameter needs to be estimated to cause reasonable amount of overlap between the classification 'bells', in order to avoid an undesirable overlap between decisions, thus worsening the generalization ability of the network. If σ is too large or too small, the corresponding probability density functions will lead to the increase in the misclassification rate. Thus, too small deviations cause a very spiky approximation, which cannot generalize well and, on the other hand, too large deviations smooth out the details. Commonly, the smoothing factor σ is chosen heuristically. Thus, the search domain is represented by the entire positive real axis $\mathbf{R}_+$, with the decision boundary continuously varying from a non-linear border if of $\sigma \to 0$, to a hyper-plane if $\sigma \to \infty$, (360). However, fortunately, PNN is not too sensitive to reasonable variations of the σ value.

Remark 5.8. Besides the above standard estimate (i.e., the Parzen-Cacoulos model), within the PNN algorithm several alternatives are used to estimate the probability density functions corresponding to the decision classes. We will present below some of the most used such estimates for a certain decision class Ω:

- $f(\mathbf{x}) = \dfrac{1}{n(2\lambda)^p} \sum\limits_{j=1}^{m} 1$, when $|x_i - x_{ij}| \leq \lambda$, $i = 1, 2, ..., p$, $j = 1, 2, ..., m$;

- $f(\mathbf{x}) = \dfrac{1}{m\lambda^p} \sum\limits_{j=1}^{m} \prod\limits_{i=1}^{p} \left[1 - \dfrac{|x_i - x_{ij}|}{\lambda}\right]$, when $|x_i - x_{ij}| \leq \lambda$, $i = 1, 2, ..., p$, $j = 1, 2, ..., m$;

- $f(\mathbf{x}) = \dfrac{1}{n(2\pi)^{p/2} \lambda^p} \sum\limits_{j=1}^{m} \prod\limits_{i=1}^{p} \exp\left[-\dfrac{1}{2} \dfrac{(x_i - x_{ij})^2}{\lambda^2}\right] = $

  $= \dfrac{1}{n(2\pi)^{p/2} \lambda^p} \sum\limits_{j=1}^{m} \exp\left[\dfrac{-\sum_{i=1}^{p}(x_i - x_{ij})^2}{2\lambda^2}\right];$

- $f(\mathbf{x}) = \dfrac{1}{n(2\lambda)^p} \sum\limits_{j=1}^{m} \prod\limits_{i=1}^{p} \exp\left[-\dfrac{|x_i - x_{ij}|}{\lambda}\right] = \dfrac{1}{n(2\lambda)^p} \sum\limits_{j=1}^{m} \exp\left[-\dfrac{1}{\lambda} \sum\limits_{i=1}^{p} |x_i - x_{ij}|\right];$

- $f(\mathbf{x}) = \dfrac{1}{n(\pi\lambda)^p} \sum\limits_{j=1}^{m} \prod\limits_{i=1}^{p} \left[1 + \dfrac{(x_i - x_{ij})^2}{\lambda^2}\right]^{-1};$

- $f(\mathbf{x}) = \dfrac{1}{n(2\pi\lambda)^p} \sum\limits_{j=1}^{m} \prod\limits_{i=1}^{p} \left[\dfrac{\sin \frac{(x_i - x_{ij})}{2\lambda}}{\frac{(x_i - x_{ij})}{2\lambda}}\right]^2;$

- $f_{kn}(\mathbf{x}) = \dfrac{1}{(2\pi)^{p/2} \sigma^p} \cdot \dfrac{1}{m} \cdot \sum\limits_{j=1}^{m} \exp\left[k^n \left(-\dfrac{d(\mathbf{x}, \mathbf{x}_j)^2}{2\sigma^2}\right)\right]$, $k \geq 2$, $n \geq 1$;

- $f_{Tr}(\mathbf{x}) = \dfrac{1}{(2\pi)^{p/2} \sigma^p} \cdot \dfrac{1}{m} \cdot \sum\limits_{j=1}^{m} \sum\limits_{k=1}^{r} \dfrac{\left(-\dfrac{d(\mathbf{x}, \mathbf{x}_j)^2}{2\sigma^2}\right)^k}{k!}$, for $r \geq 1$.

  (for details, see (359), (144)).

PNNs are seen as implementations of a statistical algorithm, known as *kernel discriminant analysis* (KDA), in which the procedures are organized into a feedforward multilayered network containing, in total, four layers:

- *Input layer*;
- *Pattern layer*;

- *Summation layer*;
- *Output layer*.

Note 5.1. Sometimes PNNs are credited with three layers only: *input/pattern layer*, *summation layer* and *output layer*, by merging the first two layers.

The fundamental PNN architecture, excluding the initial input layer, consists of nodes assigned to the three basic layers, (358), (360):

- *Pattern layer/unit.* There is only one pattern node/pattern unit for each training object (sample). Each input/pattern node forms a (dot) product of the input pattern vector $\mathbf{x}$ with a weight vector $\mathbf{w}_i$, denoted $z_i = \mathbf{x} \cdot \mathbf{w}_i^T$, and then performs a non-linear operation, that is $\exp\left[-(\mathbf{w}_i - \mathbf{x}) \cdot (\mathbf{w}_i - \mathbf{x})^T/(2\sigma^2)\right]$ (assuming that both $\mathbf{x}$ and $\mathbf{w}_i$ are normalized to unit length), before outputting its activation level to the summation node. Note that, unlike the sigmoid activation function used in MLP (in the back-propagation algorithm), the non-linear transformation used in this case is the exponential $\exp\left[(z_i - 1)/\sigma^2\right]$;
- *Summation layer/unit.* Each summation node receives the outputs from the input/pattern nodes associated with a given class and simply sums the inputs from the pattern units that correspond to the category from which the training pattern was selected, that is $\sum_i \exp\left[-(\mathbf{w}_i - \mathbf{x}) \cdot (\mathbf{w}_i - \mathbf{x})^T/(2\sigma^2)\right]$;
- *Output (decision) layer/unit.* The output nodes produce binary outputs corresponding to two different decision categories Ω_r and Ω_s, $r \neq s$, $r,s = 1,2,...,q$, by using the inequality (*classification criterion*):

$$\sum_i \exp\left[-(\mathbf{w}_i - \mathbf{x}) \cdot (\mathbf{w}_i - \mathbf{x})^T/(2\sigma^2)\right] > \sum_j \exp\left[-(\mathbf{w}_j - \mathbf{x}) \cdot (\mathbf{w}_j - \mathbf{x})^T/(2\sigma^2)\right].$$

These nodes/units have only one weight C, given by the loss (or cost) parameters, the *a priori* membership probabilities, and the number of training samples in each category. Concretely, the weight C is given by $C = -\dfrac{h_s l_s}{h_r l_r} \cdot \dfrac{n_r}{n_s}$.

This weight is determined based only on the decision's significance, so that, in case of the lack of such an information, we simply choose $C = -1$.

Technically speaking, such a PNN (training) algorithm may have the following form (see, for instance, (136).

PNN training algorithm

Input. Consider q classes of patterns/objects (p-dimensional vectors) Ω_1, $\Omega_2,..., \Omega_q$. Each decision class Ω_i contains a number of m_i vectors (or training patterns), that is $\Omega_i = \{\mathbf{x}_1, \mathbf{x}_2, ..., \mathbf{x}_{m_i}\}$.

1) For each class Ω_i, $i = 1,2,...,q$, compute the (Euclidian) distance between any pair of vectors and denote these distances by $d_1, d_2,..., d_{r_i}$, where $r_i = C_{m_i}^2 = \dfrac{m_i!}{2!(m_i - 2)!}$.

2) For each class Ω_i, $i = 1, 2, ..., q$, compute the corresponding average distances and standard deviations $D_i = \dfrac{\sum_{j=1}^{r_i} d_j}{r_i}$, $SD_i = \sqrt{\dfrac{\sum_{j=1}^{r_i} (d_j - D_i)^2}{r_i}}$.

3) (*Searching process*) Compute the "smoothing" parameter searching domain D_σ, based on the 99,7% confidence interval, given by $D_\sigma = (0, 3 \times SD)$, where $SD = \max_i \{SD_i\}$.

4) For each decision class Ω_i, $i = 1, 2, ..., q$, consider the decision functions (Parzen-Cacoulos window classifier):

$$f_{\Omega_i}(\mathbf{x}) = \frac{1}{(2\pi)^{p/2} \sigma^p} \cdot \frac{1}{m_i} \cdot \sum_{j=1}^{m_i} \exp\left(-\frac{d(\mathbf{x}, \mathbf{x}_j)^2}{2\sigma^2} \right).$$

5) (*Bayes decision rule*) In each decision class Ω_i (randomly) choose a certain vector $\mathbf{x}_i^0$. Compare $f_i(\mathbf{x}_i^0)$ and $f_j(\mathbf{x}_i^0)$, for all $i \neq j$, following the algorithm:

"IF $l_i \cdot h_i \cdot f_i > l_j \cdot h_j \cdot f_j$ (for all $j \neq i$) THEN $\mathbf{x}_i^0 \in \Omega_i$ ELSE IF $l_i \cdot h_i \cdot f_i \leq l_j \cdot h_j \cdot f_j$ (for some $j \neq i$) THEN $\mathbf{x}_i^0 \notin \Omega_i$"

6) (*Measuring the classification accuracy*) For each (fixed) decision class Ω_i consider the 3-valued logic:

"TRUE if $l_i \cdot h_i \cdot f_i > l_j \cdot h_j \cdot f_j$ (for all $j \neq i$), UNKNOWN if $l_i \cdot h_i \cdot f_i = l_j \cdot h_j \cdot f_j$ (for some $j \neq i$) and FALSE -otherwise".

Initially, each of the three variables is set to zero. Whenever a truth value is obtained, the corresponding variable is incremented with step size 1.

7) Repeat step 5 for another choice for $\mathbf{x}$ in Ω_i until all of them are chosen. Increment counter.

8) Repeat step 5 for all vectors $\mathbf{x}$ in Ω_j, for all $j \neq i$. Increment counter.

9) Obtain the classification accuracy in percentage (σ values are cached).

10) (*Estimating optimal smoothing parameter*) Choose a certain procedure to estimate the parameter σ.

11) If the current value of σ exceeds D_σ, then STOP.
Obtain the corresponding classification accuracy (σ values are cached).

12) Compute the maximum value MAX of the variable corresponding to the TRUE value.

Output. σ corresponding to MAX represents the optimal value of the "smoothing" parameter for each decision category Ω_i, $i = 1, 2, ..., q$.

Regarding the estimation procedure of the only network parameter, i.e., σ (from step 10 of the above training algorithm), we present here three variants, simple but effective, (136) (137), (138), (139), (140), (315), (147) together with the corresponding performance obtained in practical applications.

- **Incremental approach.** Divide the search domain D_σ by N dividing knots σ_1, $\sigma_2,...,\sigma_N$ into $(N + 1)$ equal sectors. Repeat step 5 by assigning to σ the values σ_k, $k = 1, 2, ..., N$.
- **Genetic algorithm approach.** Each chromosome is defined by the variable $X = (\sigma)$, the gene corresponding to the smoothing factor σ, taking its value from the value domain D_σ. A population of Y chromosomes is used. Selection is carried out by the Monte Carlo procedure. The average crossover (arithmetic recombination) $(X_1, X_2) \rightarrow \left(\dfrac{X_1 + X_2}{2}\right)$ is used to generate new chromosomes and for the mutation the following technique is applied (see also the non-uniform mutation, subsection 5.9): "assume we decide to mutate the gene σ of a chromosome. We will generate a random number, whose values are either 0 or 1. Then, the new value for the gene is determined by $\sigma \pm \delta$ (δ is a small enough value to fine tune the accuracy), "+" if 0 is generated, and "- " otherwise".
  Find the maximum of the cost function, counting the number of correct classifications.
- **Monte Carlo approach.** Generate in the search domain D_σ a number of N random dividing points $\{P_1, P_2,..., P_N\}$, uniformly distributed in D_σ. Repeat step 5 by assigning $\sigma = P_k$, $k = 1, ..., N$.

The basic PNN architecture, presented above, is suggestively illustrated in Fig. 5.19.

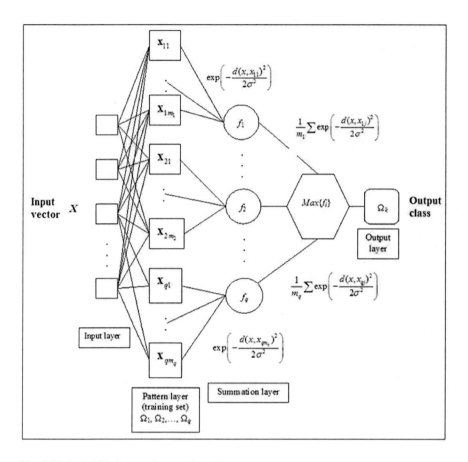

Fig. 5.19 Probabilistic neural network architecture

5.3.4   Some Neural Networks Applications

We end the presentation of this very fascinating 'world' of artificial neural networks, trying to illustrate their classification capabilities, a fundamental data mining problem, with some practical applications. It is superfluous to mention that there is a very rich literature in this field, ranging from classification, function approximation (regression analysis, time series, fitness approximation, etc.), data processing, robotics, automatic medical diagnosis, pattern recognition, etc. For this, we will select some easy to understand, but effective applications, mainly focused on the automatic medical diagnosis, a sensitive problem of our time.

Example 5.5. Iris flower classification using different NNs types. First of all, in order to illustrate the performance of NNs as classifiers, we used a classical dataset and some of the most common NNs types. Thus, we considered the well-known *Iris* flower dataset (see also Section 3.8), consisting of 50 flowers of *Setosa* type, 50 of *Versicolour* type, and 50 of *Virginica* type, and four classical NNs types: Kohonen (SOM), RBF, MLP and linear NN. RBF and MLP were considered with four different architectures. Technically, we have divided the dataset containing all the 150 flowers into two subsets: a training dataset of 80 flowers, containing 27 *Setosa* samples, 26 *Versicolour* samples, and 27 *Virginica* samples, and a testing dataset of 70 flowers, containing 23 *Setosa* samples, 24 *Versicolour* samples, and 23 *Virginica* samples. The attributes considered in the flowers classification process are the petals and sepals dimensions (i.e., length and width, respectively). The table below presents the main results obtained by applying different neural network types to classify the *Iris* flowers.

Table 5.2 NNs structure and classification performance for *Iris* flowers

NN type	Number of hidden layers/units	Testing accuracy
Kohonen (SOM)	5/-	91%
RBF (Radial Basis Function) #1	3/1	64%
RBF (Radial Basis Function) #2	3/2	91%
RBF (Radial Basis Function) #3	3/4	94%
RBF (Radial Basis Function) #4	3/10	95%
MLP (Multilayer Perceptron) #1	2/2	96%
MLP (Multilayer Perceptron) #2	2/4	96%
MLP (Multilayer Perceptron) #3	2/7	97%
MLP (Multilayer Perceptron) #4	2/8	97%
Linear model	2/-	87%

Both overall and per-class classification statistics for a MLP model are displayed in the table below. From this table we can observe: the total number of cases in each class, cases of each class that were correctly (and incorrectly) classified, and finally, cases of that class which could not be classified at all (i.e., unknown cases). In this way, more detailed information on misclassification is provided.

Table 5.3 Overall and per-class statistics

	Setosa	*Virginica*	*Versicolour*
Total	50	50	50
Correct	50	47	48
Wrong	0	3	2
Unknown	0	0	0
Correct (%)	100	94	96
Wrong (%)	0	6	4
Unknown (%)	0	0	0

Next table displays the corresponding confusion matrix.

Table 5.4 Confusion matrix

	Setosa	*Virginica*	*Versicolour*
Setosa	50	0	0
Virginica	0	47	2
Versicolour	0	3	48

The figures below illustrate the scheme of each NN used for classification of the *Iris* flowers.

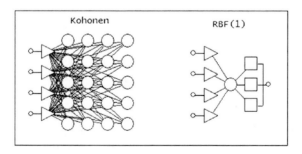

Fig. 5.20 Kohonen (SOM) and Radial Basis Function networks #1

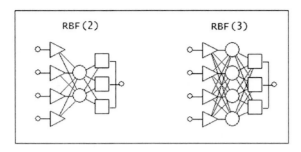

Fig. 5.21 Radial Basis Function networks (#2 and #3)

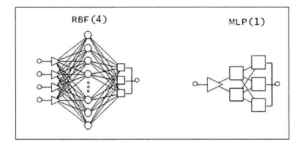

Fig. 5.22 Radial Basis Function #4 and Multi-Layer Perceptron #1

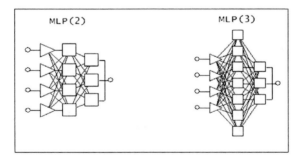

Fig. 5.23 Multi-Layer Perceptron (#2 and #3)

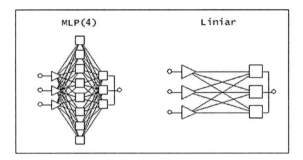

Fig. 5.24 Multi-Layer Perceptron #4 and the linear model

Example 5.6. Automatic medical diagnosis of liver diseases with PNN. The PNN model was applied to classify/diagnose a group of 299 people in four categories (decision classes) corresponding to three different liver diseases, here including the fourth class, representing the control group consisting of healthy people (137), (138), (139), (140), (141), (142), (143), (144), (145).

The four decision classes are the following:

- Chronic hepatitis (CH);
- Liver cirrhosis (LC);
- Hepatocellular carcinoma (hepatic cancer) (HCC);
- Control group (healthy people) (HP).

The dataset concerns patients from the County Emergency University Hospital of Craiova, Romania, and consisted of:

- 60 (CH) patients;
- 179 (LC) patients;
- 30 (HCC) patients;
- 30 healthy people.

Each individual in the group was mathematically represented by a vector with 15 components, each component representing a significant characteristic for the diagnosis process. Thus, an individual is identified by an object with 15 attributes: x_1 = TB (total bilirubin), x_2 = DB (direct bilirubin), x_3 = IB (indirect bilirubin), x_4 = AP (alkaline- phosphatase), x_5 = GGT (gamma-glutamyl-transpeptidase), x_6 = LAP (leucine amino peptidase), x_7 = AST (aspartate-amino-transferase), x_8 = ALT (alanine-amino- transferase), x_9 = LDH (lactic dehydrogenase), x_{10} = PI (prothrombin index), x_{11} = GAMMA, x_{12} = ALBUMIN, x_{13} = GLYCEMIA, x_{14} = CHOLESTEROL, x_{15} = Age.

PNN was trained on a training dataset of 254 individuals (85%), and tested on a dataset consisting of 45 individuals (15%). In the learning phase, in order to estimate the parameter σ, the three methods mentioned above, namely: incremental search, Monte Carlo method, and genetic algorithms-based search were used. The results were satisfactory, the accuracy obtained in the training phase was around 97%, while the testing accuracy was around 95%. There may be, obviously, different values for the classification (diagnosis) accuracy (training/testing), depending on both the used method and the choice of parameters for each method (e.g., number of search points for the incremental method, number of nodes for the Monte Carlo method, size of the chromosomes population, number of generations and genetic operators used for the genetic algorithms approach). Many other researches showed that PNNs proved to be efficacious classifiers in the automatic diagnosis in different medical branches.

The PNN algorithm has been implemented in Java (regardless of the searching method for σ). The Java programming language has been chosen because the Java technology is an object-oriented, platform-independent, multithreaded programming environment. The Standard Edition of the Java Platform was used since it is designed to develop portable, secured and high-performance applications for the widest range of desktop computing platforms possible, including Microsoft Windows, Linux, Apple Macintosh and Sun Solaris; Java applications were compiled to bytecode.

The most important issue about this implementation is that the *Java Database Connectivity* (JDBC) has been used. The (JDBC) API is the standard for database-independent connectivity between Java programming language and a wide range of databases (SQL, MS Access, spreadsheets or flat files). By using JDBC, physicians are able to query and update data in the database directly from the program.

The PNN architecture used in this study is displayed below.

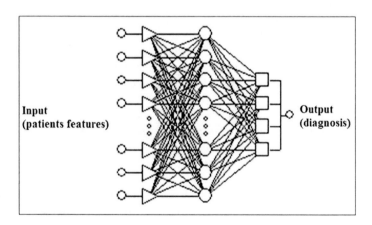

Fig. 5.25 PNN architecture

Other NNs types have been used on the same database, for comparison. The table below shows their testing performance, comparable, excepting the linear model, with that provided by PNN.

Fig. 5.26 illustrates the corresponding architectures.

Table 5.5 Comparison of different NN diagnosis accuracies

NN type	Number of hidden layers/units	Accuracy
RBF (Radial Basis Function)	3/25	92%
MLP (Multilayer Perceptron)	3/8	94%
Linear model	2/-	82%

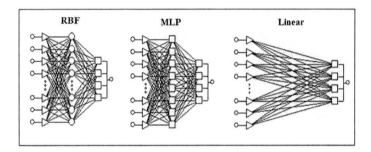

Fig. 5.26 RBF, MLP and linear model architectures

Example 5.7. Differential diagnosis of pancreatic diseases using MLP. Another way of using NNs in computer-aided diagnosis is the differential diagnosis obtained by the analysis of dynamic sequences of EUS elastography. In (331) a multilayer-perceptron with two hidden layers with different numbers of neurons in each layer and the back-propagation procedure as training algorithm has been used for the differential diagnosis of chronic pancreatitis and pancreatic cancer. The study design was prospective and included a total of 68 patients (normal pancreas -22, chronic pancreatitis -11, pancreatic adenocarcinoma -32, and pancreatic neuroendocrine tumors -3). Instead of using common numerical data format to feed NN, a new medical imaging technique was used, namely, the EUS elastography. Technically, *EUS elastography* is a newly developed imaging procedure that characterizes the differences of hardness and strain between diseased tissue and normal tissue. The elastography information is displayed in real time as a transparent color overlay in a defined region of interest, similar to color Doppler examinations. EUS elastography was performed during the EUS examinations, with 2 movies of at least 10 seconds. Each acquired movie was subjected to a computer-enhanced dynamic analysis by using a public domain Java-based image processing tool (*ImageJ*) developed at the National Institutes of Health, Bethesda, Maryland, USA. Concretely, each EUS elastography movie was converted into a numerical form, characterized by a single average hue histogram vector. Each individual value of the vector corresponded to the number of pixels of each color, in other words, to the number of pixels that correspond to the elasticity level, from 1 to 256. Thus, MLP was fed with a set of vectors that represents average hue histograms, which summarize the information provided by the EUS elastography sample movies. A very good testing performance of 95% on average was obtained, together with a high training performance, equaling 97% on average. The optimum number of hidden units in the network equals 17, with a relative high SD of 14. Thus, it is possible to have a relatively simple network structure, that is a very fast NN, with a very good performance. The area under the ROC curve (0.957) confirmed the high classification performance.

Example 5.8. Pancreatic cancer detection using committees of networks. In (151), a competitive/collaborative neural computing system has been applied to detect pancreatic cancer, using again the EUS elastography methodology. In the first stage, the computing system works in a competitive mode, that is the n initial NN algorithms will be firstly applied to the same medical database and, next, statistically evaluated (using a benchmarking process consisting of different comparison tests concerning their performances), in order to put them in a hierarchical order, depending on their classification performances. In the second stage, the computing system works in a weighted collaborative mode. Each of the k best NN algorithms is applied to new data corresponding to a previously undiagnosed patient, and an overall diagnosis is retained as the final computing system output, based on a weighted voting system (WVS). Four different NNs types were used: linear NN, MLP, RBF and PNN. They were tested on the same dataset as in the example above, consisting of average hue histogram vectors corresponding to 68 patients with different pancreatic diseases. The hierarchy of the NNs competitors (based on the testing accuracy), obtained in the competitive phase was: 2-hidden-layer MLP (95%), 1-hidden-layer MLP (91%), RBF (80%), linear NN (65%), and PNN (48%). We have to remark that the above results, obtained in the competitive phase, reflect the performances of NNs on this particular database only. It is worth to mention that the NNs performances strongly depend on the medical database in use, for instance: different diseases, same disease but different attributes types, etc. As we mention above, the previous competitive phase generated a classification (diagnosing) performance hierarchy, given as follows: MLP (2-layer), MLP (1-layer), RBF, linear NN, PNN. Taking into account the benchmarking process and the performances compared with the reported standard medical results, the first three NNs have been retained to provide the final diagnosis. A more elitist competition, focused on the inter-rater reliability and the over-learning behavior, would select the two MLP models only. The distribution of the number of votes (weights) among the three NN models is given by (default/standardized to 100): MLP (2-layer) -36 votes, MLP (1-layer) -34 votes and RBF -30 votes, with a quota q ranging between 51 and 100. For the sake of simplicity, this study dealt with equal amount of power for all voters only. Thus, a default WVS in this case might be $\{51: 36, 34, 30\}$. To highlight the effectiveness of such an approach, a concrete example consisting of three different testing cases (i.e., three new, unknown, patients) has been considered. Thus, while in the first case, all the three best NN models provided the same diagnosis, which is the real one, in the second and third cases, two of them agreed on the same diagnosis, contrary to the third one, but the WVS collaborative mechanism provided the right diagnosis. To conclude, the use of the ensemble of NN models, working both in a competitive and collaborative way, provides the most robust behavior with respect to the automatic diagnosis reliability.

Example 5.9. Modeling patient length of stay using a hybrid Self-Organizing Map (Kohonen network) and Gaussian mixture model. In (152), the SOM ability to validate the length of stay (LOS) clustering results, obtained using Gaussian mixture modeling (GMM) approach, is explored, by comparing the classification accuracy of different results. Generally speaking, the result of the clustering process may not be usually confirmed by our knowledge of the data. The self-organizing map neural network is an excellent tool in recognizing clusters of data, relating similar classes to each other in an unsupervised manner. As it is well-known, SOM is used, basically, when the training dataset contains cases featuring input variables without the associated outputs. SOM can also be used for classification when output classes are immediately available. The advantage in this case is its ability to highlight similarities between classes, thus assessing different previous classification approaches. Technically, two clustering techniques: GMM and SOM neural network are used in tandem. Recall that GMM is a probability density function comprising of m Normally distributed component functions, (383), (250). These Normally distributed components are combined together to form the overall density model, flexible enough (depending on m) to approximate almost any distribution.

Health care facilities operate administrative information systems to collect information on patient activity. In this context, patient length of stay is often used as a proxy measure of a patient's resource consumption due to the practical difficulties of directly measuring resource consumption and the easiness of calculating LOS. Grouping patients is advantageous in that it helps to simplify our view as well as improve our comprehension of the diverse patient population. Understanding the different groups of patients with regards to their LOS and predicting LOS at admission would assist hospital management and health professionals in making more informed and timely decisions on managing patients' care and planning for their discharge, and on allocating hospital resources.

This study introduces a two-phase clustering approach for deriving clinically meaningful groups of patient spells and validating the subsequently derived LOS intervals. The first phase is based on a GMM approach in order to obtain LOS-based component models of the data and the subsequent LOS intervals. Although GMM has proved to be a viable method for grouping patient spells, it is not too easy to determine whether the number of component models (clusters) derived from GMM is optimal. This has motivated the SOM application, in the second phase, to verify the optimality of the chosen clusters and to conduct a statistical analysis for further validation. This tandem hybrid model was applied on a surgical dataset which contains data that are still typically stored by hospital computerized systems. The Surgical dataset consists of 7723 records detailing the spells of patients undergoing surgery in a tertiary hospital in Adelaide, Australia between 4 February 1997 and 30 June 1998.

In this study, four different SOM/GMM hybrid models have been considered:

- SOM with GMM (5 LOS intervals): [0-2 days];[3-5 days];[6-13 days];[14-36 days]];[37+ days];
- SOM with GMM (5 LOS intervals): [0-3 days];[4-6 days];[7-14 days];[15-37 days];[38+ days];

- SOM with GMM (4 LOS intervals): [0-3 days];[4-9 days];[10-28 days];[29+ days];
- SOM with GMM (4 LOS intervals): [0-3 days];[4-8 days];[9-37 days];[38+ days].

The SOM application to assess the optimal LOS intervals has provided the following hierarchy:

1. LOS intervals: [0-3 days];[4-8 days];[9-37 days];[38+ days];
2. LOS intervals: [0-2 days];[3-5 days];[6-13 days];[14-36 days]];[37+ days];
3. LOS intervals: [0-3 days];[4-9 days];[10-28 days];[29+ days];
4. LOS intervals: [0-3 days];[4-6 days];[7-14 days];[15-37 days];[38+ days].

The subsequent statistical analysis has proved that SOM, as applied using the previous GMM clustering, significantly discriminates between the LOS intervals. Thus, the hybrid two-stage clustering system can be used to assess and validate the GMM results regarding the LOS grouping, and to choose the optimal number of LOS components (clusters). To conclude, SOM, as an unsupervised neural network, uses its self-organizing capability of classifying objects to assess the GMM performances by comparing its classification with those inferred from the GMM clustering process. This represents a reliable and a viable way of verifying the optimal LOS clustering among the feasible alternatives generated by GMM.

Example 5.10. Real-life applications of NNs models. It is obvious to say that NN-based solutions are extremely efficient in terms of running speed and computational resources, and, in many complex real-life problems, neural networks provide performance that is difficult to be reached by other technologies. The area of NNs applications is very wide, here we just mention some of them. Thus, NNs have been used in weather forecasting - (312), (246), (117), stock market - (252), (213), fraud detection - (289), (42), pattern recognition - (266), (34), signal processing - (419), (244), agricultural production estimates - (280), (332), and some particular applications, such as: prediction of dew point temperature - (342), fault diagnosis in analog circuits - (102), pooled flood frequency analysis - (344), etc.

Example 5.11. Computer-aided (medical) diagnosis with NNs models. We now briefly present other NNs concrete applications in the medical field. Thus, the use of computer technology within medical decision support is now widespread and pervasive across a wide range of medical areas. Computer-aided diagnosis (CAD), as a relatively recent interdisciplinary methodology, interprets medical images and other specific measurements using the computer technology in order to obtain automatic differential diagnosis of lesions. Many recent studies focus on applications of different classification algorithms on a wide range of diseases. In this regard, the comparison of the performances of automated diagnosis systems is essential - (52), (371), (255), (352). The use of neural computing techniques in the CAD field is now sufficiently widespread - (281), (216), (59), (185), (226), (245), (225), (351), (122), (83), (54), (58), (256), (211),(28), (27), (148), (149), (150), (153) since the ability of NNs to learn from input data with or without a teacher makes them very flexible and powerful in medical diagnosis.

5.3.5 Support Vector Machines

A special class of universal networks of feedforward type is represented by the so-called *support vector machines* (SVMs), extensively used in pattern classification problems and non-linear regression. SVMs are based on the *statistical learning theory* (SLT), and among their promoters we can include: Boser, Guyon, Vapnik, (38), (391), (393), Cortes (68).

Conceptually, SVM is a linear machine, equipped with special features, and based on the *structural risk minimization* (SRM) method and the *statistical learning theory*. Consequently, SVM can provide a good generalization performance in pattern recognition problems, without incorporating problem-domain knowledge, which gives it a unique feature among other learning machines.

From the previous presentation of NNs one can derive the following dilemma: "The perceptron, or a more complex single-layer neural network, despite their simple and efficient learning algorithm, have a limited power of classification, because they learn only the linear decision boundaries in the input space. Multi-layer networks have, however, a much higher classification power, but, unfortunately, they are not always easily trained because of the multitude of local minima and the high dimensional weights space". A solution to this dilemma may come from SVMs or, more generally, *kernel machines*, that have efficient training algorithms and can represent, at the same time, complex non-linear boundaries.

Without going into details, we illustrate the problem of separating points that are not linearly separable (i.e., points that cannot be totally classified using a separating hyperplane) by the following very simple example, shown in the figure below. The basic idea drawn from this figure can be summarized by the *kernel trick*, i.e., the way to solve a non-linear separation problem by mapping the original non-linearly separable points into a higher-dimensional space, where a linear classifier is subsequently used. Therefore, a linear classification in the new space is equivalent to non-linear classification in the original space.

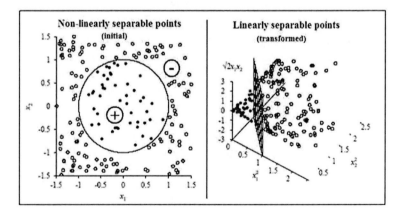

Fig. 5.27 *Kernel trick* paradigm

Thus, the figure on the left shows the two-dimension input space, in which an object is denoted by $\mathbf{x} = (x_1, x_2)$. Let us note that there are two classes of objects:

- The "positive" ones ($d = +1$), which are located inside the circle of equation $x_1^2 + x_2^2 = 1$;
- The "negative" ones ($d = -1$), which are located outside the circle.

As we can easily observe, there is not a linear separation between the two sets of objects, instead there is a circular one. But, if we map the two-dimensional input space ($\mathbf{x}$) into another three-dimensional space ($\mathbf{z}$), by an application $z = g(\mathbf{x})$, given by:

- $z_1 = g_1(\mathbf{x}) = x_1^2$,
- $z_2 = g_2(\mathbf{x}) = x_2^2$,
- $z_3 = g_3(\mathbf{x}) = \sqrt{2}x_1 x_2$,

we observe, in the right figure, that in the new three-dimensional space ($\mathbf{z}$), the transformed objects are now linearly separable. If we project this new space onto the first two axes, we get a detailed 'image' of this method, illustrated in the figure below, in which the separator is a line, and the closest points -the *support vectors*- are marked with little circles.

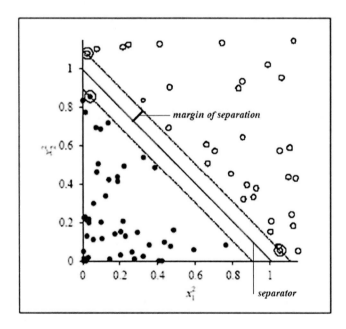

Fig. 5.28 Linear separation obtained by the *kernel trick*

The *margin of separation*, which is represented in the figure above by half of the bandwidth centered on the linear separator (the central line), defines a sort of 'distance' between the "positive" objects and the "negative" objects. In other words, the margin of separation measures the separation between the linear separator (in our case the line and, generally, a hyperplane) and the closest point of the training dataset, thus separating the points.

Generalizing, if the initial space is mapped into a space with sufficiently high dimension, then such non-linearly separable objects will become linearly separable.

Remark 5.9. Regarding the ease to find a linear separator between points in space ($\mathbf{z}$), it is worth mentioning that its equation in a n-dimensional space is defined by n parameters, and so, the *overfitting* phenomenon will occur when n is approximately equal to the number N of points to be classified. Therefore, kernel machines usually search the *optimal* linear separator, i.e., the one that has the largest margin (distance) between "positive" and "negative" patterns. One can show (*computational learning*) that this separator has suitable properties regarding the robust generalization concerning new patterns.

The general scheme of a kernel machine is displayed in the figure below.

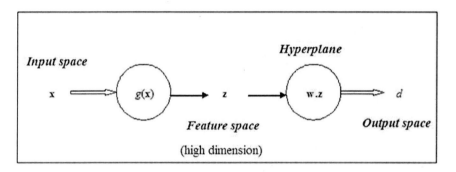

Fig. 5.29 Scheme of a *kernel machine*

Conceptually, SVMs (kernel machines, in general) operate based on two steps, illustrated in the figure above:

- A non-linear mapping of the input space into a high-dimensional *feature space*, that is hidden from both sides (i.e., input and output);
- The construction of an (optimal) separation hyperplane for the features obtained in the first step.

Remark 5.10. Theoretically speaking, the first step is based on Cover's theorem regarding the *separability of patterns*, (70), while the second step is based on structural risk minimization.

A key idea in building a SVM is represented by the use of the *inner-product kernel* between a support vector and an arbitrary vector from the input space. In what follows we will briefly present the basic concept of the inner-product kernel (for details, see (181)). Thus, let us denote by $\mathbf{x}$ a certain vector from the input space, and by $g(\mathbf{x}) = \{g_j(\mathbf{x}), j = 1, 2, ..., m_1\}$ a set of non-linear transformations from the input space $\mathbf{x}$ to the feature space $\mathbf{z}$, where m_1 is the dimension of the feature space.

Given the linear transformation g (*a priori* defined), we define a hyperplane, seen in this context as a decision surface, by:

$$\sum_{j=1}^{m_1} w_j \cdot g_j(\mathbf{x}) + b = 0, \tag{5.45}$$

where $\mathbf{w} = \{w_j, j = 1, 2, ..., m_1\}$ represents a weight vector, connecting the feature space $\mathbf{z}$ to the output space $\mathbf{d}$, b being a bias. Denoting $g_0(\mathbf{x}) \equiv 1$ and $w_0 = b$, we may write:

$$\sum_{j=1}^{m_1} w_j \cdot g_j(\mathbf{x}) = 0. \tag{5.46}$$

Thus, $g(\mathbf{x}) = (g_0(\mathbf{x}), g_1(\mathbf{x}), ..., g_{m_1}(\mathbf{x}))$ represents the image (transformation) of the input vector in the feature space, weighted by $\mathbf{w} = (w_0, w_1, ..., w_{m_1})$, and the decision surface in the feature space is then given by:

$$\mathbf{w} \cdot g^T(\mathbf{x}) = 0. \tag{5.47}$$

Denoting by $\mathbf{x}_i, i = 1, 2, ..., N$, an input training vector, we may build the inner-product kernel defined by:

$$K(\mathbf{x}, \mathbf{x}_i) = g(\mathbf{x}) \cdot g^T(\mathbf{x}_i) = \sum_{j=0}^{m_1} g_j(\mathbf{x}) \cdot g_j(\mathbf{x}_i), \ i = 1, 2, ..., N, \tag{5.48}$$

kernel used in the process of finding the optimal separation hyperplane, highlighting the fact that explicit knowledge about the feature space is not at all involved in this process.

As a computationally efficient procedure to obtain the optimal hyperplane, we may mention the *quadratic optimization*, (181).

Next, we present in short the principles underlying the construction and operation of SVMs. Thus, let us consider the training dataset $T = \{\mathbf{x}_i, d_i; i = 1, 2, ..., N\}$, where $\mathbf{x}_i$ represents an input pattern and d_i the corresponding output.

We first assume that the two classes $d_i = +1$ ("positive" patterns) and $d_i = -1$ ("negative" patterns) are linearly separable (see Fig. 5.30).

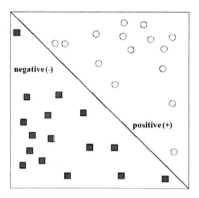

Fig. 5.30 Linearly separable patterns

Then, the equation of a separation hyperplane (decision surface) is given by:

$$\mathbf{w} \cdot \mathbf{x}^T + b = 0, \tag{5.49}$$

where $\mathbf{x}$ is an input vector, $\mathbf{w}$ is an adjustable weight, and b is a bias. Next, we have:

- $\mathbf{w} \cdot \mathbf{x}_i^T + b \geq 0$ *if* $d_i = +1$,
- $\mathbf{w} \cdot \mathbf{x}_i^T + b < 0$ *if* $d_i = -1$.

Recall that, for a known weight vector $\mathbf{w}$ and a bias b, the separation given by the hyperplane, defined by the above equation, and the closest point from the two regions is just the *margin of separation*. Under these circumstances, the goal of SVM is to find that hyperplane which maximizes the margin of separation. In this case, the decision surface thus found is called *optimal hyperplane*. The figure below illustrates the way of choosing the optimal hyperplane.

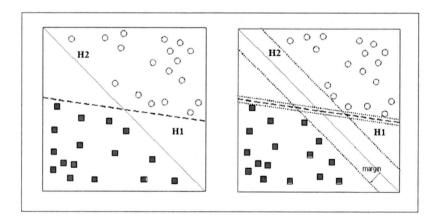

Fig. 5.31 Choosing the optimal hyperplane (H2)

As it easy to see, the optimal hyperplane is H2 (the largest margin of separation).

Let $\mathbf{w}_0$ and b_0 denote the values corresponding to the optimal hyperplane, given by:

$$\mathbf{w}_0 \cdot \mathbf{x}^T + b_0 = 0. \tag{5.50}$$

Then, the *discriminant function*, defined by:

$$g(\mathbf{x}) = \mathbf{w}_0 \cdot \mathbf{x}^T + b_0, \tag{5.51}$$

(analytically) measures the distance from vector $\mathbf{x}$ to the optimal hyperplane, (88), (89), (181). If we denote by r the desired distance, then:

$$g(\mathbf{x}) = \mathbf{w}_0 \cdot \mathbf{x}^T + b_0 = r \parallel \mathbf{w}_0 \parallel,$$

or:

$$r = \frac{g(\mathbf{x})}{\parallel \mathbf{w}_0 \parallel}.$$

The problem at hand is to estimate the parameters $\mathbf{w}_0$ and b_0, corresponding to the optimal hyperplane, based on the training dataset T. We can see that (Fig. 5.32) the parameters $\mathbf{w}_0$ and b_0 satisfy the following two conditions:

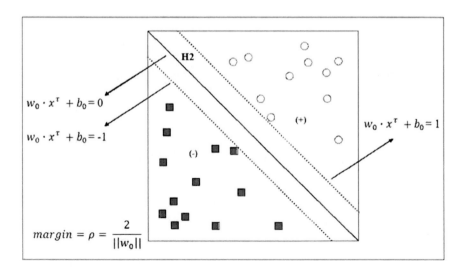

Fig. 5.32 Optimal hyperplane and corresponding margin of separation

$$\mathbf{w}_0 \cdot \mathbf{x}_i^T + b_0 \geq 1, \;\; if \; d_i = +1,$$

$$\mathbf{w}_0 \cdot \mathbf{x}_i^T + b_0 \leq -1, \;\; if \; d_i = -1.$$

The particular points $(\mathbf{x}_i, d_i)$ which belong to the two lines defined by the above equations:

$$\mathbf{w}_0 \cdot \mathbf{x}_i^T + b_0 = \pm 1, \tag{5.52}$$

are called *support vectors*, hence the name "support vector machine". These vectors are, thus, placed on the decision boundaries, separating the two categories, and they are therefore the most difficult to classify. In other words, based on them we are able to concretely separate the two categories (vectors 'supporting' the optimal decision process).

As one can easily see, the value of the margin of separation is obtained by maximizing the 'distance':

$$\rho = \frac{2}{\| w \|},$$

which is equivalent to minimizing the function:

$$L(\mathbf{w}) = \frac{\| \mathbf{w} \|^2}{2},$$

(i.e., minimizing the Euclidean norm of the weight vector $\mathbf{w}$) with the constraint:

$$f(\mathbf{x}_i) = \begin{cases} +1, & if \; \mathbf{w} \cdot \mathbf{x}_i^T + b \geq 1 \\ -1, & if \; \mathbf{w} \cdot \mathbf{x}_i^T + b \leq -1. \end{cases}$$

The solution to this problem may be obtained by using, for instance, *quadratic programming*, (181).

Remark 5.11. The optimal hyperplane is defined by the optimal weight $\mathbf{w}_0$, which provides the maximum possible separation between the two classes ("positive" and "negative").

More computational details are to be found, for instance, in (181), (419), (378).

Remark 5.12. SVMs are efficient and elegant learning machines, seen as an approximate implementation of the SRM technique, that is rooted in *Vapnik-Chervonenkis*

(VC) dimension theory, (389). It is well-known that the VC *dimension* of a learning machine determines the way in which a nested structure of approximation functions is used under SRM. It is also known that the VC dimension for a set of separating hyperplanes is equal to the input dimension plus one. Thus, to use SRM, we need to construct a set of separating hyperplanes of varying VC dimension, in such a way that the empirical risk (training error) and the VC dimension are simultaneously minimized. This approach is based on Vapnik's theorem, presented below. First, denote by $\mathbf{R}^n$ the n-dimensional Euclidean space, and by H_ρ the set of linear classifiers that separate $\mathbf{R}^n$ using hyperplanes of thickness ρ; let $H_{\rho+}$ be the set of linear classifiers with thickness greater than or equal to ρ.

Theorem 5.4. *(Vapnik 1982, (390)). Let $X_r = \{x_1, x_2, ..., x_k\} \subset \mathbf{R}^n$ denote a set of points contained within a sphere of radius r. The VC dimension of $H_{\rho+}$, restricted to X_r, satisfies the inequality:*

$$VC\dim\left(H_{\rho+}\right) \leq \min\left(\left[\frac{4r^2}{\rho^2}\right], n\right) + 1,$$

where $[\cdot]$ represents the nearest integer function.

In our context, the equivalent of the above theorem asserts that: "Let D be the smallest diameter containing all training input vectors $x_1, x_2, ..., x_N$. The VC dimension boundary for the set of possible separating hyperplanes is given by the inequality:

$$h \leq \min\left(\left[\frac{D^2}{\rho_0^2}\right], m_0\right) + 1,$$

where h is the VC dimension, $\rho_0 = \dfrac{2}{\|w_0\|}$ is the margin of separation between the two decision classes, and m_0 represents the input space dimension".

This result allows us to control the VC dimension h (hyperplane complexity) independently of the input space dimension m_0, if the margin of separation is properly chosen. Thus, whatever the learning task, SVM provides a method for controlling the model complexity without taking into account the dimensionality. For technical details, see (391), (393), (181), (419).

We talked so far about the simple linear separability between decision classes. What happens however in the general case of non-linearly separable patterns, as shown in Fig. 5.33?

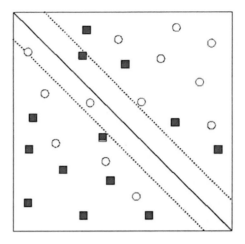

Fig. 5.33 Non-linearly separable patterns

Unlike the previous situation, given such a training dataset, it is no more possible to construct a separating hyperplane, without encountering classification errors.

Without going into technical details, we only mention that in this case we consider a set of additional non-negative scalar variables $\{\xi_i, i = 1, 2, ..., N\}$, called *slack variables*, measuring the deviation of a point from the ideal condition of pattern separability. In this case, the goal is to minimize the cost function:

$$L(\mathbf{w}, \xi) = \frac{\|\mathbf{w}\|^2}{2} + C \cdot \sum_{i=1}^{N} \xi_i, \tag{5.53}$$

with the constraints:

$$f(\mathbf{x}_i) = \begin{cases} +1, & if \ \mathbf{w} \cdot \mathbf{x}_i^T + b \geq 1 - \xi_i \\ -1, & if \ \mathbf{w} \cdot \mathbf{x}_i^T + b \leq -1 + \xi_i, \end{cases}$$

where C is called the *regularization* parameter, being selected by the user in one of the two ways:

- *Experimental* determination, via the standard use of a training/testing dataset;
- *Analytical* determination, via the estimation of the VC dimension (see VC dimension boundary) and then by using bound on the generalization performance of the machine based on the VC dimension.

(for technical details see, for instance, (181), (419)).

Finally, we shall outline three types of SVMs, starting from the kernel types taken into consideration (*idem*).

1. *Polynomial learning machine*, with the inner product kernel given by:

$$K(\mathbf{x}, \mathbf{x}_i) = \left(\mathbf{x} \cdot \mathbf{x}_i^T + 1\right)^p,$$

where power p is specified *a priori* by the user;

2. *Radial-basis function network* (RBFN), with the inner product kernel given by:

$$K(\mathbf{x}, \mathbf{x}_i) = \exp\left(-\frac{1}{2\sigma^2}\|\mathbf{x} - \mathbf{x}_i\|^2\right),$$

where the smoothing parameter σ^2, common to all the kernels, is specified *a priori* by the user (see also subsection 5.3.2);

3. *Two-layer perceptron* (i.e., MLP with a single hidden layer), with the inner product kernel given by:

$$K(\mathbf{x}, \mathbf{x}_i) = \tanh\left(\beta_0 \cdot \mathbf{x} \cdot \mathbf{x}_i^T + \beta_1\right),$$

where some values of the parameters β_0 and β_1 satisfy Mercer's theorem.

Remark 5.13. For each of the three above NNs one can use the SVM learning algorithm to implement the learning process. Thus, for the RBFN type of SVM, the number of radial-basis functions and their centers are automatically determined by the number of support vectors and their values. For the MLP type of SVM, the number of hidden neurons and their weights are automatically determined by the number of support vectors and their values. For more technical details, see (181).

Although SVMs are a special type of linear learning machine, we have considered them as belonging to the NNs domain, because their architecture follows the same 'philosophy', as we can easily see from the diagram illustrated in Fig. 5.34 (*idem*).

Example 5.12. XOR problem, (57), (181). In order to concretely illustrate the way of using SVMs, we consider the classical XOR problem (*exclusive OR problem*). Starting from the meaning of "exclusive OR" in natural language, and passing through the logical operation "*exclusive disjunction*" of a pair of propositions (p, q), we may consider the formal situation when "p is true or q is true, but not both". An example of such a situation is illustrated by the following statement: "*You may go to work by car or by subway/tube*". Recall that such a problem involves non-linear separability, and cannot be solved by the use of a single-layer network.

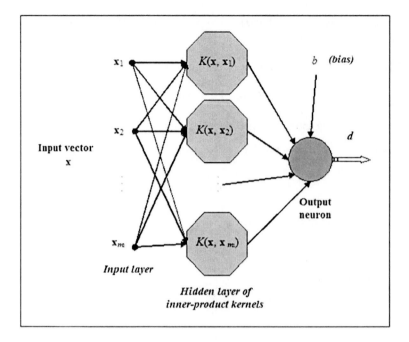

Fig. 5.34 SVM architecture

A summary of the XOR problem is presented in Table 5.6 below

Table 5.6 XOR problem

Input vector $\mathbf{x}$	Desired output d
(-1, -1)	-1
(-1, +1)	+1
(+1, -1)	+1
(+1, +1)	-1

Fig. 5.35 illustrates the XOR problem.

If we denote $\mathbf{x} = (x_1, x_2)$ and $\mathbf{x}_i = (x_{i_1}, x_{i_2})$, then the inner-product kernel, given by:

$$K(\mathbf{x}, \mathbf{x}_i) = \left(\mathbf{x} \cdot \mathbf{x}_i^T + 1 \right)^2, \tag{5.54}$$

is expressed as:

$$K(\mathbf{x}, \mathbf{x}_i) = 1 + x_1^2 x_{i_1}^2 + 2x_1 x_2 x_{i_1} x_{i_2} + x_2^2 x_{i_2}^2 + 2x_1 x_{i_1} + 2x_2 x_{i_2}. \qquad (5.55)$$

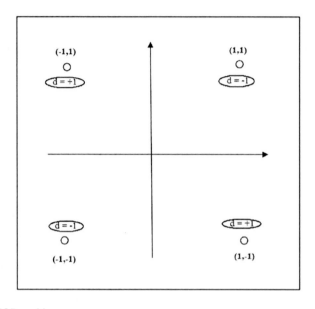

Fig. 5.35 XOR problem

The image $\mathbf{z}$ of the input vector $\mathbf{x}$, induced in the feature space, is given by:

$$\mathbf{z} = g(\mathbf{x}) = \left(1, x_1^2, \sqrt{2}x_1 x_2, x_2^2, \sqrt{2}x_1, \sqrt{2}x_2\right).$$

Similarly,

$$\mathbf{z}_i = g(\mathbf{x}_i) = \left(1, x_{i_1}^2, \sqrt{2}x_{i_1} x_{i_2}, x_{i_2}^2, \sqrt{2}x_{i_1}, \sqrt{2}x_{i_2}\right).$$

Denoting by K the matrix $[K(\mathbf{x}_i, \mathbf{x}_j)]$, where $K(\mathbf{x}_i, \mathbf{x}_j)$ represents the inner-product kernel, then:

$$K = \begin{bmatrix} 9 & 1 & 1 & 1 \\ 1 & 9 & 1 & 1 \\ 1 & 1 & 9 & 1 \\ 1 & 1 & 1 & 9 \end{bmatrix}.$$

Using the quadratic optimization with respect to the *Lagrange multipliers*, (181), we obtain the optimum weight vector:

$$\mathbf{w}_0 = \left(0, 0, -\frac{1}{\sqrt{2}}, 0, 0, 0\right),$$

and the optimal hyperplane equation:

$$\mathbf{w}_0 \cdot g^T(\mathbf{x}) = \left(0, 0, -\frac{1}{\sqrt{2}}, 0, 0, 0\right) \cdot \left(1, x_1^2, \sqrt{2}x_1 x_2, x_2^2, \sqrt{2}x_1, \sqrt{2}x_2\right)^T = 0, \quad (5.56)$$

which reduces to the equation $x_1 \cdot x_2 = 0$.

Example 5.13. Real-life applications of SVM models. SVMs are popular for real-world classification due to several advantages over classical paradigms. From the performance perspective, they are one of the most powerful classifiers. Their working is independent of the number of features involved in a decision problem and thus the "curse of dimensionality" is clearly handled. From the computational point of view, SVMs provide a fast training and they are very useful for a fast insight into the capability of a robust learning technique, when facing an explicit problem to be solved. The domain list of SVM applications is extremely large and varied. Nevertheless, here are some examples: text categorization, (208), (368), (369); image retrieval, (384); 3D object recognition, (304); time series prediction, (273); analysis of microarray gene expression data, (45); face detection, (288); protein subcellular localization prediction, (195); drug design, (48); cancer classification, (162); liver fibrosis grade in chronic hepatitis C, (416), (206), (370).

Remark 5.14. 1) Apart from the use in pattern recognition, SVMs are also utilized for nonlinear regression, under the name of *support vector regression* -SVR, see (85), (392). We present here only the fundaments of this SVM version. Thus, given a (non-linear) regression problem, described by the following regression equation:

$$d = f(\mathbf{x}) + v, \quad (5.57)$$

where d is the scalar dependent variable, $\mathbf{x}$ is the vector independent variable (predictor), f is a scalar-valued non-linear function defined by the conditional expectation $E[D|\mathbf{x}]$, with D a random variable with realization denoted by d, and v represents the 'noise'. Then, based on a training dataset $T = \{\mathbf{x}_i, d_i; i = 1, 2, ..., N\}$, we have to estimate the dependence of d on $\mathbf{x}$ (i.e., the function f and the distribution of the noise v). Denoting by y the estimate of d, and by $g(\mathbf{x}) = \{g_j(\mathbf{x}), j = 0, 1, 2, ..., m_1\}$ a set of non-linear basis functions, consider the expansion of y in terms of $g(\mathbf{x})$, given by:

$$y = \sum_{j=0}^{m_1} \mathbf{w}_j \cdot g_j(\mathbf{x}) = \mathbf{w} \cdot g^T(\mathbf{x}),$$

where, as before, $\mathbf{w}$ represents the weight vector, and the bias b equals the weight $\mathbf{w}_0$. To construct a SVR for estimating the dependence of d on $\mathbf{x}$, an ε *-insensitive loss function*, (391), (393), given by:

$$L_\varepsilon(d, y) = \begin{cases} |d - y|, & |d - y| \geq \varepsilon \\ 0, & |d - y| < \varepsilon, \end{cases}$$

is used. Thus, the issue to be solved lies in minimizing the empirical risk:

$$R = \frac{1}{N} \sum_{i=1}^{N} L_{\varepsilon}(d_i, y_i),$$

with the constraint:

$$\|\mathbf{w}\|^2 \leq c_0,$$

where c_0 is a constant.

Considering two sets of non-negative slack variables $\{\xi_i, i = 1, 2, ..., N\}$ and $\{\xi'_i, i = 1, 2, ..., N\}$, we have now to minimize the cost function:

$$L(\mathbf{w}) = \frac{\|\mathbf{w}\|^2}{2} + C \cdot \sum_{i=1}^{N} (\xi_i + \xi'_i), \tag{5.58}$$

with the following constraints:

$$d_i - \mathbf{w} \cdot g^T(\mathbf{x}_i) \leq \varepsilon + \xi_i, \ i = 1, ..., N,$$

$$\mathbf{w} \cdot g^T(\mathbf{x}_i) - d_i \leq \varepsilon + \xi'_i, \ i = 1, ..., N,$$

$$\xi_i \geq 0, \ i = 1, ..., N,$$

$$\xi'_i \geq 0, \ i = 1, ..., N.$$

(for computational details, see (181)).

2) The standard optimization technique of SVM, based on Lagrange multipliers, may seem quite complicated in terms of completely understanding its philosophy, the corresponding computation methodology, and its correct implementation. In this context, a new approach has been developed (see, for instance (368), (369)), called *evolutionary support vector machine* -ESVM, which provides a simple alternative to the standard SVM optimization by using evolutionary algorithms (see Section 5.9). ESVM inherits the SVM classical approach regarding the learning process, but the estimation of the decision function parameters is performed by an evolutionary algorithm. Thus, the parameters to be estimated are either of the separation hyperplane, in the classification case, or of the function defining the connection between the predictor and dependent variables, in the regression case. In this context, ESVM can always determine the learning function parameters, which is often impossible with the classical technique. Moreover, ESVM obtains the coefficients directly from the evolutionary algorithm, and can make reference to them at any time within a

computer run. ESVMs were validated on real-world problems of classification (e.g., spam detection, diagnosis of diabetes mellitus, liver degree prediction in chronic hepatitis C, *Iris* plant recognition, soybean diseases diagnosis), and regression (e.g., the Boston real estate problem), the results, (368), (369), (370), demonstrating the functionality of this hybrid technique.

We present here only one concrete application of ESVMs, namely the prediction of the degree of hepatic fibrosis in patients with chronic hepatitis C, (370). Thus, hepatic fibrosis, the principal pointer to the development of a liver disease within chronic hepatitis C, can be measured through several stages -from F_0 (no fibrosis) to F_4 (cirrhosis). The correct evaluation of its degree, based on recent different non-invasive procedures, is of current major concern. One of the latest methodologies for assessing it is the *Fibroscan* (http://www.echosens.com/) and the effect of its employment is impressive. However, the complex interaction between its stiffness indicator and the other biochemical and clinical examinations towards a respective degree of liver fibrosis is hard to be manually discovered. Hence, the ESVM approach is applied with the purpose of achieving an automated learning of the relationship between the medical attributes and fibrosis levels, investigating both the direct multifaceted discrimination into all five degrees of fibrosis and the slightly less difficult common separation into solely three related stages. The resulting performance proved a significant superiority over the standard SVM classification. What is more, a feature selection genetic algorithm was further embedded into the method structure, in order to dynamically concentrate search only on the most relevant attributes. Finally, an additionally resulting formula is helpful in providing an immediate calculation of the liver stage for new cases, while establishing the presence/absence and comprehending the weight of each medical factor with respect to a certain fibrosis level. All these therefore confirm the promise of the new ESVM methodology towards a dependable support within the particular field of medical decision-making and beyond.

We end the brief overview concerning SVMs, indicating some useful links regarding SVMs and their implementation.

- *SVM* - General presentation, http://www.support-vector-machines.org/.
- *SVM^{light}* - implementation of Support Vector Machines in C, for classification and regression problems, ranking problems, etc. Author: Thorsten J. - Cornell University, http://svmlight.joachims.org/.
- *mySVM* - C++ implementation of SVM classification and regression, http://www-ai.cs.uni-dortmund.de/SOFTWARE/MYSVM/index.html.
- *BSVM* - implementations of multi-class classification, SVM regression, http://www.csie.ntu.edu.tw/c̃jlin/bsvm/.
- *SVM in R* - SVM implementation in *R*, http://cran.r-project.org/web/packages/e1071/index.html.
- MATLAB *SVM Toolbox* - SVM classification and regression, http://www.isis.ecs.soton.ac.uk/resources/svminfo/.

Besides the references mentioned in the text, the interested reader in SVMs may also consider (49), (73), (336), (364).

5.4 Association Rule Mining

Basically, *association rule learning* is a well-known method in data mining for discovering interesting relations between variables in large databases - see, for instance, (299), (4), (5).

An association rule can be seen as an implication of the form $X \rightarrow Y$, where X and Y are distinct items or itemsets (collections of one or more items), X being the rule *antecedent* and Y being the rule *consequent*. In other words, a rule antecedent is the portion of a conditional rule that needs to be satisfied in order that the rule consequent is true. It is an unsupervised data mining technique looking for connections between items/records belonging to a large dataset. A typical and widely-used example of association rule mining is *market basket analysis*. Market basket analysis tries to identify customers, purchasing certain grouped items, providing insight into the combination of products within a customer's 'basket'. For example, in retail, market basket analysis helps retailers to understand the purchase behavior of customers. On the other hand, using the bar-code scanners information, a supermarket database consists of a large number of transaction records, listing all items bought by a customer on a single purchase transaction. Based on association rule mining, managers could use the rules discovered in such a database for adjusting store layouts, for cross-selling, for promotions, for catalog design, for identifying the customer segments based on their purchase pattern. As a classical example, besides the famous *"beer-diaper"* case:

$$\{X \rightarrow Y\} \Leftrightarrow \{\textbf{diaper}, milk\} \rightarrow \{\textbf{beer}\}$$

Let us also mention the *"Amazon effect"*, when, looking, for instance, for a book on data mining, a list of potentially interesting books on neighboring subjects (e.g., data analysis, statistical analysis, statistical learning, etc.) is presented (list built based on a profile of what other "similar" customers have ordered).

Example 5.14. We will illustrate the way of mining association rules by a very simple example. Thus, let us consider a list of d items {A, B, C, D, E, F, G}, bought by ten customers (see Fig. 5.36).

Starting from this list, we create a matrix displaying the frequency of occurrence of pairs of items.

Thus, for example, item A occurred in 60% of purchases, item E has never been purchased together with item A, while item A has been purchased together with item F in 30% of cases.

Customer	Items						
	A	B	C	D	E	F	G
#1	×		×			×	
#2	×	×		×			
#3		×	×			×	
#4		×		×		×	
#5	×						×
#6					×	×	
#7	×		×				
#8			×		×		
#9	×	×	×	×		×	
#10	×			×		×	

Fig. 5.36 List of bought items

Table 5.7 Frequency of occurrence of pairs of items

	A	B	C	D	E	F	G
A	6	2	3	2	0	3	1
B	2	4	2	3	0	3	0
C	3	2	5	1	1	2	0
D	2	3	1	4	0	3	0
E	0	0	1	0	2	1	0
F	3	3	2	3	1	6	0
G	1	0	0	0	0	0	1

Remark 5.15. In practical applications, association rules discovery needs several hundred/thousand transactions before being considered statistically significant, and datasets often contain thousands or millions of transactions (e.g., chain of supermarkets). Let us note the complexity of such a task since the process of finding all frequent itemsets in a database is difficult, involving the search of all possible itemsets. Thus, the set of possible itemsets has $2^d - 1$ elements, in the above case $2^7 - 1$ elements.

To select interesting rules from the set of all possible rules, we need some 'measures' assessing the effectiveness of the association rule process: support, confidence and lift. Thus, the *support* represents the number of transactions that include all items in the antecedent and consequent parts of the rule (transactions that contain both X and Y), being sometimes expressed as a percentage. The *confidence* represents the ratio of the number of transactions that include all items in the consequent as well as the antecedent to the number of transactions that include all items in the antecedent. Finally, the *lift* represents the ratio of the confidence of the rule and the expected confidence of the rule.

We illustrate the above notions with the following purchasing situation. Consider the case of selling two beverage types: beer and whiskey in 500,000 transactions:

- 20,000 transactions concern whiskey (4% of all transactions);
- 30,000 transactions concern beer (6% of all transactions);
- 10,000 transactions concern both whiskey and beer (2% of all transactions).

Then, we have:

- The support is given by 10,000/500,000 = 2%;
- For confidence we consider the following two scenarios:

    - The association rule *"When people buy whiskey, buy also beer"* has a confidence of 10,000/20,000 = 50%;
    - The association rule *"When people buy beer, buy also whiskey"* has a confidence of 10,000/30,000 = 33%.

Note that the two rules have the same support (2%).

If there is no more information about other transactions, we can make the following statement from the available data:

- Customers buy whiskey 4% of the time;
- Customers buy beer 6% of the time.

The two rates, 4% and 6% respectively, are called the *expected confidence* to buy whiskey or beer, regardless of other shopping.

Since the confidence of the rule *"buying whiskey & beer"* is 50%, while the expected confidence to buy beer is 6%, then the lift of the rule *"buying whiskey & beer"* is 8.33 (0.5/0.06).

The rule *"buying whiskey & beer"* may be expressed in term of lift by the following assertion: *"Customers buying whiskey are 8.33 times more tempted to buy beer at the same time with whiskey"*. Thus, the interaction between whiskey and beer is very strong for those people.

To not be suspected that we focused on a specific type of customers (alcohol beverage drinkers), we can consider instead of beer and whiskey, flour and butter in the pastry industry, or laptops and printers in the IT products sales.

We can define the process of discovering association rules as follows: *"Given a set of transactions, discover all possible rules when both the support and the confidence are equal or greater than some predefined thresholds"*. The way these rules are discovered depends on the chosen procedure, but the idea behind this methodology may be summarized in the following two steps:

1. Find those itemsets whose occurrences exceed a predefined threshold in the database, itemsets called *frequent* or *large itemsets*;
2. Generate association rules from those large itemsets with the constraints of minimal confidence.

Regarding the latter step, one can show, (378), that the total number R of rules that can be extracted from a dataset containing a number d of items is given by:

$$R = 3^d - 2^{d+1} + 1,$$

so there is a sufficient large computational complexity in this process. For instance, in the case above, for $d = 2$, we obtain two rules. If instead, we deal with $d = 10$ different items, then we have to handle the situation of 57,002 rules!

We end this subsection by presenting some frequently used algorithms in association rule discovery:

- *A priori* algorithm, proposed by Rakesh Agrawal and Ramakrishnan Srikant, (5), being considered the best-known algorithm to mine association rules;
- *FP-growth* (frequent pattern growth) algorithm, proposed by Jiawei Han, Jian Pei, Yiwen Yin, and Runying Mao, (167);
- *ECLAT* (Equivalence Class Clustering and Bottom-up Lattice Traversal), proposed by Mohammed Javeed Zaki, Srinivasan Parthasarathy, Mitsunori Ogihara, Wei Li, (418).

More information on association rules discovery can be found in (3), (420), (378).

5.5   Rule-Based Classification

Rule-based classification basically means the process of using a training dataset of labeled objects from which classification rules are extracted for building a classifier (i.e., a set of rules, used in a given order during the prediction process, to classify new (unlabeled/unseen) objects). In short, we classify objects by using a collection of "*If...Then...*" type rules. A rule is an implication of the form $X \rightarrow Y$, where X represents the rule *antecedent*, or *condition* (LHS: *left-hand-side*), consisting of a conjunction of attributes values, and Y represents the rule *consequent* (RHS: *right-hand-side*), representing the class label.

We further present the rule-based classification terminology:

- A rule R *covers* an object **x** if the attributes of **x** satisfy the condition of the rule R;
- The *coverage* of a rule R represents the fraction of objects that satisfy the antecedent of R;
- The *accuracy* of the rule R represents the fraction of objects that satisfy both the antecedent and consequent of R;
- The *length* of R represents the number of descriptors.

Example 5.15. Let us consider again the example presented in Chapter 4, related to customer profile regarding the place where he/she goes shopping (see Table 5.8). We can get the following two rules.

Table 5.8 Training dataset regarding the shopping behavior

Taxable income	Car ownership	Marital status	Buy from shop
125,000	Yes	Single	NO
100,000	No	Married	NO
70,000	No	Single	NO
120,000	Yes	Married	NO
95,000	No	Divorced	YES
60,000	No	Married	NO
220,000	Yes	Divorced	NO
85,000	No	Single	YES
75,000	No	Married	NO
90,000	No	Single	YES

- Rule R_1: (*Taxable income* < 95,000) $\wedge$ (*Car ownership* = No) $\wedge$ (*Marital status* = Married) $\rightarrow$ (*Buy from shop* = NO), with objects:

  - x_1: (*Taxable income* = 90,000) $\wedge$ (*Car ownership* = No) $\wedge$ (*Marital status* = Single);
  - x_2: (*Taxable income* = 85,000) $\wedge$ (*Car ownership* = No) $\wedge$ (*Marital status* = Single);
  - x_3: (*Taxable income* = 75,000) $\wedge$ (*Car ownership* = No) $\wedge$ (*Marital status* = Married);

  Only object x_3 is covered by R_1.
- Rule R_2: (*Marital status* = Married) $\rightarrow$ (*Buy from shop* = NO). Then:

  - Coverage of R_2 = 40%;
  - Accuracy of R_2 = 100%;
  - Length of R_2 = 1.

If, for instance, R: (*Marital status* = Single) $\rightarrow$ (*Buy from shop* = NO), then the coverage is 40% again, but the accuracy is now 50%.

Example 5.16. We present now a simple and well-known example of rule-based classification regarding the taxonomy of some creatures, based on certain attributes (see, for instance, (378)). The training dataset contains 20 samples consisting of 4 attributes and the corresponding class labels (see Fig. 5.37).

From this table we can extract rules such that:

- R_1: (*Blood type* = warm) $\wedge$ (*Give Birth* = no) $\rightarrow$ Birds;
- R_2: (*Give Birth* = no) $\wedge$ (*Can Fly* = no) $\rightarrow$ Reptiles.

Name	Blood Type	Give Birth	Can Fly	Live in Water	Class
human	warm	yes	no	no	mammals
python	cold	no	no	no	reptiles
salmon	cold	no	no	yes	fishes
whale	warm	yes	no	yes	mammals
frog	cold	no	no	sometimes	amphibians
komodo	cold	no	no	no	reptiles
bat	warm	yes	yes	no	mammals
pigeon	warm	no	yes	no	birds
cat	warm	yes	no	no	mammals
leopard shark	cold	yes	no	yes	fishes
turtle	cold	no	no	sometimes	reptiles
penguin	warm	no	no	sometimes	birds
porcupine	warm	yes	no	no	mammals
eel	cold	no	no	yes	fishes
salamander	cold	no	no	sometimes	amphibians
gila monster	cold	no	no	no	reptiles
platypus	warm	no	no	no	mammals
owl	warm	no	yes	no	birds
dolphin	warm	yes	no	yes	mammals
eagle	warm	no	yes	no	birds

Fig. 5.37 Extracting rules from training dataset

Observe that the two rules above cover all 'objects' in this dataset signifying *birds* (i.e., {pigeon, penguin, owl, eagle}), and reptiles (i.e., {python, komodo, turtle, gila monster}), respectively. The corresponding accuracy equals 100%.

Next, if we consider the rule:

- R_3: (*Give Birth* = no) $\land$ (*Can Fly* = yes) $\to$ Birds, then we see that 3 of 4 birds in the dataset fit the antecedent of the rule, excepting the penguin.

If we now consider the rule:

- R_4: (*Can fly* = no) $\land$ (*Live in Water* = sometimes) $\to$ {amphibians, reptiles, birds},

we see that its consequent consists of three different 'objects', so it is not helpful for a good classification.

Regarding the main characteristics of rule-based classifiers, let us mention two types of constituent rules:

- *Mutually exclusive rules*, when: (*a*) the rules are (mutually) independent of each other, (*b*) no two rules are triggered by the same object, and (*c*) every object is covered by at most one rule;
- *Exhaustive rules*, when: (*a*) the corresponding rule-based classifier accounts for every possible combination of attributes values, and (*b*) each object is covered by at least one rule.

We see that the rule-based classifier regarding mammals, given by the following set of rules:

- R_1: (*Blood type* = cold) → Non-mammal,
- R_2: (*Blood type* = warm) ∧ (*Give Birth* = no) → Non-mammal,
- R_3: (*Blood type* = warm) ∧ (*Give Birth* = yes) → Mammal,

is both mutually exclusive and exhaustive. In this context, let us however remark the bizarre case of the platypus (*Ornithorhynchus anatinus*), which does not obey this classification rules (warm blood, but egg-laying, even it is a mammal).

Concerning the process of building classification rules, we can mention the two possible approaches:

1. *Direct method*, consisting in extracting rules directly from data. Examples of such algorithms are: RIPPER, (66), CN2, (61), Holte's 1R, (190), Boolean reasoning, (346), (14);
2. *Indirect method*, consisting in extracting rules from other classification models, e.g., decision trees, neural networks, etc. In this context, let us mention the C4.5 algorithm, (310).

For a comparison between the two approaches above, see (378), where C4.5 rules are confronted with RIPPER rules, using the above training dataset. Concerning the indirect method, let us illustrate below the way of using a decision tree to extract rules. For this purpose, let us reconsider the example regarding the shopping place (i.e., buying from shop), presented in Chapter 4. The corresponding decision tree is illustrated below (Fig. 5.38).

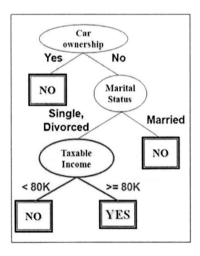

Fig. 5.38 Decision tree graph

From this decision tree we can extract the following rule-based classifier:

- R_1: (*Car ownership* = Yes) → NO;
- R_2: (*Car ownership* = No) ∧ (*Marital status* = Single/Divorced) ∧ (*Taxable income* < 80,000) → NO;

- R_3: (*Car ownership* = No) ∧ (*Marital status* = Single/Divorced) ∧ (*Taxable income* > 80,000) → YES;
- R_4: (*Car ownership* = No) ∧ (*Marital status* = Married) → NO.

with mutually exclusive and exhaustive rules; more details in (378).

Finally, from the foregoing brief presentation, it results that the main advantages of using rule-based classifiers are (*idem*):

- High expressiveness and performance, analogous to that of decision trees;
- Easy to interpret;
- Easy to generate;
- High classification speed of new objects.

Much more technical details concerning this topic can be found in (267), (378), (410), (166).

5.6 *k*-Nearest Neighbor

Starting from the "*duck test*" seen as a funny motto introducing the inductive reasoning, "*If it looks like a duck, swims like a duck, and quacks like a duck, then it probably is a duck*", we can talk about a well-known classification method, *k*-nearest neighbor, intuitively based on this idea.

In the pattern recognition area, the "*k*-Nearest Neighbor" (*k*-NN) algorithm represents that classification method, in which a new object is labeled based on its closest (*k*) neighboring objects. To better understand what *k*-nearest algorithm really means, let us carefully look at the figure below.

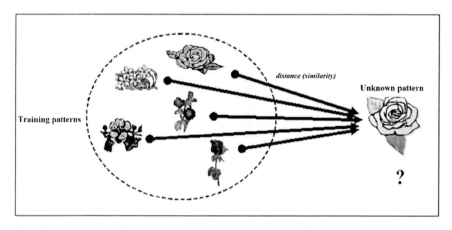

Fig. 5.39 Illustration of the *k*-nearest algorithm (flower recognition)

In principle, given a training dataset (left) and a new object to be classified (right), the "distance" (referring to some kind of similarity) between the new object and the training objects is first computed, and the nearest (most similar) *k* objects are then chosen. To construct the algorithm, we need the following items (algorithm input):

- A set of stored records (training dataset);
- A distance (metric) to compute the similarity between objects;
- The value of k, i.e., the (necessary) number of objects (records) belonging to the training dataset, based on which we will achieve the classification of a new object.

Based on these three requirements, a new (not yet classified) object will be classified by performing the following steps:

- Compute the distance (similarity) between all the training records and the new object (naive approach);
- Identify the k nearest objects (most similar k neighbors), by ordering the training objects taking into account the computed distances in the first step;
- Assign the label which is most frequent among the k training records nearest to that object ("majority voting").

To conclude, the k-nearest neighbor algorithm is amongst the simplest of all machine learning algorithms, since it simply consists in classifying an object by the majority vote of its neighbors.

Remark 5.16. 1) The naive approach of this algorithm is computationally intensive, especially when the size of the training dataset grows. To avoid this situation, many nearest neighbor algorithms have been proposed over the years, generally seeking to reduce the number of distance evaluations and thus becoming more tractable.
2) A drawback of the classical "majority voting" methodology is that the classes with the more frequent objects tend to dominate the decision concerning the classification of a new object. The alternative may consist in considering a "weighted voting system", i.e., weighting somehow each of the k nearest neighbors (e.g., by choosing the weight $w = 1/d^2$, where d represents the distance between the new object and the corresponding neighbor).

Fig. 5.40 synthetically presents the algorithm for $k = 1, 2, 3$ neighbors of a new object -the center of the circle.

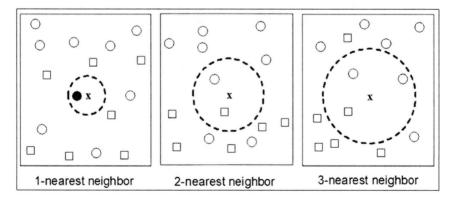

| 1-nearest neighbor | 2-nearest neighbor | 3-nearest neighbor |

Fig. 5.40 Different k-nearest neighbors cases ($k = 1, 2, 3$)

Remark 5.17. 1) The best choice of k depends upon the data. Thus, larger values of k reduce the effect of noise (outliers) on the classification, but make boundaries between classes less distinct. A good selection of the parameter k can be performed by using the *cross-validation* method. Technically, we divide the given dataset into a number of r randomly drawn, disjointed subsets, and for a fixed value of k we apply the k-nearest algorithm to make predictions on the rth subset, and evaluate the error one cycle. At the end of the r cycles, the computed errors are then averaged to yield a measure of how well the algorithm predicts new objects. The above steps are then repeated for various k and the value achieving the highest classification accuracy is then selected as the optimal value for k. The special case where the class is predicted to be the class of the closest training object (i.e., when $k = 1$) is called the *nearest neighbor* algorithm.

2) The accuracy of the k-NN algorithm can be severely degraded by the presence of noisy or irrelevant features, or if the feature scales are not consistent with their importance. Accordingly, features may have to be previously scaled to prevent confusion due to the 'domination' by a certain feature. Concerning the problem of selecting or scaling features to improve the classification accuracy, let us only mention the use of evolutionary algorithms and the mutual information of the training data with the training classes -see, for instance, (353), (278), (123).

3) The k-NN algorithm is easy to implement, but on the other hand, it is a 'lazy' classifier, especially in the presence of large training sets. Recall that, unlike the *eager learners*, where the model tries to generalize the training data before receiving tasks, the *lazy learners* delay the generalization of the training data until a task is received by the model. Seen as an *instance-based learner*, its negative points can be summarized in the following:

- It does not build models explicitly like decision trees or rule-based classifiers;
- The classification process requires long runtime (especially in the naive approach case) and is relatively computationally 'expensive';
- The prediction is based on local information, therefore it is likely to be influenced by extreme values/outliers.

As we mentioned above, to avoid the above drawbacks, various improvements to the k-NN method have been proposed (e.g., *proximity graphs*, (385), *SVMs*, (362), *adaptive distance measure*, (397), distance functions based on *Receiver Operating Characteristics*, (173), etc.).

Finally, we mention a well known variant of the k-NN algorithm, of 'nearest neighbor' type (i.e., $k = 1$), namely PEBLS (*Parallel Examplar-Based Learning System*), (69), having the following characteristics:

- It can handle both continuous and categorical attributes;
- Each record is assigned a weight.

Example 5.17. We will exemplify the PEBLS algorithm by reconsidering again the example regarding the shopping behavior presented in Chapter 4 (see also (378)).

Table 5.9 Training dataset regarding the shopping behavior

Taxable income	Car ownership	Marital status	Buy from shop
125,000	Yes	Single	NO
100,000	No	Married	NO
70,000	No	Single	NO
120,000	Yes	Married	NO
95,000	No	Divorced	YES
60,000	No	Married	NO
220,000	Yes	Divorced	NO
85,000	No	Single	YES
75,000	No	Married	NO
90,000	No	Single	YES

If in terms of continuous attributes, the usual metric is given by the Euclidean distance, in terms of categorical attributes, one uses the formula:

$$d(X_1, X_2) = \sum_i \left| \frac{n_{1i}}{n_1} - \frac{n_{2i}}{n_2} \right|,$$

where X_l, $l = 1, 2$, are the categorical attributes, and n_{lj}, n_l represent the corresponding frequencies. For example, regarding the marital status, with the corresponding distribution tabulated below (Fig. 5.41)

Class	Marital status		
	Single	Married	Divorced
Yes	2	0	1
No	2	4	1

Fig. 5.41 "Marital status" statistics

the corresponding distances are given by:

- $d(Single, Married) = |2/4 - 0/4| + |2/4 - 4/4| = 1$;
- $d(Single, Divorced) = |2/4 - 1/2| + |2/4 - 1/2| = 0$;
- $d(Married, Divorced) = |0/4 - 1/2| + |4/4 - 1/2| = 1$.

Next, regarding the attribute "Car ownership", with the corresponding distribution tabulated below (Fig. 5.42)

Class	Car ownership	
	Yes	No
Yes	0	3
No	3	4

Fig. 5.42 "Car ownership" statistics

we have:

- $d(Car\ ownership = Yes,\ Car\ ownership = No) = |0/3 - 3/7| + |3/3 - 4/7| = 6/7$.

On the other hand, regarding the distance between records X and Y, the following formula is used:

$$d(X,Y) = w_X \cdot w_Y \cdot \sum_i d^2 (X_i, Y_i),$$

where w_X and w_Y are the corresponding weights, given by:

$$w_X = \frac{Number\ of\ times\ X\ is\ used\ for\ prediction}{Number\ of\ times\ X\ predicts\ correctly},$$

$$w_Y = \frac{Number\ of\ times\ Y\ is\ used\ for\ prediction}{Number\ of\ times\ Y\ predicts\ correctly}.$$

For those interested in technical details concerning the k-NN algorithm, we refer, besides the references already mentioned, to (77), (341), (378).

5.7 Rough Sets

Rough sets (RS) were proposed by Zdzislaw Pawlak in the early 1980's, (291), in an attempt to mathematically tackle the vagueness concepts, their main purpose being the process of automated transformation of data into knowledge. We can thus see RS as a mathematical approach to imperfect knowledge. It is worth mentioning in this context the fuzzy set theory, proposed by Zadeh, (417). Returning to our subject, let us note that Pawlak showed that the principles of learning from examples (using a training dataset, as we saw in the previous sections) can be formulated in the context of this approach. We very briefly present below the principles underlying this concept and some data mining applications.

The RS concept can be generally defined by means of two topological notions, namely *interior* and *closure*, called *approximations*. The basic idea in the RS theory consists in the fact that, based on a set of objects, a set of attributes and decision values, one can create rules for finding the upper and the lower approximation, and the boundary region of that set of objects. These rules once built, a new object can be easily classified into one of the regions. We shortly outline below the 'idea' behind the RS approach.

- The first step of the RS approach consists in transforming the given dataset into an *information table* (i.e., objects with attributes, and attribute values for each object). In an information table, each row represents an object, and every column represents an attribute. Like for any other classical decision problem, usually one of the attributes will be chosen as the *decision attribute* (dependent attribute), while the remaining attributes will represent attributes based on which a decision is made (independent/predictive), called *conditional attributes*.
- The second step of the process is the formation of *equivalence classes*. They are actually groups of objects for which all the values of the conditional attributes are the same for each object, and therefore we cannot distinguish between them (*indiscernible objects*) using the available attributes.
- The third step of the process consists in building the *discernibility matrix*, whose elements are the attributes that distinguish between equivalence classes.
- One builds two approximations: (*a*) *lower approximation* and (*b*) *upper approximation* of a set X of objects. Thus, the lower approximation of X is a collection of objects that can be classified for sure as members of X (with maximum certainty), while the upper approximation of X is a collection of objects that can be classified as potential elements of X (uncertain membership).
- One considers the *boundary region* of X, containing objects that cannot be classified with any certainty neither in X, nor outside of X.
- Finally, based on *reducts*, one builds the rules. Considering now a new object (i.e., with no decision value/unknown label), and using the such built decision rules, it will be thus classified (i.e., a decision value will be assigned to it).

Remark 5.18. 1) Not all attributes may be required for building the equivalence classes. Therefore, we consider a *'reduct'*, consisting of sufficient information necessary to discern a class of objects through the other classes.
2) The classes for which there is more than one value of the decision attribute are called *vague classes*.

In what follows, we will present in mathematical (formalized) terms some basic notions from the Rough Sets' world.

An *information system* (IS) means an information table in which each row represents an object/event/instance/case, while each column represents an attribute (observation/variable/property, etc.), which will have a value attached to each object (measured for each object). The attributes are the same for each object, only their values are different. Table 5.10 represents such an information system.

In this IS, the attribute *"Taxable income"* represents the decision attribute (de-pendent), while the attributes *"Marital status"*, *"Education"* and *"Gender"* are con-ditional attributes (independent). The aim of the study is the rules discovery, based on which one can predict the taxable income of a new person, taking into account its conditional attributes values {Marital status, Education, Gender}.

Table 5.10 An example of information system

#	Marital status	Education	Gender	Taxable income
1	Single	Good	Male	High
2	Single	Good	Male	Medium
3	Married	Good	Female	Medium
4	Divorced	Poor	Male	Low
5	Divorced	Poor	Male	Low

More formally, an information system represents a pair $I = (U,A)$, where:

- U is a non-empty finite set of objects called the *universe*;
- A is a non-empty finite set of attributes;
- A function $a : U \rightarrow V_a$, for every $a \in A$; the set V_a is called the *value set* of a.

Remark 5.19. 1) Function a is not injective, so there may be different objects with the same value for attribute a.
2) In the case of supervised learning, i.e., where there is a decision attribute, the information system is called *decision system*. Thus, a decision system is any infor-mation system of the form $I = (U, A \cup \{d\})$, where $d \notin A$ is the decision attribute. In our example, since we have considered *"Taxable income"* as decision attribute, this information system is actually a decision system.

The following definition introduces the concept of *indiscernibility relation*. Such a relationship exists between certain objects when all the corresponding values of a given attribute are the same, so they cannot be distinguished taking into account that attribute. Formally, if o_1 and o_2 are two different objects, such that:

$$a_i(o_1) = a_i(o_2),$$

we say that o_1 and o_2 are *indiscernible objects* in relation to attribute $a_i \in A$.

Thus, for the above decision system (or *decision table*), we see that we obtain the following three groups of people with similar values of the conditional of attributes $\{1, 2\}$, $\{3\}$, and $\{4, 5\}$.

Starting from the three above groups of (indiscernible) people, we therefore may consider three (equivalence) classes, presented in Table 5.11.

Table 5.11 Corresponding equivalence classes

#	Marital status	Education	Gender
1	Single	Good	Male
2	Married	Good	Female
3	Divorced	Poor	Male

Assuming that the definition of the equivalence relation is already known, we proceed to introducing the notion of *indiscernibility relation*. Formally, if $I = (U, A)$ represents an information system and $B \subseteq A$ is a certain subset of attributes, then B induces an equivalence relation $IND(B) \subseteq U \times U$, defined by:

$$IND(B) = \{(x, y) \in U \times U | \forall a \in B, a(x) = a(y)\}, \tag{5.59}$$

called the *B-indiscernibility relation*.

If $(x, y) \in IND(B)$, then objects x and y are *indiscernible* from each other by attributes from B. The equivalence classes of the B-indiscernibility relation are denoted by $[x]_B$, representing a partition of the set U. Formally, the partition of U generated by $IND(B)$ is given by:

$$U/IND(B) = \otimes \{a \in B | U/IND(\{a\})\}, \tag{5.60}$$

where:

$$A \otimes B = \{X \cap Y | \forall X \in A, \forall Y \in B, X \cap Y \neq \emptyset\}.$$

Remark 5.20. In the case of a decision system, let us divide the set A of attributes into two distinct subsets: P and D, where P is the set of predictive (conditional) attributes, and D consists of the decision attribute d. Note, in context, that there are cases where the set D contains more than one decision attribute. Starting from the definition of the B-indiscernibility relation, we can define, in particular, the partitions $U/IND(P)$ and $U/IND(D)$, called *prediction equivalence classes* and *decision equivalence classes*, respectively.

If the set $U = \{o_1, o_2, ..., o_n\}$ represents the universe of a decision system, then the *discernibility matrix* is a $n \times n$ symmetric matrix, defined by:

$$m_{ij} = \{a \in P | a(o_i) \neq a(o_j) \land (d \in D, d(o_i) \neq d(o_j))\}, \tag{5.61}$$

for $i, j = 1, 2, ..., n$.

Thus, m_{ij} is the set of all attributes that classify the objects o_i and o_j into different decision classes, i.e., 'discern' objects o_i and o_j.

For an information system $I = (U, A)$, the discernibility matrix is given by:

$$m_{ij} = \{a \in A | a(o_i) \neq a(o_j)\}, \tag{5.62}$$

for $i, j = 1, 2, ..., n$.

Example 5.18. Let $U = \{o_1, o_2, ..., o_8\}$ be the universe of objects, $P = \{a_1, a_2, a_3, a_4\}$ be the set of the predictive attributes, and $D = \{d\}$ be the set of the decision attributes (here, the classical case of one decision attribute), with the corresponding decision table given below (Table 5.12).

Table 5.12 Example of a decision table

#	a_1	a_2	a_3	a_4	d
o_1	1	2	1	3	1
o_2	1	1	1	1	0
o_3	2	1	1	2	0
o_4	3	3	1	1	1
o_5	3	2	1	1	0
o_6	3	3	1	3	1
o_7	1	3	0	2	1
o_8	2	1	0	3	0

Then, the corresponding discernibility matrix is given by (Table 5.13):

Table 5.13 Discernibility matrix

#	o_1	o_2	o_3	o_4	o_5	o_6	o_7	o_8
o_1	$\emptyset$							
o_2	a_2, a_4	$\emptyset$		a_1, a_2		a_1, a_2, a_4	a_2, a_3, a_4	
o_3	a_1, a_2, a_4	$\emptyset$	$\emptyset$	a_1, a_2, a_4		a_1, a_2, a_4	a_1, a_2, a_3	
o_4	$\emptyset$			$\emptyset$				
o_5	a_1, a_4	$\emptyset$	$\emptyset$	a_2	$\emptyset$	a_2, a_4	a_1, a_2, a_3, a_4	
o_6	$\emptyset$			$\emptyset$		$\emptyset$		
o_7	$\emptyset$			$\emptyset$		$\emptyset$	$\emptyset$	
o_8	a_1, a_2, a_3	$\emptyset$	$\emptyset$	a_1, a_2, a_3, a_4	$\emptyset$	a_1, a_2, a_3	a_1, a_2, a_4	$\emptyset$

Remark 5.21. Every discernibility matrix (uniquely) defines a *discernibility function* f, which is a Boolean function of m Boolean variables $a_1^*, a_2^*,..., a_m^*$, corresponding to the attributes $a_1, a_2,..., a_m$, given by:

$$f(a_1^*,...,a_m^*) = \bigwedge \left\{ \bigvee m_{ij}^* | 1 \leq j \leq i \leq n, m_{ij} \neq \emptyset \right\},$$

where $m_{ij}^* = \{a^* | a \in m_{ij}\}$.

Let us now define the notion of a rough set. But first, we introduce the two approximations (lower and upper) of a given set, already mentioned at the beginning of this subsection.

Let $I = (U,A)$ be an information system and let $B \subseteq A$ and $X \subseteq U$. Then, based on knowledge extracted from B, we can build two approximations of the set X and, in relation to these sets, we can get a clue regarding the membership of certain elements of U to X. In other words, X can be approximated using only information contained in B, by using these B-approximation sets of X. Formally, we have:

- $\underline{B}X = \{x \in U | [x]_B \subseteq X\}$ -the *B-lower approximation* of X, which represents, as we already mentioned, the set of objects of U which can be surely (i.e., with maximum confidence) classified as members of X;
- $\overline{B}X = \{x \in U | [x]_B \cap X \neq \emptyset\}$ -the *B-upper approximation* of X, which represents the set of objects that can be classified as possible elements of X (i.e., with uncertainty).

Next, the set:

$$BN_B(X) = \overline{B}X - \underline{B}X,$$

will be referred to as the *B-boundary region* of X, consisting of those objects that cannot decisively be classified into X on the basis of knowledge in B.

We are now in the position to introduce the notion of a rough set. Thus, a set is said to be *rough* (respectively *crisp*) if the boundary region is non-empty (respectively empty).

Remark 5.22. Sometimes, the pair $(\underline{B}X, \overline{B}X)$ is known as *rough set*.

Example 5.19. Let us consider the following decision table (adapted from (219)), where the conditional attributes are A_1 and A_2, and the decision attribute is A_3.

If we choose $B = A = \{A_1, A_2\}$ and $X = \{o | A_3(o) = 1\}$, we then obtain the approximation regions:

$$\underline{B}X = \{o_1, o_6\},$$

and

$$\overline{B}X = \{o_1, o_3, o_4, o_6\},$$

Table 5.14 Decision table

#	A_1	A_2	A_3
o_1	16-30	50	1
o_2	16-30	0	0
o_3	31-45	1-25	0
o_4	31-45	1-25	1
o_5	46-60	26-49	0
o_6	16-30	26-49	1
o_7	46-60	26-49	0

and therefore, since $BN_B(X) = \{o_3, o_4\} \neq \emptyset$, it follows that X is rough since the boundary region is not empty.

We show in the figure below a suggestive illustration of the two approximations of a set.

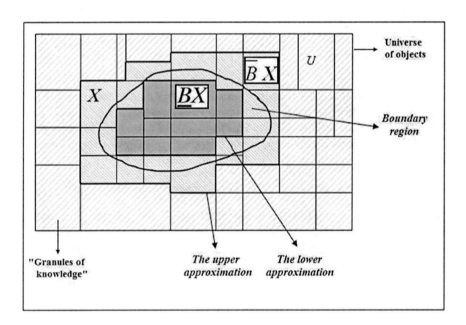

Fig. 5.43 Illustration of the main characteristics of rough sets

Remark 5.23. We can define the *approximation accuracy* (accuracy of the rough-set representation of a set X) using the formula:

$$\alpha_B(X) = Card(\underline{B}X)/Card(\overline{B}X), \tag{5.63}$$

where $Card(Y)$ denotes the cardinality of the set Y. If X is rough with respect to B, then $0 \le \alpha_B(X) < 1$, otherwise (i.e., $\alpha_B(X) = 1$) X is crisp with respect to B.

We previously mentioned that not all the attributes are necessary for the formation of equivalence classes, and therefore a 'reduct' is considered, consisting of sufficient information necessary to discern objects in a class from objects belonging to other classes. Formally, given an information system $I = (U,A)$, a *reduct* of A is a minimal set of attributes $B \subseteq A$, such that $IND(B) = IND(A)$. In other words, a reduct is a minimal set of attributes from A that preserves the partitioning of the universe, and hence performs the same classification as the whole set of attributes A.

One can show that there is more than one reduct of an information system, so we need to deepen this problem. In this context, one of the main ideas is to consider as relevant attributes those in the core of the information system, i.e., attributes that belong to the intersection of all reducts of the information system. Technically, the *core* of the information system is given by:

$$CORE(I) = \bigcap RED(A), \tag{5.64}$$

where $RED(A)$ is the set of all reducts of A.

Example 5.20. Let us consider the following decision system concerning certain symptoms possibly indicating flu, together with the corresponding decision (Table 5.15), (291).

Table 5.15 Decision table

Patient	Headache	Muscle-pain	Temperature	Flu
X_1	No	Yes	High	Yes
X_2	Yes	No	High	Yes
X_3	Yes	Yes	Very high	Yes
X_4	No	Yes	Normal	No
X_5	Yes	No	High	No
X_6	No	Yes	Very high	Yes

In this case, we can consider two reducts with respect to flu:

- Reduct 1 = {*Headache, Temperature*}, with the corresponding table:

Table 5.16 Decision table corresponding to Reduct 1

Patient	Headache	Temperature	Flu
X_1	No	High	Yes
X_2	Yes	High	Yes
X_3	Yes	Very high	Yes
X_4	No	Normal	No
X_5	Yes	High	No
X_6	No	Very high	Yes

- Reduct 2 = {*Muscle-pain, Temperature*}, with the corresponding table:

Table 5.17 Decision table corresponding to Reduct 2

Patient	Muscle-pain	Temperature	Flu
X_1	Yes	High	Yes
X_2	No	High	Yes
X_3	Yes	Very high	Yes
X_4	Yes	Normal	No
X_5	No	High	No
X_6	Yes	Very high	Yes

As we stated above, instead of using the initial decision table (Table 5.15), we can use either Table 5.16, or Table 5.17.

Next, it is easy to see that the core of this information system is given by:

$$CORE(I) = \{Temperature\}.$$

To conclude, it is obvious to realize that the reducts can be used to build minimal decision rules. Once the reducts have been computed, the rules are easily constructed. Thus, RS generates a collection of "*IF...THEN...*" decision rules that are used to classify the objects in decision tables. These rules are generated from the application of reducts to the decision table, looking for instances where the conditionals match those contained in the set of reducts and reading off the values from the decision table. If the data is consistent, then all objects with the same conditional values as those found in a particular reduct will always map to the same decision value. In many cases though, the decision table is not consistent, and instead we must contend with some amount of indeterminism. In this case, a decision has to be made regarding which decision class should be used when there are more than one matching conditioned attribute values. Discussions of these ideas are found, for instance, in (347).

For example, from Table 5.16 and 5.17 we can get very simple rules, like these: "IF {*Headache* = No ∧ *Temperature* = High} THEN {*Flu* = Yes}", "IF {*Muscle-pain* = Yes ∧ *Temperature* = Normal} THEN {*Flu* = No}", etc.

Finally, let us focus, in short, on the decision-making aspect of the RS approach. Let $I = (U, A \cup \{d\})$ be a given decision system. The cardinality of the image $d(U) = \{k | d(x) = k, x \in U\}$, denoted $r(d)$, is called the *rank* of d. Then, the decision d determines the partition:

$$CLASS_I(d) = \{X_1, X_2, ..., X_{r(d)}\}, \tag{5.65}$$

of the universe U, where $X_k = \{x \in U | d(x) = k\}$, $1 \leq k \leq r(d)$. This partition is called the *classification of objects in I determined by the decision d*, and the set X_k is called the k-th decision class of I.

We illustrate in Fig. 5.44 such a classification, comprising three decision classes.

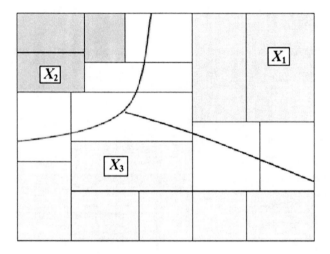

Fig. 5.44 Decision classes of a decision system

Partially paraphrasing the information system concept, we can say that there is a whole 'universe' regarding the Rough Sets, our arguments including even the existence of an *International Rough Set Society* -http://www.roughsets.org/; http://roughsets.home.pl/www/.

Rough sets applications are very diverse, here we mention only a few of them:

- First of all, let us underline that the RS theory has proved to be useful in data mining. See, for instance, (234), (223), (276), (300), (301), (302), (237), (296).
- RS have been successfully applied in advanced medical research: treatment of duodenal ulcer, (109), image analysis, (271), (174), (176), (339), breast cancer

detection, (376), (175), (314), (148), hepatic cancer detection, (141), (314), heart diseases, (120), (55), diagnosis of biliary cirrhosis, (315), Parkinson's disease, (316), etc.

- RS have also been used in economics, finance and business, (348), (379), (224), (345), (128), (412), (56), etc.

Note also that the RS theory has been extended to the fuzzy sets field, by considering the concept of *fuzzy-rough sets* (fuzzy equivalence classes) -see, for instance (86), (87), (303), (311), (64).

For those who wish to apply RS in different areas of research, we recommend using, among other software, a specialized *Rough Set Toolkit* called ROSETTA (http://www.lcb.uu.se/tools/rosetta/). As the authors specify, "ROSETTA is a toolkit for analyzing tabular data within the framework of rough set theory. ROSETTA is designed to support the overall data mining and knowledge discovery process: From initial browsing and preprocessing of the data, via computation of minimal attribute sets and generation of if-then rules or descriptive patterns, to validation and analysis of the induced rules or patterns".

More details about ROSETTA can be found in (282), (283), (220). The figure below illustrates a screenshot of this software.

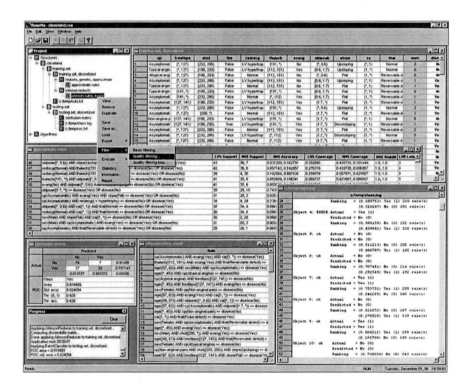

Fig. 5.45 ROSETTA screenshot

5.8   Clustering

As we showed in the introductory chapter, the cluster analysis, also known as clustering, means finding groups (clusters) of objects, based on their similarity (some sort of resemblance), so that within each group there is a great similarity, while the groups are different enough from one another. Fig. 5.46 illustrates the 'philosophy' behind the clustering procedure.

In terms of machine learning, the clustering process is a form of unsupervised learning.

While we briefly explained above what the clustering process is, let us now see what the cluster analysis is not, (378):

- Conventional supervised classification, based on complete information (i.e., objects with both predictive attributes and class labels). Here, a classification function is learned from (or fitted to) training objects. It is then tested on testing objects (e.g., decision and classification trees, decision graphs, supervised neural networks, etc.) -see, for instance, (43), (284), (205), (322);
- Simple segmentation of objects, based on certain rules, that do not directly involve similarity -see, for instance, (201), (37), (249);
- Results of a query, following an external specification -see, for instance, "*Query-based classification* -SAP NetWeaver 7.0 EHP - http://help.sap.com/, *query-based learning*, (200), (327);
- Graph partitioning, based on the same size of component pieces, but with few connections between them -see, for instance, (103).

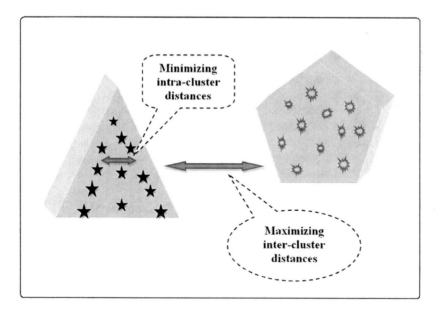

Fig. 5.46 Clustering process

In principle, the clustering methodology involves two distinct approaches:

- *Hierarchical clustering*;

- *Non-hierarchical/partitional/flat* clustering.

Hierarchical clustering reveals successive clusters using previously established clusters, thus building a hierarchy of clusters (producing a *dendrogram* = a tree diagram) and not merely a simple partition of objects. The number of clusters is not required as an input condition of the algorithm, while a certain condition can be used to end it (e.g., a predetermined number of clusters). Thus, one can obtain any desired number of clusters by 'cutting' the dendrogram at the proper level.

We will mention here three hierarchical clustering types:

1. *Agglomerative* (*bottom-up*), in which pairs of objects/clusters are sequentially connected to produce larger clusters. The method consists of:

   - Place each object in its own cluster (i.e., each object is considered as a separate cluster);

   - Merge at each step the closest clusters until only one is obtained or, alternatively, the stop condition is met.

2. *Divisive* (*top-down*), in which all objects are initially placed in a single cluster (the root), and then successively divided (split) into separate groups. The method consists of:

   - One starts with a single (the largest) cluster (the root) containing all objects;

   - At each step divide (split) a cluster in smaller clusters, until each cluster contains one point, or some other stop condition is met.

3. *Conceptual*, consisting in finding clusters that share some common property or represent a particular concept, generating a concept description for each generated class. One incrementally builds a structure out of the data by trying to subdivide a group of observations into subclasses. The result is a hierarchical structure known as the concept hierarchy -see, for instance, (261), (262), (231), (110), (111). Among the conceptual clustering algorithms we can mention, for instance, CLUSTER/2, (262) and COBWEB, (110), (111).

The clustering process depends on the type of similarity chosen for the objects segmentation. Consequently, they can be divided in several ways, taking into account the kind of similarity between them.

We synthetically illustrate below two different ways of clustering objects, depending on two different types of similarity.

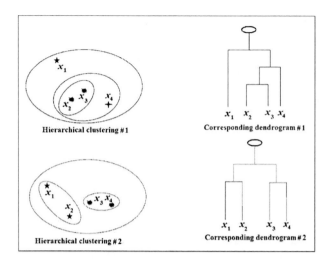

Fig. 5.47 Examples of hierarchical clustering of the same objects

The non-hierarchical clustering, that is the partitional clustering, consists in the initial split of objects into non-overlapping subsets (clusters), so each object belongs to just one cluster, as shown in Fig. 5.48.

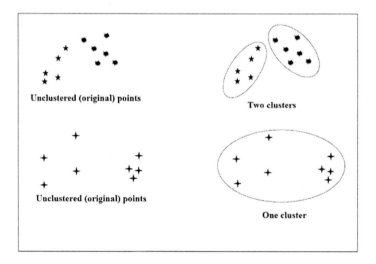

Fig. 5.48 Examples of partitional clustering

The clustering process basically involves three main steps:

1. Defining a similarity measure;
2. Defining a criterion for the clusters building process;
3. Building an algorithm to construct clusters based on the chosen criterion.

A clustering algorithm aims to identify natural groups of objects in a given set and, as such, it needs to measure the degree of similarity between objects, based on a certain criterion. The first action to take is therefore to consider an appropriate measure, corresponding to the intrinsic nature of data, and intended to assess a certain 'distance' (dissimilarity) between objects.

Another important aspect of the cluster analysis is the way to validate the clustering structure, built based on a certain algorithm, (164), (378). Although it is difficult to consider an objective function in this case, we will mention some important criteria in this direction:

- *External validation*, which consists in comparing the obtained clustering with other segmentation/classification approaches (e.g., using statistical techniques);
- *Internal validation*, which evaluates the clustering result of an algorithm using only quantities and features inherent to the dataset, i.e., without respect to external information (e.g., using *sum of squared error* - SSE);
- *Relative validation*, which compares two different clustering models or clusters (e.g., statistical testing, using SSE, etc.).

The main steps performed in a cluster analysis consist of:

- Data preparation - collecting and arranging data for the clustering process;
- The choice of a similarity measure - establishing how to compute the 'resemblance' distance between objects;
- Prior knowledge - the use of available knowledge concerning the given field, which can help in preparing the data and choosing an appropriate similarity measure;
- Effectiveness of the cluster structure - the construction quality and the time affected for it.

Summarizing the above considerations, the main points to be considered in the clustering process are the following:

- *Problem formulation* - the selection of objects for clustering;
- *The choice of a similarity measure* - the selection of an appropriate 'distance' between the objects to be clustered, based on the proposed criterion;
- *Selection* of the clustering model;
- *Selection* of the number of clusters (or the STOP condition), as the case may be;
- *Graphic illustration and clusters interpretation* (drawing conclusions);
- *Assessing the validity and robustness* of the model using various methods, such as:

  - Repeating the process using other similarity measures corresponding to the context;
  - Repeating the process using other appropriate clustering techniques;
  - Repeating the process several times, but ignoring at each iteration one or more objects.

We have focused so far on the problem concerning the measurement of the 'distance' (similarity) between objects, trying thus to solve the problem of minimizing the intra-cluster distance. On the other hand, we must solve the problem of maximizing the inter-cluster distance, in other words, we need to define a 'distance' between two clusters (i.e., inter-clusters similarity). This problem relates to the hierarchical clustering approach. There are several ways to solve this problem, the most known approaches are the following:

- *Single linkage (nearest neighbor)* - the distance between two clusters is determined by the distance of the two closest objects (nearest neighbors) in the different clusters;
- *Complete linkage (furthest neighbor)* - the distances between clusters are determined by the greatest distance between any two objects in the different clusters (i.e., by the "furthest neighbors");
- *Unweighted pair-group average (group average)* - the distance between two clusters is calculated as the average distance between all pairs of objects in the two different clusters;
- *Weighted pair-group average* - identical to the unweighted pair-group average method, except that in the computations, the size of the respective clusters (i.e., the number of objects contained in them) is used as a weight;
- *Unweighted pair-group centroid* - the distance between two clusters is determined as the distance between the corresponding centroids (the *centroid* of a cluster is the average point in the multidimensional space defined by the dimensions, i.e., the center of gravity for the respective cluster);
- *Weighted pair-group centroid (median)* - identical to the previous one, except that weighting is introduced into the computations to take into consideration differences in cluster sizes;
- *Ward's method*, (394) - distinct from all other methods, using an analysis of variance approach to evaluate the distances between clusters (i.e., attempting to minimize the *Sum of Squares* (SS) of any two (hypothetical) clusters that can be formed at each step).

We illustrate in the figures below four of the criteria presented above.

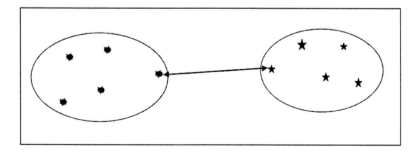

Fig. 5.49 Single linkage

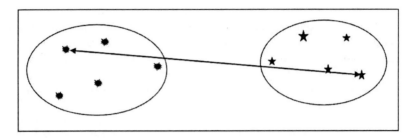

Fig. 5.50 Complete linkage

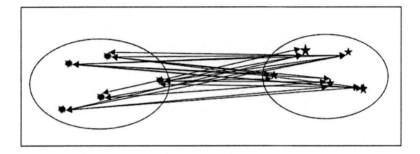

Fig. 5.51 Unweighted pair-group average (group average)

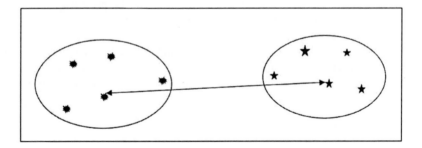

Fig. 5.52 Unweighted pair-group centroid (distance between centroids)

Next, we will present some basic issues concerning the measurement of the similarity between objects, the first effective step in the clustering process. In the context of the clustering methodology, the similarity measure indicates how similar two objects are. Often, however, instead of using the similarity, we can consider the dissimilarity since it is more appropriate to the idea of measuring the distance between objects. Irrespective of the way of comparing two objects, the issue of choosing a specific measure essentially depends on the problem itself, on our aim, and on the expected results.

Usually, it is required for such a measure to have certain properties, depending on the specific problem to which it is applied. In principle, a measure of similarity is a function $d : D \times D \to \mathbf{R}_+$ applied on a set of objects D, and having certain specific properties. Conceptually, we can say that $similarity = distance^{-1}$, and therefore, the term of measure of dissimilarity, regarded as the distance between two objects, can be used as well. However, the established term is *measure of similarity*. Seen as a distance function, such a measure should possess the basic properties of a metric, namely:

- *Non-negativity* and *identity of indiscernibles*: $d(A,B) \geq 0$, $d(A,B) = 0 \Leftrightarrow A = B$;

- *Symmetry*: $d(A,B) = d(B,A)$;

- *Sub-additivity/triangle inequality*: $d(A,B) + d(B,C) \geq d(A,C)$.

Remark 5.24. Some of the above properties are not always required. For example, the symmetry property (e.g., in image processing there are cases in which, almost paradoxically, the image of a child is considered more similar to the image of one of his (her) parent than *vice versa*) is not satisfied (case sometimes known as quasi-metric). The same is the case for the triangle inequality. Consider, for example, the vectors $\mathbf{x}_1 = (a,a,a,a)$, $\mathbf{x}_2 = (a,a,b,b)$ and $\mathbf{x}_3 = (b,b,b,b)$. Intuitively, if we can say that the distances between $\mathbf{x}_1$ and $\mathbf{x}_2$, as well as of $\mathbf{x}_2$ and $\mathbf{x}_3$, are small enough, the distance between the first and last can be considered very high in some circumstances (i.e., when measuring the similarity), exceeding the sum of the first two distances. A very suggestive illustration of this 'strange' mathematical situation is the well-known (similarity) comparison between human being, centaur and horse.

Let us mention, in this context, other possible properties of a similarity measure:

- *Continuity properties*, often encountered in pattern recognition: *perturbation robustness*, *crack robustness*, *blur robustness*, *noise* and *occlusion robustness*.

- *Invariance properties*. A similarity measure d is invariant to a transformation group G, if for every $g \in G$, $d(g(A),g(B)) = d(A,B))$. For example, in pattern recognition, the measure is usually invariant to *affine* transformations.

Remark 5.25. The choice of a measure of similarity must be always in accordance with the type of available data (e.g., numerical, categorical, rank, fuzzy, etc.).

We present below some of the most popular similarity measures, which are applied in almost all cases. But first, we have to specify that, in order to measure the similarity between two objects/instances, we will consider them as vectors: $\mathbf{x} = (x_1, x_2, ..., x_n)$, $\mathbf{y} = (y_1, y_2, ..., y_n)$, having, for the sake of simplicity, the same dimension n. Note that, in some instances, one may consider (with some modifications) vectors with different dimensions.

1) *Minkowski* distance:

$$d_p(\mathbf{x},\mathbf{y}) = \left(\sum_{i=1}^{n} |x_i - y_i|^p \right)^{1/p}, \quad p \in \mathbf{N}. \tag{5.66}$$

Note that a generalization of the Minkowski distance is the *power distance*, given by:

$$d_{p,r}(\mathbf{x},\mathbf{y}) = \left(\sum_{i=1}^{n} |x_i - y_i|^p \right)^{1/r}, \tag{5.67}$$

and used when one may want to increase or decrease the progressive weight that is placed on dimensions on which the respective objects are very different. Thus, parameter p controls the progressive weight that is placed on differences on individual dimensions, and parameter r controls the progressive weight that is placed on larger differences between objects.

Remark 5.26. 1) For $p = 1$ one obtains the *Manhattan distance* (or *city block*, or *taxicab*, or L_1 *distance*):

$$d_{cb}(\mathbf{x},\mathbf{y}) = \sum_{i=1}^{n} |x_i - y_i|. \tag{5.68}$$

This is the same problem as getting from corner A to corner B in a rectilinear street map, hence the name "city-block". This distance is simply the sum of difference across dimensions. However, the effect of single large differences (outliers) is dampened (since they are not squared).

2) For $p = 2$ one obtains the well-known *Euclidean distance* (or L_2 *distance*, or even *crow flies* distance), probably the most commonly chosen type of distance, especially in case of numerical data. Basically, it is the geometric distance in the multidimensional space, and is given by:

$$d_E(\mathbf{x},\mathbf{y}) = \left(\sum_{i=1}^{n} |x_i - y_i|^2 \right)^{1/2}. \tag{5.69}$$

Note that the Euclidean distance is computed from raw data, and not from standardized data, and can be greatly affected by differences in scale among the dimensions from which the distances are computed.

3) For $p = \infty$ one obtains the *Chebychev distance*, given by:

$$d_C(\mathbf{x},\mathbf{y}) = \max_i |x_i - y_i|. \tag{5.70}$$

Note that this distance may be appropriate in cases when one wants to define two objects as "different" if they are different on anyone of their dimensions.

4) The Manhattan distance for binary vectors becomes the *Hamming distance*. Note that this distance is defined as the number of bits which differ between two binary

strings, i.e., the number of bits which need to be changed to turn one string into the other.

2) *Cosine* distance/measure:

$$d_c(\mathbf{x}, \mathbf{y}) = \frac{\mathbf{x} \cdot \mathbf{y}^T}{\|\mathbf{x}\|_E \cdot \|\mathbf{y}\|_E}, \tag{5.71}$$

i.e., the cosine of the angle between two vectors $\mathbf{x}$ and $\mathbf{y}$ (see the *dot* product). Note that this similarity measure is often used to compare documents in text mining, where the (attribute) vectors $\mathbf{x}$ and $\mathbf{y}$ are usually the term frequency vectors of the documents.

3) *Tanimoto* distance/measure:

$$d_T(\mathbf{x}, \mathbf{y}) = \frac{\mathbf{x} \cdot \mathbf{y}^T}{\mathbf{x} \cdot \mathbf{x}^T + \mathbf{y} \cdot \mathbf{y}^T - \mathbf{x} \cdot \mathbf{y}^T}. \tag{5.72}$$

4) *Jaccard* index (or similarity coefficient) and *Jaccard* distance, used in Statistics to measure the similarity/dissimilarity between two sample sets. Thus, the *Jaccard index* between two sample sets, $J(A,B)$, is the ratio of the size of their intersection to the size of their union:

$$J(A,B) = \frac{\|A \cap B\|}{\|A \cup B\|}. \tag{5.73}$$

Then, the *Jaccard dissimilarity distance* between A and B is given by $[1 - J(A,B)]$.

5) *Pearson'r* distance (or *correlation coefficient* measure):

$$r(\mathbf{x}, \mathbf{y}) = \frac{\sum_{i=1}^{n} (x_i - \bar{x})(y_i - \bar{y})}{\sqrt{\sum_{i=1}^{n} (x_i - \bar{x})^2 \sum_{i=1}^{n} (y_i - \bar{y})^2}}. \tag{5.74}$$

6) *Mahalanobis* distance/measure. The Mahalanobis measure is generally given by:

$$d_M(\mathbf{x}, \mathbf{y}) = \sqrt{(\mathbf{x} - \mathbf{y}) \cdot B \cdot (\mathbf{x} - \mathbf{y})^T}, \tag{5.75}$$

where B is any symmetric and positive-definite matrix. Note that, for two random vectors $\mathbf{x}$ and $\mathbf{y}$, with the same distribution and covariance matrix $cov(D)$, the Mahalanobis measure is given by:

$$d_M(\mathbf{x}, \mathbf{y}) = \sqrt{(\mathbf{x} - \mathbf{y}) \cdot cov(D)^{-1} \cdot (\mathbf{x} - \mathbf{y})^T}. \tag{5.76}$$

Remark 5.27. If the covariance matrix is the identity matrix, the Mahalanobis distance reduces to the Euclidean distance.

7) *Fuzzy* extensions of classical similarity measures. The fuzzy extensions of the traditional similarity measures are used as ready-built tools for comparing vectors or matrices whose elements take values in the interval [0, 1]. Thus, let **x** and **y** be two vectors so that their components x_i and y_i belong to the interval [0, 1]; such vectors are called *fuzzy vectors* -see, for instance, (343). Note in this context that a *crisp vector* can be seen as a special case of fuzzy vector, when x_i belongs to the set $\{0, 1\}$. Next, let us denote:

$$s(x_i, y_i) = \max\{\min\{x_i, y_i\}, \min\{1 - x_i, 1 - y_i\}\},$$

often used to define different fuzzy similarity measures, starting from the corresponding classical cases. Thus, for example, the *fuzzy Minkowski* measure is given by the following formula:

$$d_F^p(\mathbf{x}, \mathbf{y}) = \left(\sum_{i=1}^n s(x_i, y_i)^p\right)^{1/p}. \tag{5.77}$$

Remark 5.28. 1) *Weighted similarity measures.* Weights can be incorporated in any similarity measure, in order to rank the importance of each attribute taking into account the context. For example, a *weighted Minkowsky* measure has the form:

$$d_{p,\alpha}(\mathbf{x}, \mathbf{y}) = \left(\sum_{i=1}^n \alpha_i |x_i - y_i|^p\right)^{1/p}, \quad \alpha_i > 0, \quad \sum_{i=1}^n \alpha_i = 1, \tag{5.78}$$

where α_i represents the weight corresponding to the attribute x_i.

2) *Mixed-weighted similarity measures.* In many real applications, there are situations in which objects/instances are represented as mixed vector, in which the components have different meanings, thus having different natures (e.g., numeric, categorical, ranks, fuzzy, etc.). For instance, if we are situated in the medical field, when the vector **x** may mathematically represents a patient, then some attributes are numerical (e.g., age, weight, height, cholesterol level, blood pressure, etc.), others could be categorical (e.g., gender, place of residence, disease type), others may represent ranks (e.g., disease stage, tumor size), others may be considered as fuzzy attributes (e.g., risk factors: smoking, alcohol consumption, etc.), etc. In meteorological studies, for instance, we have to deal like in the previous case with numerical attributes (e.g., temperature, pressure), percentage/rank attributes (e.g., humidity, wind strength, hurricane category), categorical (e.g., location), etc. It is obvious that in these cases we have to use special similarity measures, which take into account the specificity of the targeted domain. A natural way to solve such problems is to consider a mixed similarity measure (especially designed to quantify the type of data), and weighted at the same time (to quantify the significance of each attribute). Specifically, consider two objects to be compared, having the following form:

$$\mathbf{x} = \left(\left(x_1^1, x_2^1, ..., x_{k_1}^1 \right), \left(x_1^2, x_2^2, ..., x_{k_2}^2 \right), ..., \left(x_1^s, x_2^s, ..., x_{k_s}^s \right) \right),$$

$$\mathbf{y} = \left(\left(y_1^1, y_2^1, ..., y_{k_1}^1 \right), \left(y_1^2, y_2^2, ..., y_{k_2}^2 \right), ..., \left(y_1^s, y_2^s, ..., y_{k_s}^s \right) \right),$$

where there are s different types of data, with the corresponding dimensions k_1, $k_2,..., k_s$. First, we will put the s sequences into hierarchical order depending on their contextual importance, by considering certain weights α_j, $\sum_j \alpha_j = 1$. Then, we will consider for each particular sequence a certain similarity measure, appropriate to that case. If we denote the component sequences by:

$$x_j = \left(x_1^j, x_2^j, ..., x_{k_j}^j \right),$$

$$y_j = \left(y_1^j, y_2^j, ..., y_{k_j}^j \right),$$

then the mixed-weighted similarity measure has the form:

$$d(\mathbf{x}, \mathbf{y}) = \sum_{j=1}^{s} \alpha_j \cdot d_j (x_j, y_j). \tag{5.79}$$

3) *Features standardization.* The process of standardizing the characteristics (attributes) is an important issue when we measure the (similarity) distance between objects. In principle, there is no general valid solution to this problem, but everything depends on the specific situation. For example, we have the transformation $x_i \rightarrow \dfrac{x_i - \overline{x_i}}{SD(x_i)}$, so that each component (attribute) has the *mean* = 0 and *variance* = 1, or the transformation $x_i \rightarrow \dfrac{x_i - \overline{x}}{SD(x)}$, so that each vector (object) has *mean* = 0 and *variance* = 1.

4) We mentioned above that we should take into account the nature of the data when choosing a suitable similarity measure. Thus, for categorical (ordinal) data, we mention two classical approaches:

- Conversion of the original ordinal data into numerical form, considering a normalized scale [0, 1], where *min* = 0 *max* = 1, the rest of data being properly interpolated (e.g., the individual's body weight categories using BMI (body mass index): 0.0 = *severely underweight*, 0.3 = *underweight*, 0.5 = *normal*, 0.7 = *overweight*, 0.9 = *obese class I*, 1.0 = *obese class II-III*).
- The use of a (symmetric) similarity matrix, where the elements (entries) belonging to the main diagonal equal 1 (similarity), or 0 (dissimilarity). Considering the example before, for instance, we have (hypothetical figures):

	severely underweight	underweight	normal	overweight	obese class I	obese class II-II
severely underweight	1	0.8	0.7	0.5	0.2	0
underweight	0.8	1	0.9	0.7	0.3	0.1
normal	0.7	0.9	1	0.7	0.3	0.2
overweight	0.5	0.7	0.7	1	0.5	0.3
obese class I	0.2	0.3	0.3	0.5	1	0.8
obese class II-III	0	0.1	0.2	0.3	0.8	1

Fig. 5.53 Example of a similarity matrix

For other categorical data, either binary rules are usually considered (e.g., IF x_i = y_i for all i, THEN *similarity* = 1, ELSE 0), or a certain semantic property (e.g., $d(wood, iron) = \alpha |wood\ density - iron\ density|$, $d(Joan, Mary) = \alpha |Joan's\ age - Mary's\ age|$, $d(Paris, London) = \alpha |Paris\ population - London\ population|$, or $d(Paris, London) = \alpha |cost\ of\ living\ in\ Paris - cost\ of\ living\ in\ London|$, where α is a certain parameter), or a certain similarity matrix is used like before.

5.8.1 Hierarchical Clustering

Once a similarity distance selected, based on which we can compare the objects, the second step consists in partitioning them using a particular methodology. As we said at the beginning, the result could be represented in the classical manner, as a dendrogram (tree diagram). The difference between the two main hierarchical clustering methods consists in the manner of building the tree: bottom-up - *agglomerative* clustering, and top-down - *divisive* clustering.

The hierarchical clustering model lies in iteratively grouping objects using a particular method of "amalgamation/linkage" (see also the way of defining the distance between clusters), such as:

- *single linkage* (nearest neighbor);
- *complete linkage* (furthest neighbor);
- *average linkage* (unweighted pair-group average)/(weighted pair-group average);
- *centroid method* (unweighted pair-group centroid)/(weighted pair-group centroid (median));
- *Ward's method.*

We schematically illustrate in the figures below the hierarchical clustering method for partitioning certain car brands, taking into account the following characteristics (attributes): price, acceleration time (0-100 km/h or 0-62 mph), braking (stopping) distance, top speed, and fuel consumption. It is obvious that this car grouping should be taken purely illustrative.

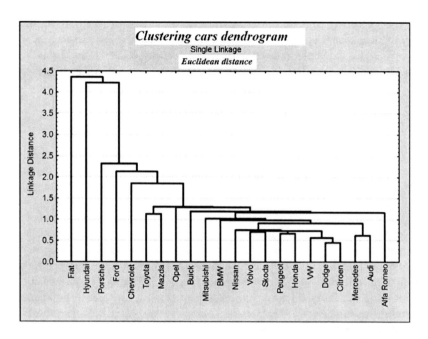

Fig. 5.54 Dendrogram of cars clustering (vertical tree diagram -Euclidean distance)

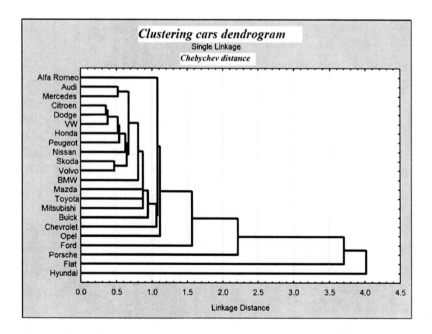

Fig. 5.55 Dendrogram of cars clustering (horizontal tree diagram -Chebychev distance)

Note. As it is easy to see from the two figures, the first dendrogram represents a vertical tree diagram, the similarity is given by the Euclidean distance and the method is based on single linkage. In the second case, the Chebychev distance was chosen, while the single linkage method was used again, thus obtaining a horizontal tree diagram.

5.8.2   Non-hierarchical/Partitional Clustering

Non-hierarchical clustering (or partitional clustering) is often known as clustering of k-means type. This model is totally different as methodology from the hierarchical clustering technique, in this case assuming an *a priori* knowledge of the number of clusters. The issue in this context is to create an algorithm such that an exact and previous stipulated number of clusters, as distinct as possible, have to be obtained. Although, in principle, the k-means clustering method produces exactly k clusters which divide the initial set of objects in groups as distinct as possible, the open problem of estimating the optimal number of clusters leading to the (near) best separation of objects is still unsolved. We mention here the most commonly used method for solving this issue, based on the *cross-validation* technique, in order to automatically determine the number of clusters in the dataset (see also k-nearest neighbor). Another simple approach is to compare the results of multiple runs with different k classes and choose the best one according to a given criterion (e.g., finding the solution that minimizes the *Schwarz criterion* - also related to BIC). Computationally, we may consider this method as the "reverse" of the *analysis of variance* (ANOVA) technique. Thus, the model starts with k random clusters, and then moves objects between those clusters with the goal to:

- minimize variability within clusters;
- maximize variability between clusters.

Thus, while the significance test in ANOVA evaluates the between-group variability against the within-group variability when computing the significance test for the hypothesis that the means in the groups are different from each other, in k-means clustering the model 'tries' to move objects in and out of clusters, to get the most significant ANOVA results - for other technical details, see *"Cluster Analysis"*/StatSoft electronic Statistics textbook http://www.statsoft.com/textbook/cluster-analysis/.

We now present the general scheme of the k-means algorithm.

k-means algorithm

1. Select k points at random as cluster centers.

2. Assign instances to their closest cluster center according to some similarity distance function.

3. Calculate the *centroid* or mean of all instances in each cluster (this is the *'mean'* part of the algorithm).

4. Cluster the data into k groups where k is predefined.

5. GOTO the step 3. Continue until the same points are assigned to each cluster in consecutive rounds.

Technically speaking, the algorithm steps are the following:

1. Suppose there are N data points $\mathbf{x}^l = (x_1{}^l, x_2{}^l, ..., x_n{}^l)$ in total, $l = 1, 2, ..., N$;
2. Find a set of k representative vectors $\mathbf{c}_j$, where $j = 1, 2, ..., k$;
3. Partition the data points into k disjoint subsets S_j, containing N_j data points, in such a way as to minimize the *sum-of-squares* clustering function given by:

$$J = \sum_{j=1}^{k} \sum_{l \in S_j} \left\| \mathbf{x}^l - \mathbf{c}_j \right\|^2, \tag{5.80}$$

where $\mathbf{c}_j$ is the mean of the data points in set S_j, given by:

$$\mathbf{c}_j = \frac{\sum_{l \in S_j} \mathbf{x}^l}{N_j}. \tag{5.81}$$

Note. There is a major problem in using the *k-means* method, namely the need to define the 'mean', which implies the fact that it is practically inapplicable to non-numeric data (e.g., categorical data).

Remark 5.29. Let us also mention in context three related algorithms to cluster analysis:

1. *Expectation Maximization* (EM algorithm) used for data clustering in machine learning and computer vision - see, for instance, (254), (274);
2. *Quality Threshold* clustering, seen as an alternative method of partitioning data, and not requiring an *a priori* number of clusters - see, for instance, (183);
3. *Fuzzy c-means* clustering, as its name indicates, assumes that each point has a degree of belonging to clusters rather than completely belonging to just one cluster - see, for instance, (32), (192).

Finally let us briefly review both the "white balls" and the "black balls" corresponding to these two major types of clustering.

- *Hierarchical clustering*:

    (+) intuitive and easily understandable. It can produce an ordering of the objects, which may be informative for data display;

    (+) smaller clusters are generated, which may be helpful for knowledge discovery;

    (-) sensitivity to noise and outliers;

    (-) unwanted persistence of the starting clusters;

(-) difficulty in handling differently sized clusters and convex shapes;

(-) difficulty in handling large datasets.

- *Partitional clustering*:

  (+) well suited to generating globular clusters (i.e., clusters which are very roughly spherical or elliptical - convex);

  (+) may be computationally faster than hierarchical clustering for large number of variables (if k is small);

  (+) may produce tighter clusters than hierarchical clustering, especially if the clusters are globular;

  (-) the *a priori* fixed number of clusters can make it difficult to choose the optimal k (see also the comments referring to the estimation of the number of clusters);

  (-) has problems when clusters are of differing (sizes, densities, non-globular shapes);

  (-) has problems when data contains outliers;

  (-) different initial partitions can result in different final clusters.

Judging from the above aspects, we conclude that it is wiser to use, if possible, both methods simultaneously. For example, we can first use the hierarchical clustering to determine a suitable number of clusters, and then use the partitional clustering to improve the results by reconfiguring the irrelevant clusters.

For more details on clustering techniques, see (105), (323), (212), (121), (1).

Example 5.21. We present an application of the k-means algorithm to increase the performance of a probabilistic neural network (PNN), used in the liver cancer diagnosis, (138). Thus, starting from a group of 299 subjects, divided as follows:

- 60 individuals with chronic hepatitis (CH),
- 179 individuals with liver cirrhosis (LC),
- 30 individuals with hepatocellular carcinoma (hepatic cancer) (HCC),
- 30 healthy people (control group) (HP),

the PNN algorithm was first applied to the initial group to classify individuals into the four above categories. Based on the patients' classification thus obtained, one can establish the standard patient profile corresponding to each disease, and, secondly, one can create an intelligent diagnosis system that can help physicians to take an optimal diagnosis decision. Note that for this classification, 15 significant medical characteristics have been used, features which are highly correlated with the diagnosis (see also subsection 5.3.3 concerned with PNN applications). Since the main PNN drawback consists in computation problems when handling large datasets, the k-means algorithm was consequently used to reduce the number of required data (i.e., patients in this context) for this study, in order to increase the computation speed of the neural network. In this respect, a number of 250, 200, 150 and 100 clusters has been chosen. Thus, instead of all the 299 individuals being classified, a number of 250, 200, 150 or 100 'virtual' individuals has been classified.

We called them 'virtual' individuals since they represent the clusters' centroids, and not real people. Practically, unlike when using the initial dataset of 299 (real) individuals, each of the N clusters ($N = 250, 200, 150, 100$) were represented by their centroids, regarded as 'virtual patients'. For searching the parameter σ both the genetic algorithms (GA) and the Monte Carlo (MC) method were used. The results of this experiment are shown in the table below.

Table 5.18 Mixed PNN/k-means approach in hepatic diseases

Number of patients (real or virtual)	Training accuracy (%)		Testing accuracy (%)	
	GA	MC	GA	MC
299	87	89	82	86
250	88	90	80	83
200	91	92	76	79
150	93	93	74	74
100	94	95	68	67

From the table above we see that reducing the computational effort (and hence, increasing the corresponding processing speed) by 16% (the case of 250 individuals) did not involve a significant decrease in classification accuracy. On the other hand, when we deal with fewer (virtual) individuals to be classified, the training accuracy apparently increases while the testing accuracy decreases. We say the training accuracy 'apparently' increases, because we no longer deal with real people, but with 'fictitious' people -the clusters centroids seen as virtual people. Finally, it is noted that a reduction of up to 33% (200 individuals) is still reasonable in practical terms (76-79% testing accuracy).

Example 5.22. While the above example illustrates the applicability of the clustering technology in computer-aided medical diagnosis, the following example demonstrates its applicability for patient management, (100), (101). Thus, understanding the types of patients treated and their resulting workload, could achieve better use of hospital resources. Patients typically vary in a multitude of ways, some of which include their diagnosis, severity of illness, medical complications, speed of recovery, resource consumption, length of stay (LOS) in hospital, discharge destination, and social circumstances. Such heterogeneity in patient populations, coupled with the uncertainty inherent within health care systems (e.g., emergency patients arriving at random), makes it complicated to plan for effective resource use. The difficulties caused by heterogeneous patient populations have generated a need to group the population into a set of comprehensible and homogeneous groups. Many benefits and uses of grouping patients have been identified and recognized as being vital for improving the planning and management of hospitals and health facilities alike. Grouping patients is advantageous in that it helps to simplify our view as well as improve our comprehension of the diverse patient population. Clustering algorithms have also been used to better understand the relationships between data when the groups are neither known nor cannot be predefined. In this context, LOS is often used as a proxy measure of a patient' resource consumption because of the practical difficulties of directly measuring resource consumption and the easiness of calculating LOS. Grouping patient spells according to their LOS has proved to be a challenge in health care applications due to the inherent variability in the LOS distribution.

This study deals with three clustering techniques, *Gaussian mixture model* (GMM), *k-means* algorithm and the *two-step clustering* algorithm, applied on a stroke dataset containing data that are typically stored by hospital computerized systems. The stroke dataset originates from the *English Hospital Episode Statistics* (HES) database and concerns all finished consultant episodes of stroke patients, aged 65 and over discharged from all English hospitals between April 1st 1994 and March 31st 1995. Regarding the use of the *k*-means algorithm, in order to evaluate the appropriateness of the data segmentation and thus the number of clusters, the algorithm has been run for $k = 2$ to 6, and an analysis of variances, comparing the within-cluster variability (small if the classification is good) and between-cluster variability (large if the classification is good) has also been performed. A standard between-groups analysis of variances (F-values) was performed to evaluate the intra-cluster/inter-cluster variability. Consequently, a near optimal number of clusters ranges from 4 and 5 (p-level < 0.01). Table 5.19 presents the parameters derived from the five-cluster run.

Table 5.19 LOS parameters derived from the five-cluster model for the stroke dataset

	Components			
Clusters #	1st 2nd	3rd	4th 5th	
Mean (days)	9.1 69	607	1,689 3,452	
Standard deviation (days)	8.6 36.7	225.9	371.8 804.3	
Mixing coefficient (%)	94.0 5.8	0.1	>0.1 > 0.1	

Example 5.23. Another application of the *k*-means algorithm refers to the cancer treatment segmentation, (135). Thus, four classical types of treatment have been considered:

- *C1: Chemotherapy (CT)*;
- *C2: Chemotherapy (CT) + Hormonotherapy (HT)*;
- *C3: Chemotherapy (CT) + Radiotherapy (RT) + Curietherapy (QT)*;
- *C4: Chemotherapy (CT) + Radiotherapy (RT) + Curietherapy (QT) + Hormonotherapy (HT)*.

To estimate the optimal treatment, each patient was mathematically equated with a vector with four components. The first three components represent domain-specific predictor attributes: average tumor diameter, age and disease stage, while the last component represents the corresponding treatment type. By this patients' segmentation method, based on the treatment type, one can obtain a specific technique to correlate a certain patient with a certain treatment type, helping the physician to decide the appropriate treatment of each patient.

5.9   Genetic Algorithms

The evolutionary simulations by using computers (John von Neumann's high-speed computer at the Institute for Advanced Study in Princeton, New Jersey) seem to have been initiated in 1954 by Barricelli, focused on the simulation of the evolution of the ability to play a simple game, (22). Later on, Fraser (1957), (115), deals with computer simulation of genetic systems using the existing computers at that time -the so-called *'automatic digital computers'*. We may mention from this pioneering period Box's research on optimization of industrial productivity, (39), and Bremermann with optimization through evolution and recombination, (44). Starting from these first initial steps and many others, genetic algorithms have been recognized as a separate domain in the context of optimization methods. This recognition started with the researches regarding *cellular automata* from 60's, undertaken by John Holland and his students at the University of Michigan, the theoretical approach coming soon through the publication of his book *"Adaptation in Natural and Artificial Systems"* in 1975, (189). Let us mention, within the context, their extended presentation due to Goldberg, (129), (131), covering all of the important topics in the field, described in a very 'friendly' way. The development, more theoretical indeed, of the genetic algorithms continued thereafter, culminating with *"The First International Conference on Genetic Algorithms"*, held in 1985 at the University of Illinois, after which, simultaneously with the exponential growth of the computing power, the genetic algorithms have known a real practical development, mentioning here the first commercial software for personal computers -*Evolver* (Axcelis, Inc. / Palisade Corporation) - appeared in 1990.

The genetic algorithms are based on the modern theory of evolution, with roots both in the natural selection principle, developed by Darwin in his famous book from 1859 *"Origin of species"*, (76), full title *"On the Origin of Species by Means of Natural Selection, or the Preservation of Favoured Races in the Struggle for Life"*, which states that the evolution is based on natural selection, and in the Mendel's genetics, who has revealed that the hereditary factors which are transferred from parents to children have a discreet nature (*"Versuche uber Pflanzenhybride"* -*"Research about plant hybrids/Experiments in plant hybridization"*, paper presented on February 8th and March 8th, 1865 at Brunn Natural History Society).

The *genetic algorithms* (GAs) represent an identification technique of approximating solutions for optimization and search problems, being a particular class of *evolutionary algorithms* (EAs). They are placed in the wider domain of the *global search heuristics*, and can be viewed, in principle, as:

- problem solvers;
- competent basis for machine learning;
- computational models for innovation and creativity;
- computational philosophy.

GA, considered as the most popular case of EAs, represents a metaheuristic optimization algorithm, based on a population of potential solutions and using specific mechanisms inspired by the biological evolution (natural genetics), such as: individuals (chromosomes), reproduction, mutation, recombination, selection, survival of the fittest. We recall here some basic terms related to the biological evolution, some of which being found in the EAs vocabulary:

- *Chromosomes*, representing the carriers of genetic information. Chromosomes are linear structures with the components represented by *genes*, which control the hereditary characteristics of parents (functional units of heredity, encoding phenotypic characteristics).
- *Phenotype*, representing all the observable characteristics of an organism, such as shape, size, color, and behavior, that result from the interaction of its genotype (total genetic inheritance) with the environment. The common type of a group of physically similar organisms is sometimes also known as the phenotype (Encyclopedia Britannica. 2010. Encyclopedia Britannica Online. 17 Jun. 2010/ http://www.britannica.com/EBchecked/topic/455632/phenotype). In EAs terminology, the meaning of a particular chromosome is seen as its phenotype.
- *Genotype*, representing the genetic constitution of an organism. The genotype determines the hereditary potentials and limitations of an individual, comprising the entire complex of genes inherited from both parents (Encyclopedia Britannica. 2010. Encyclopedia Britannica Online. 17 Jun. 2010/http://www.britannica.com/ EBchecked/topic/229258/genotype). In EAs terminology, a single chromosome is seen as a genotype, representing a potential solution to a problem.
- *Natural selection* is the process that results in the adaptation of an organism to its environment by means of selectively reproducing changes in its genotype, or genetic constitution. Evolution often occurs as a consequence of this process. Darwin saw natural selection as the mechanism by which advantageous variations were passed on to later generations and less advantageous traits gradually disappeared (Encyclopedia Britannica. 2010. Encyclopedia Britannica Online. 17 Jun. 2010/ http://www.britannica.com/EBchecked/topic/406351/natural-selection).
- *Evolution* states that animals and plants have their origin in other preexisting types, and that the distinguishable differences are due to modifications in successive generations, representing one of the keystones of modern biological theory (Encyclopedia Britannica. 2010. Encyclopedia Britannica Online. 17 Jun. 2010/http://www.britannica.com/EBchecked/topic/197367/evolution).

The idea behind EAs is simple, regardless of their type: given a population of individuals, under the environmental pressure, which can only support a limited number of them, a selection will be involved, which in turn will involve an increase in adaptation of the selected individuals to the environmental conditions. Given an evaluation (performance) function which has to be maximized, one randomly creates a set of candidate solutions (i.e., elements belonging to the domain of the function), and then one applies the evaluation function as an abstract *fitness* measure. Further, based on this adequacy (the highest adequacy = the best individual = the fittest), one selects some of the best candidates to obtain a new generation, by applying the recombination/crossover and/or the mutation. Recombination/crossover is an operator applied to two selected candidates (parent chromosomes), forming two similar offspring, by swapping corresponding parts of the parents. Mutation applies to a single candidate, arbitrarily altering one or more parts, resulting in a new candidate. Based on these mechanisms of biological origin, one obtains a set of new candidates ready to compete with the old ones to form the new generation, i.e., *survival of the fittest in the struggle for life* - see Charles Darwin/Herbert Spencer. This process of biological inspiration will iterate until 'good enough' individuals will be obtained, or a STOP condition will be met.

EAs are of several types, although all are based on the idea exposed above, the differences consisting of certain technical details concerning their implementation, the history of occurrence and type of problems to which they apply. We listed below the main types of EAs.

- *Genetic algorithms*, representing, as we mentioned before, the most popular type of EAs. In the GAs case, one seeks the candidate solution of a problem in the form of a finite string of numbers (i.e., sequence over a finite alphabet). Traditionally, the GAs components are binary of 0s and 1s, although the best representations are usually those that reflect something about the problem being solved. This metaheuristic technique is routinely used to generate useful solutions to optimization and search problems, recombination representing a primary variation operator, mutation representing the secondary variation operator, 'parents' being randomly selected, biased according to fitness, while the 'survivors' selection is made using different methods (e.g., age-based replacement, random replacement, fitness-based replacement, elitist replacement, replacing the worst, etc.) -see, for instance, (129), (130), (131), (260), (268), (180).
- *Evolutionary programming* (EP), invented by Lawrence J. Fogel (1960) while serving at the National Science Foundation (NSF), was initially based on experiments in which *finite state machines* (FSMs) represented individual organisms in a population of problem solvers. EP was then extended to use arbitrary data representations and be applied to generalized optimization problems see, for instance, (80), (97).
- *Genetic programming* (GP) evolves computer programs, traditionally represented in memory as tree structures. Specifically, GP iteratively transforms a population of computer programs into a new generation of programs by applying analogs of naturally occurring genetic operations. The genetic operations include crossover, mutation, reproduction, gene duplication, and gene deletion -see, for instance, (222), (21), (227).
- *Evolution strategy* (ES) uses natural problem-dependent representations, and primarily mutation and selection as search operators. Note that the operators are applied in a loop, and an iteration of the loop is called a *generation* -see, for instance, (313), (337), (33).
- *Learning classifier systems* (LCS), introduced by Holland (1976), are rule-based systems, where the rules are usually in the traditional form of *"IF state THEN action"*. They are close to reinforcement learning and GAs. LCS is a broadly-applicable adaptive system that learns from external reinforcement and through

an internal structural evolution derived from that reinforcement. LCS show great promise in the area of data mining, modeling and optimization, control, etc. Examples of LCS applications: *Armano* - NXCS experts for financial time series forecasting, *Smith* - fighter aircraft LCS, *Hercog* - traffic balance using classifier systems in an agent-based simulation, *Takadama* - exploring organizational-learning orientated classifier system in real-world problems, etc. -see, for instance, (188), (46), (47).

As it has been seen above, there are two fundamental components in this area underlying an evolutionary system:

1. Variation operators (recombination and mutation);
2. The selection process.

The general scheme of an EA (pseudo-code) can be summarized in the following algorithm, (97):

BEGIN

    INITIALISE population with random candidate solutions;

    EVALUATE each candidate;

    REPEAT UNTIL (*TERMINATION CONDITION* is satisfied) DO

        1. *SELECT* parents;

        2. *RECOMBINE* pairs of parents;

        3. *MUTATE* the resulting offspring;

        4. *EVALUATE* new candidates;

        5. *SELECT* individuals for the next generation;

    OD

END

We finish this brief introduction to the EAs field, citing the following particularly suggestive assertion: *"Computer programs that "evolve" in ways that resemble natural selection can solve complex problems even their creators do not fully understand"* (John H. Holland - http://econ2.econ.iastate.edu/tesfatsi/holland.GAIntro.htm).

5.9.1   Components of GAs

We will briefly present the main components of the GAs architecture, which are obviously the same for EAs, and which we already mentioned in the introduction above:

- *Representation* (definition of individuals);
- *Evaluation (fitness)* function;

- *Population*:

  - Parent selection mechanism;
  - Survivor selection mechanism.

- *Variation operators* (recombination and mutation);
- *Parameters* that GAs use (population size, probability of applying variation operators, etc.).

Note that each of these components has to be specified, together with the initialization and termination procedures.

A. *Representation* (*definition of individuals*)

We begin by defining the GA 'universe', based on the dual entity *solution/chromosome*, seen as the consequence of the initial duality, given by "real-world/GA world". Thus, in the context of the real-world problems to be solved, the *candidate solution* (possible solution) of the original problem is equivalent to the *phenotype*, representing a point in the space of possible solutions of the original problem, i.e., the phenotype space. On the other hand, in the GA 'world', a possible/candidate solution (an individual $\sim$ the encoding of the original object), is called *chromosome*, equivalent to the *genotype*, representing a point in the evolutionary search space, i.e., the genotype space. The term '*representation*' refers, in principle, to the mapping of the phenotype onto the genotype (encoding process), or to the reverse decoding process, although it may relate to the structure of the data from the genotype space. For instance, if we have to find the maximum of a given function using GA, and we decide to represent the possible solutions in the binary code, then a phenotype may be the number 54, while the corresponding genotype is 0110110.

The first type of representing chromosomes in GA is the *binary representation* (i.e., the genotype is given by a bit string/array of bits/binary vector, e.g., 10011, 11000, 10000, etc.), an example of practical application being the Boolean decision problems. Although it is the simplest possible representation of chromosomes, it can effectively cover only a narrow range of issues. Therefore, the next step up on the 'stair' of representation types is the *integer representation*, covering an area much wider of applications, such as finding optimal values for variables with integer values (e.g., categorical values, image processing parameters, etc.). In this case, the chromosomes are represented by vectors with integer components (e.g., (1,2,3,4), (14,23,15,7)). However, many real-world problems occur as real valued problems, and, in this case, the genotype of the candidate solutions will be naturally represented by vectors of real components $\mathbf{x} = (x_1, x_2, ..., x_p)$, each gene x_k being a real number (e.g., (0.1, 2.3, 4.1, 7.5), (2.5, 1.3, 3.0, 4.7), etc.). This is called the *floating point representation* (or the *real-coded representation*). In a case where we should decide the order in which the sequence of events may occur, it is natural to choose the representation given by permutations of integers (*permutation representation*), where, unlike the integer representation, a number cannot appear more than once (e.g., (1,2,3,4), (2,1,4,3), etc.). We should note here that the permutations choice depends on the problem, that is there are problems where there is an intrinsic order, and there are other situations where the order is not so important. For example, if we consider the problem of optimizing a production process, and a chromosome contains four genes, which represent the components of a product, then the order in which they are built is important for the final assembly, so we have different representations for chromosomes (1,2,3,4) or (2,1,4,3). In other cases the order may not be so important. For instance, in the well-known *Traveling Salesman Problem* (TSP), we can consider that the starting point is not important, so sequences of cities such as (C,A,B,D) and (A,D,B,C) are equivalent.

Initialization is a simple process within the GAs, by which we practically create an *initial* population of chromosomes, each having a binary/integer/floating point, etc. representation. Usually, the initial (first) population is randomly seeded, observing that some heuristics can be chosen at this step, in order to better fit to the problem being solved - solutions (individuals) chosen in areas where optimal solutions are likely to occur, for instance -see also (97). The population size depends on the nature of the problem, ranging from several hundred to several thousand chromosomes (possible solutions).

If the initialization is the beginning of building a GA, the end of the process is determined by the *termination condition* (STOP condition). In principle, one can consider two termination conditions:

1. The case, very rare, when the problem has a known optimal fitness level (coming, for instance, from a known optimum of the given objective function), in which case the reaching of this level will stop the searching process.
2. The usual case, based on the stochastic nature of GAs, implying no guarantees to reach an optimum, in which case one can choose different termination conditions, determined by the following considerations:

   • The time allowed for the computer to run expires;
   • The total number of fitness evaluations reaches a given (pre-determined) level;
   • The fitness improvement is limited for a given period of time (or number of generations), by a certain level (flat improvement curve);
   • The population diversity is limited by a given threshold;
   • The combination of the above.

In conclusion, the completion of the searching process is obtained by either the reaching of an optimum (if known, which is unfortunately very unlikely) or a STOP condition is satisfied.

B. *Evaluation (fitness) function*

Once the representation of solutions chosen, and the initial population defined, we have to handle the issue of evaluating the individuals' quality. In natural (biological) conditions, this role is assumed by the environment, which 'selects' the best individuals for survival and reproduction (*natural selection principle* - Darwin: "the best ones should survive and create new offspring"). In the GAs 'universe', this role is assumed by the *evaluation function* (*fitness function*), which can be seen as an

objective function (for the original problem), serving to quantify the degree of optimality of a solution (chromosome), thus defining the requirements for adaptation
to the 'environment'. It represents the basis for selection, generating therefore subsequent 'improvements' of the individuals. For this reason, it is always problem dependent. Because in many problems it is difficult or even impossible to define such
a function, the related field of *interactive genetic algorithms* has been developed,
which uses the human evaluation (*aesthetic selection*) too (e.g., colors, images, music, etc. - probably rooted in the famous assertion "*De gustibus et coloribus non est
disputandum - There is no arguing about tastes and colors*"). Let us note that the
evaluation function is nonnegative and, assuming that the population contains n individuals x_1, x_2,..., x_n, one can define the *overall performance* (*total fitness*) by the
formula $F = \sum_{i=1}^{n} f(x_i)$. Let us also mention that the speed of execution of GAs is
highly correlated with the fitness function, due to the fact that a GA must be iterated
many times to obtain good enough solutions, and so the choice of the fitness function represents a difficult task in real-world applications. Thus, there are cases when
a fixed fitness function is used, while there are cases when it changes during the
search process. We finally remark that the choice of an effective fitness function becomes difficult in the *multi-criteria/multi-objective optimization* problems, in which
one can consider, for example, the *Pareto optimality*, or in optimization problems
with constraints, in which one considers, for example, *penalty functions*.

Very often, the problems solved by using GAs refer to optimization problems
(e.g., finding the extrema (minima/maxima) of a function in a given domain), case in
which the fitness function is the same as the objective function. We illustrate below
this situation by two very simple classical examples - see, for instance, (129), (260).
In the first case, the objective function to be maximized is given by $f : [0, 31] \rightarrow \mathbf{R}_+$,
$f(x) = x^2$ (a parabola). We use a 5-bit binary representation (e.g., for $x_1 = 13$, the
corresponding representation is 01101, $x_2 = 8$ is equivalent to 01000, while $x_3 =$
19 is represented by 10011), naturally choosing the objective function as evaluation
function. Obviously, from the three above chromosomes, x_3 is chosen as the best
solution, since the function f reached a local maximum for it (see Fig. 5.56).

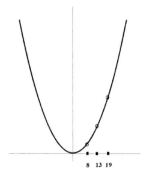

8 13 19

Fig. 5.56 Graph of the function $f(x) = x^2$, with three chromosomes

In the second case, the objective function is given by $f : [-1,2] \rightarrow \mathbf{R}$, $f(x) = x \cdot \sin(10 \cdot \pi \cdot x) + 1$, which has to be maximized as well. We consider here a 22-bit binary representation, and choose, for instance, three chromosomes: $x_1 = 0.637197 \sim 1000101110110101000111$, $x_2 = -0.958973 \sim 0000001110000000001000$, and $x_3 = 1.627888 \sim 1110000000111111000101$ (for technical details related to this real numbers/binary conversion see (260)). After evaluating the three chromosomes above, the best has proved to be x_3, since we obtain the highest value of f for it (see Fig. 5.57).

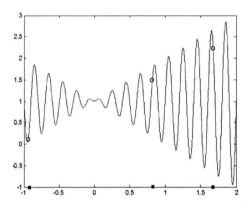

Fig. 5.57 Graph of the function $f(x) = x \cdot \sin(10 \cdot \pi \cdot x) + 1$, with three chromosomes

C. *Population*

The role of the *population* is to hold the representation of all possible solutions, being a set of multiple copies of genotypes. Thus, while the individuals are static objects, not changing in time, the population evolves, representing the core of the evolution process. The diversity found in a population illustrates the number of different solutions present at a time, different measures of this diversity being considered (e.g., number of different fitness values, number of different phenotypes/genotypes, entropy, etc.).

The process of defining a population can be very simple - just specifying the number of individuals in its composition (i.e., population size) -or more complicated- defining a spatial structure, related to a distance, or defining a neighborhood relation between its members.

While the variation operators act on the one or two individuals in the population (mutation/recombination), the selection operators (parents and survivors selection) act on the population, taking into account the entire current population, the selection (choice) being made relative to the existing individuals at that moment (e.g., usually the best individuals are selected for giving birth to a new generation, while the worst individuals are selected for replacement).

A large population is useful for a better exploration of the space of all possible solutions, especially when we are more interested in the global optimum than in local optima. At the same time, a large population implies larger computational costs (runtime and memory). Let us mention here that, usually, the population size is constant during the evolutionary process, a choice of 100 individuals being frequent in many cases.

C_1. Parents selection mechanism

One of the main aspects regarding the *'evolution'* concept which characterizes GAs refers to the *parent/mating* selection process, allowing the best individuals from the current population to generate a new population. By *'parent'* we mean an individual selected from the current population in order to be transformed (by the variation operators) with the aim of creating offspring (the new generation). Together with the survivors' selection mechanism, representing the other aspect of the overall selection process, the parents' selection represents the 'engine' of the individuals' evolution from one generation to another. In principle, the parents' selection is typically probabilistic, allowing the best individuals to be involved in producing the next generation. Without going into details, we further present some of the most popular types of parents' selection:

- *Fitness proportional selection* -FPS, (189), represents a selection strategy that considers an individual's survival probability as a function of its fitness score. The simplest form of FPS is known as the *roulette-wheel selection*, because this technique is somehow similar to a roulette wheel in a casino, each chromosome in the population occupying an area on the roulette wheel proportional to its fitness. Conceptually, each member of the population has a section of an imaginary roulette wheel allocated, but, unlike a real roulette wheel in a casino, the sections have different sizes, proportional to each individual's fitness. The wheel is then spun and the individual associated with the winning section is selected. In this case, the probability that an individual $\mathbf{x}_i$ will be selected is given by $p_i = \dfrac{f(\mathbf{x}_i)}{F}$. Fig. 5.58 illustrates this selection procedure.

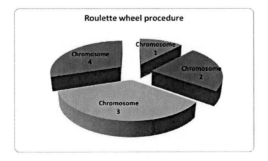

Fig. 5.58 Illustration of the roulette ('exploded') wheel procedure

The pseudo-code for the (Monte Carlo) roulette wheel algorithm is given below, (97). The algorithm is applied to select k individuals from a set of n parents. Assuming some order over the population, $[\mathbf{x}_1, \mathbf{x}_2, ..., \mathbf{x}_n]$, a list of values $[a_1, a_2, ..., a_n]$ are computed by using the formula $a_i = \sum_1^i p_j$, where p_j is defined by the selection distribution.

```
BEGIN
    set current_member = 1;
    WHILE (current_member ≤ n) DO
        Pick a random value r uniformly from [0, 1];
        set i = 1;
        WHILE (a_i < r) DO
            set i = i + 1;
        OD
        set mating_pool [current_member] = parents [i];
        set current_member = current_member + 1;
    OD
END
```

Note that the roulette wheel algorithm does not give a particularly good sample of the required distribution. Thus, for more than one sample that has to be drawn from the (uniform) distribution, the use of of the *stochastic universal sampling* -SUS, (20), is preferable. Thus, SUS is a development of the fitness proportional selection, which exhibits no bias and less variance than repeated calls to the roulette wheel algorithm. Basically, the idea is to make a single draw from the (uniform) distribution, and use this to determine how many offspring to assign to all parents. The pseudo-code for SUS is given below, (97).

```
BEGIN
    set current_member = i = 1;
    Pick a random value r uniformly from [0, 1/n];
    WHILE (current_member ≤ n) DO
        WHILE (r < a_i) DO
            set mating_pool [current_member] = parents [i];
            set r = r + 1/n;
            set current_member = current_member + 1;
        OD
```

```
        set i = i + 1;
    OD
END
```

- *Rank-based selection* (*ranking selection*), in which one first ranks the population and then every chromosome receives fitness from this ranking. Thus, each chromosome is selected based on a probability allocated in accordance with its rank. The way of 'converting' the individual rank from the overall hierarchy into the probability of selection is arbitrary, and can be achieved, for example, by linear or exponential ranking. Note that the ranking selection tends to avoid premature convergence by tempering selection pressure for large fitness differentials that occur in early generations;

- *Tournament selection* is especially used in cases where it is difficult to obtain information about the entire population (e.g., the case of very large population). In this case, the tournament type selection considers the competitors only (hence the term 'tournament', i.e., selection through competition or contest). Thus, no comprehensive information about the population is needed, but just the direct comparisons of any two chromosomes (binary tournament) and the selection of the best 'knight'. In general, however, one can choose a larger number of competitors at once, for example k competitors, thus defining the tournament size. Note that the tournament selection is efficient to code, works on parallel architectures, and allows the selection pressure to be easily adjusted. A pseudo-code for the selection of n parents has the following form (97).

```
BEGIN
    set current_member = 1;
    WHILE (current_member ≤ n) DO
        Pick k individuals randomly, with or without replacement;
        Select the best of these k comparing their fitness values;
        Denote this individual as i;
        set mating_pool [current_member] = i;
        set current_member = current_member + 1;
    OD
END
```

The probability p of selecting the most efficient competitor of the tournament may be chosen deterministically, namely $p = 1$ (*deterministic tournament*), but there are probabilistic versions with $p < 1$ (*stochastic tournament*). For other details concerning the tournament selection see, for instance, (264).

C_2. Survivor selection mechanism

The survivor selection mechanism (replacement), also known as the environment-based selection, is the process by which individuals can be distinguished based on their quality. Basically, this selection mechanism is similar to the parent selection, but it operates on another level of the evolutionary cycle. Thus, after selecting parents and obtaining their descendants, we need a mechanism for selecting those individuals who will breed the next generation, because the whole population remains, in principle, constant during the evolutionary process. The decision in this regard is done through a fitness-based process involving, in essence, each individual, usually favoring those with higher qualities, although the concept of *age* is often considered. Note that, unlike the mechanism of parent selection, which is typically stochastic, the mechanism of survivor selection is sometimes deterministic.

Next, we present some classical strategies for selecting the survivors, proposed over time.

- *Age-based strategy (replacement)*, used in the simple GA schemes, is based on the idea that, instead of using the fitness of each individual, one can consider its survivorship in the system for a while. In the classical case when the number of parents n is the same with the number of offspring m, each individual survives for just one cycle, the parents being eliminated and replaced by their offspring. Note that this strategy can be also implemented when the number of offspring m is less than the number of parents n, or in the extreme case when a single offspring is created and introduced in the population in each cycle. There is also the variant of randomly selecting a parent for replacement (for details, see (97)).
- *Fitness-based strategy (replacement)* refers to the selection techniques involving the individuals' assessment, by which n elements are chosen of the total population of $n + m$ parents and offspring, giving birth to the next generation, the methods mentioned in Section C_1 remaining valid in this case.

Remark 5.30. 1) When addressing the concept of GA, one first talks about the way of representing potential solutions in order to obtain a population of individuals, which will change over time, producing offspring that will inherit some of their characteristics, differing however more or less from them. Thus, new potential solutions will be subjected to the evaluation process. Secondly, one talks about the survival of the competitors in the selection process, enabling the 'best' of them for reproduction, yielding offspring for the next generation. For the latter aspect, we mention two population models known in literature.

1. *Generational model*, in which each generation starts with a population of size n, from which a *mating pool* of n parents is selected. Next, by applying the variation operators, a same number $m = n$ offspring is created and then evaluated. After each cycle, the whole population is replaced by its offspring, forming thus the *"next generation"*, hence the model name.
2. *Steady-state model*, in which, unlike the previous one, the entire population is no longer replaced at once, but a number of m ($< n$) old individuals are replaced by m new individuals, namely the offspring; the replacement rate $r = m/n$ of

the population is called *generational gap*. We mention here the algorithm GENI-TOR, developed by Whitley, (399), (400). This algorithm is specifically designed to allocate reproductive trials according to rank and the worst m members of the population are selected for replacement (usually by choosing $m = 1$).

2) In this context, we also mention the *elitist strategy* (*elitism*), where the best chromosome (or a few best chromosomes) is copied to the population in the next generation, the rest are being chosen in the classical way. Note that elitism can very rapidly increase the performance of GA, because it prevents losing the best found solution so far. This strategy is usually applied in conjunction with age-based and stochastic fitness-based strategies.

D. *Variation operators (recombination, mutation)*

The role of variation operators is to create new individuals, based on those already existing. After selecting, in one way or another, the individuals who will give birth to a new generation, the next step in the evolutionary cycle consists in the specific mechanisms by which this is achieved. For each offspring who will be created, we need either a pair of parents having been previously selected for 'reproduction' or a single individual selected for undergoing a change in its structure (i.e., a mutation). A new descendant/offspring (i.e., a new solution), whether obtained from two parents through the recombination operator, or from a single individual through the mutation operator, will 'inherit' some of the characteristics of their ancestors, being more or less different than them. Afterwards, new parents will be selected, and in this way new generations are produced, so that the average performance will be improved from one generation to another, because by the selection mechanism the 'best' individuals are allowed to reproduce.

Let us recall that the variation operators depend on the chromosome representation. During the years a rich literature has appeared dealing with this fundamental issue in the GAs field. Here we mainly used for this brief presentation (97) and (19).

D_1. *Recombination (crossover)*

In the GAs field, the *recombination/crossover* operator is a binary (stochastic) variation operator, which is usually applied to two parents (chromosomes) to produce one or two offspring which will inherit combined characteristics from both parents. It is considered by many as the most important way to create diversity, being thus placed ahead of mutation. The principle behind it, based on the biological evolutionary metaphor, is simple, stating that by mating two individuals with different but good characteristics (for the proposed goal), their offspring will likely inherit traits at least as good, obtained by combining their features in a successful manner. The recombination operator is, in general, a stochastic operator, applied according to a *crossover!rate* p_c, usually chosen in the interval [0.5, 1], thus determining the chance that the crossover operator will be applied to a selected pair (of parents). The recombination mechanism is simple: in principle, two parents are selected at the same time with a random number from the interval [0, 1), which is compared with the crossover rate p_c. Afterwards, if the random number is lower than p_c, then two offspring will be created from the two parents by crossover, otherwise they will be created by simply copying the parents (i.e., asexually). Thus, the descendants will be new individuals, with characteristics inherited from parents by recombination, or they will be *clones* (of their parents). Note that, generally speaking, one can choose p parents to give birth to q offspring, but the crossing scheme $p = 2$, $q = 2$ is however predominant.

We briefly present below some of the most popular recombination methods, depending on the way in which the chromosome representation has been chosen, since the recombination operators are representation dependent.

- *Recombination operators for binary representation*, basically used for the standard recombination scheme $p = q = 2$, although generalizations have been proposed:

  - *One-point crossover*, (189), (79), in which one chooses a random number from the set $\{1, 2, ..., l - 1\}$, where l is the length of a chromosome, and the two parents (i.e., the corresponding chromosome sequences) are split at this point, thus creating the two offspring by exchanging the corresponding segments obtained by this operation. We illustrate this process, considering the parents (chromosomes): $(a, a, a, |b, b, a, b)$ and $(b, b, a, |b, a, a, b)$, and $k = 3$. The corresponding offspring in this case have the following structure: $(a, a, a, |b, a, a, b)$ and $(b, b, a, |b, b, a, b)$.

  - *N-point crossover*, seen as a natural generalization of the above schema, in which the chromosome representation (sequence) is split into more than two contiguous segments. Basically, N random numbers are chosen from the set $\{1, 2, ..., l - 1\}$, the parents are split at these points, and the offspring are obtained by alternatively coupling segments from the two parents. For example, starting from the two above chromosomes, and choosing $N = 2$ crossover points $k_1 = 3$ and $k_2 = 5$, we can get the next two offspring: $(a, a, a, |b, a, |a, b)$ and $(b, b, a, |b, b, |a, b)$.

  - *Uniform crossover*, (2), (377), which, unlike the previous ones, that split the parents into one or more separate parts, treats each gene individually, randomly choosing from which parent it will be inherited. Concretely, one generates a sequence of l random numbers (recall that l is the encoding length), uniformly distributed on the interval $[0, 1]$. Thereafter, for the first offspring and for each position of the encoding sequence, the corresponding value is compared with a fixed parameter p (usually equal to 0.5), and if the value is below p the gene is inherited from the first parent, otherwise from the second parent. The second offspring will be naturally created by using the inverse mapping (i.e., one chooses the corresponding value from the other parent). For instance, for the previous parents (chromosomes), let us consider the random sequence $[0.47, 0.53, 0.64, 0.27, 0.31, 0.78, 0.83]$ drawn uniformly from $[0, 1]$, and $p = 0.5$. Then, the first offspring is given by $(a, |b, a, |b, b, |a, b)$, while the second by $(b, |a, a, |b, a, |a, b)$.

- *Recombination operators for integer representation* are usually the same as for binary representation.
- *Recombination operators for floating-point representation* may be either those presented above for binary/integer representations, known in this case as *discrete-recombination*, or may be the class of *arithmetic (intermediate) recombination*. Concretely, in the discrete-recombination case, for two parents **x** and **y**, an offspring **z** will have the value for the ith gene either x_i or y_i with equal likelihood. On the other hand, in the arithmetic recombination case, an offspring **z** will have the value for the ith gene given by the convex combination, with $z_i = \alpha \cdot x_i + (1 - \alpha) \cdot y_i$, $\alpha \in [0, 1]$. Parameter α is usually 0.5 (intrinsic value/*uniform arithmetic recombination*), i.e., **z** is the middle point of the line segment $[\mathbf{x}, \mathbf{y}]$, although, theoretically, it should be randomly chosen in the interval $[0, 1]$. We briefly present below three types of such arithmetic recombination -see also (260).

  - *Simple arithmetic recombination*, in which one chooses a crossover point k from the set $\{1, 2, ..., l-1\}$, and the offspring are created as follows: the first 'child' will take the first k genes from the first parent, the remaining values being computed using the formula $z_i = \alpha \cdot x_i + (1 - \alpha) \cdot y_i$, $i = k+1, k+2, ..., l$, while the second 'child' is created in the same way, but with parents reversed. For illustration, let $\mathbf{x} = (1.2, 2.5, 3.4, 4.5)$ and $\mathbf{y} = (0.3, 2.1, 7.6, 0.5)$ be the parents, and $k = 2$, $\alpha = 0.5$. Then, the two offspring are given by: $(1.2, 2.5, |5.5, 2.5)$, $(0.3, 2.1, |5.5, 2.5)$.
  - *Single arithmetic recombination*, in which one chooses a crossover point k in the set $(1, 2, ..., l-1)$, and offspring are constructed as follows: the first 'child' will have the first $(k-1)$ genes inherited from the first parent, the kth gene is computed using the formula: $z_k = \alpha \cdot x_k + (1 - \alpha) \cdot y_k$, and the remaining genes are again inherited from the first parent, while the second 'child' is obtained similarly, but choosing the parents in reverse order. Considering again the previous case, we get: $(1.2, |2.3, |3.4, 4.5)$, $(0.3, |2.3, |7.6, 0.5)$.
  - *Whole arithmetic recombination*, representing the most commonly used recombination operator, in which the replacement of genes is complete (complete change), made using the following formulas: $z_i = \alpha \cdot x_i + (1 - \alpha) \cdot y_i$, $i = 1, 2, ..., l$, for the first 'child', and $z_i' = \alpha \cdot y_i + (1 - \alpha) \cdot x_i$, $i = 1, 2, ..., l$, for the second 'child'. For the previous case, we obtain: $(0.75, 2.3, 5.5, 2.5)$, $(0.75, 2.3, 5.5, 2.5)$.

- *Recombination operators for permutation representation* are a more special case, because here a certain order matters, in principle, therefore it is not indicated to use the arbitrary mix of genes, the basic idea being to preserve in offspring the common information held by both parents. The most common methods used in this case are the following -for more details, see (97).

  - *Partially mapped crossover* -PMX, proposed by Goldberg and Lingle, (130), to solve the TSP problem, has been thereafter used for *adjacency-type problems*. In principle, PMX transmits information regarding both the order and the values from parents to offspring. A portion of one parent's string is mapped onto a portion of the other parent's string (i.e., *partial mapped*-PM), and the remaining information is exchanged. A version of PMX, due to Whitley, (405), (97), is presented here: (*i*) choose two crossover points at random and copy the segment between them from the first parent into the first offspring; (*ii*) starting from the first crossover point look for elements in that segment of the second parent that have not been copied; (*iii*) for each of these elements (e.g., i) look in the offspring to see what element (e.g., j) has been copied in its place from the first parent; (*iv*) place i into the position occupied by j in the second parent; (*v*) if the place occupied by j in the second parent has already been filled in the offspring by an element k, put i in the position occupied by k in the second parent; (*vi*) having dealt with the elements from the crossover segment, the rest of the offspring can be filled from the second parent, and the second offspring is created analogously with the parental roles reversed. A slight variation of this scheme is illustrated below. Thus, let us consider, for instance, the parents $\mathbf{x} = (1,2,3|5,4,6,7|8,9)$, and $\mathbf{y} = (4,5,2|1,8,7,6|9,3)$, in which figures may represent labels of cities for the TSP case. Two crossover points are chosen at random, say 3 and 7 (marked with vertical bars). First, we obtain the following *proto*-offspring: $(*,*,*,1,8,7,6,*,*)$ and $(*,*,*,5,4,6,7,*,*)$, copying the sequence between the crossover points from the first parent into the second offspring, and the sequence from the second parent into the first offspring, obtaining at the same time the correspondence between the elements of the two sequences, given by: $1 \sim 5$, $8 \sim 4$, $7 \sim 6$ and $6 \sim 7$. Then, we copy into the two offspring the elements of the two parents in the vacant places, being careful not to repeat the values. If a value is repeated, it will be replaced with the value provided by the above correspondence. In our case, we obtain: $(*,2,3,1,8,7,6,*,9)$ for the first offspring, and $(*,*,2,5,4,6,7,9,3)$ for the second offspring, here the sign '$*$' indicating that an automatic copy would have generated an overlapping (e.g., value 1 for the first offspring, copied in the first place, would have repeated the existing value 1). Thus, following the above rule, since $1 \sim 5$, we copy 5 on the first place, and regarding the sign '$*$' on the 8-th place, instead of copying 8 we copy 4, because we have the correspondence $8 \sim 4$. Finally, the two offspring are: $(5,2,3,1,8,7,6,4,9)$ and $(8,1,2,5,4,6,7,9,3)$.

– *Edge crossover* -EX, proposed by Whitley et al., (402), (403), to solve TSP, is a kind of crossover operator that considers two chromosomes as parents in order to produce one child. The aim of EX is to inherit as much edges as possible from parents to child. Here, the creation of progeny is based on considering, as far as possible, only edges that are present in one or more parents. To accomplish this task, one uses the so-called *edge/adjacency table*, which contains, for each element of the permutation, the list of the other items connected to it in the two parents (i.e., its 'neighbors'). Let us mention that improved variants of this procedure have appeared over the years, (248), (93), (277), but the most commonly used version is the so-called *edge-3 crossover*, described below, (405), (97). Thus, (*i*) construct edge/adjacency table; (*ii*) pick an initial element at random and put it in the offspring; (*iii*) set the variable *current_element = entry*; (*iv*) remove all references to *current_element* from the table; (*v*) examine list for *current_element*: (*a*) if there is a common edge (indicated by the sign '+'), pick that to be next element; (*b*) otherwise pick the entry in the list which itself has the shortest list; (*c*) ties are split at random; (*vi*) in the case of reaching an empty list, the other end of the offspring is examined for extension, otherwise a new element is chosen at random. To better understand this algorithm, we present the following example regarding TSP. Thus, let the two parents be $\mathbf{x} = (g, d, m, h, b, j, f, i, a, k, e, c)$ and $\mathbf{y} = (c, e, k, a, g, b, h, i, j, f, m, d)$, where the letters a, b,..., m, represent 12 cities taken into consideration. In this case, the edge/adjacency table is shown below (recall that '+' indicates a common edge).

Table 5.20 Edge/adjacency table

City	Edges	City	Edges
$a \rightarrow$	$+k, g, i$	$g \rightarrow$	a, b, c, d
$b \rightarrow$	$+h, g, j$	$h \rightarrow$	$+b, i, m$
$c \rightarrow$	$+e, d, g$	$i \rightarrow$	h, j, a, f
$d \rightarrow$	$+m, g, c$	$j \rightarrow$	$+f, i, b$
$e \rightarrow$	$+k, +c$	$k \rightarrow$	$+e, +a$
$f \rightarrow$	$+j, m, i$	$m \rightarrow$	$+d, f, h$

One choose at random, say city 'a', as the first element of the sequence. Thereafter, city 'k' is chosen, since edges $\{a, k\}$ appear in both parents. Next, the city 'e' is chosen from the list of edges of 'k', because it is the only one remaining in this list. We repeat the procedure, thus obtaining the partial sequence (a, k, e, c), noting that from this time we have no longer a deterministic choice to continue expanding the sequence. City 'c' has as edges the cities 'd' and 'g', each of them having two unused edges. We will randomly choose city 'd', for instance, to continue the process of building the offspring. The deterministic construction continues until position 7, thus obtaining the partial sequence (a, k, e, c, d, m). At this position we have to choose again at random, between 'f' and 'h'. We choose, for example, the city 'h'. Next, following the deterministic procedure, we obtain the sequence $(a, k, e, c, d, m, h, b, g)$, point from which we can no longer continue like this, because there are no remaining edges in the list of 'g' (situation called '*recombination failure*' in the edge-3 crossover). Thus, when a failure occurs at both edges of a partial sequence (sub-tour), edge-3 crossover starts a new partial sequence, using the so-called *terminal*. Assume, as in our case, that the partial sequence that has been constructed so far has edges lacking a live terminal by which to continue (a terminal is considered as a *live terminal* if there are still entries in its list, otherwise it is called a *dead terminal*). Since only one sequence has been constructed, and since every city has initially at least two edges in the edge/adjacency table, there must be edges inside the current sequence, representing possible edges to the terminal cities of the partial sequence. Returning to our case, we will consider the element 'a' (which also represented the starting point), which still has an unused entry in its list. The procedure chosen in this case is to reverse the partial sequence, in order to continue with 'a'. Starting from the reversed partial sequence (g,b,h,m,d,c,e,k,a), we add the city 'i' remaining in the list of 'a'. Continuing in the usual way, and no longer having failures, we obtain the desired offspring having the form $(g,b,h,m,d,c,e,k,a,i,f,j)$.

- *Order crossover* -OX, was proposed by Davis, (78), to solve order-based permutation problems. In this case, a randomly chosen segment of the first parent is copied into the offspring, but aiming at the same time to transmit information about the relative order from the second parent. We shortly present the main steps of this procedure, (97). First, choose two crossover points at random and copy the segment between them from the first parent into the first offspring. Second, starting from the second crossover point in the second parent, copy the remaining unused elements into the first offspring in the order that they appear in the second parent, treating this sequence as a torus (i.e., simplistically explained, like eating clockwise a (torus-shaped ring) doughnut, jumping portions already ate). Third, create the second offspring in an analogous way, with the parent roles reversed. We illustrate this recombination type by a TSP example, where $\mathbf{x} = (j,h,d,|e,f,i,|g,c,b,a)$ and $\mathbf{y} = (h,g,e,|b,c,j,|i,a,d,f)$, with the two crossover points 3 and 6. Thus, the corresponding sequence from the first parent which is copied in the first offspring is (e,f,i). First, we get the first offspring having the form $(*,*,*,e,f,i,*,*,*,*)$. Completing then "tail-to-head" the first offspring by copying unused elements from the second parent, the sequence of the first offspring is thus obtained $(b,c,j,|e,f,i,|a,d,h,g)$. The second offspring is given by $(e,f,i,|b,c,j,|g,a,h,d)$.

- *Cycle crossover* -CX, (285), which aims to preserve as much information as possible about the absolute position in which elements occur. Thus, the crossover is performed under the constraint that each element must come from one parent or the other, by transferring element cycles between parents, in this way the absolute positions of the elements of permutations are preserved. The main steps of this procedure, (285), (97), (19), are the following: (*i*) start with the first unused position and value of the first parent; (*ii*) look at the value in the same position in the second parent; (*iii*) go to the position with the same value in the first parent; (*iv*) add this value to the cycle; (*v*) repeat steps 2-4 until one arrives at the first value of the first parent. Basically, the process continues as long as we do not use a value already copied, which is the time for ending the cycle (started with the first value from the first parent). In this case, the remaining positions of the first offspring will be filled with values having the same positions in the second parent. We illustrate below this recombination operator by considering the parents $\mathbf{x} = (1,2,3,4,5,6,7,8,9)$ and $\mathbf{y} = (4,1,2,8,7,6,9,3,5)$, where figures may represent labels of cities in TSP. We start with the first value of $\mathbf{x}$, which is copied into the first offspring in the first position $(1,*,*,*,*,*,*,*,*)$, and we seek then the value in $\mathbf{y}$ in the same position (i.e., the first), that is 4, which will be copied into the offspring in the 4-th position $(1,*,*,4,*,*,*,*,*)$, because this is its position in $\mathbf{x}$. Next, in the same position (i.e., the 4-th) in $\mathbf{y}$ we find the value 8, so we have $(1,*,*,4,*,*,*,8,*)$, because in the first parent, 8 occupies the 8-th position. The process continues with $(1,*,3,4,*,*,*,8,*)$ and $(1,2,3,4,*,*,*,8,*)$. At this point, the selection of 2 will involve the selection of 1, which has already been copied, therefore we have to end the cycle. Therefore, we use the second parent for filling the free positions, so we obtain $(1,2,3,4,7,6,9,8,5)$. Similarly, starting from $\mathbf{y}$, one obtains $(4,1,2,8,5,6,7,3,9)$.

- *Multi-parent recombination* extends the variation operators using two parents (crossover operators), or one parent (mutation operators), by considering schemes involving 3, 4 or more parents, schemes which have no longer an equivalent in biology, but are mathematically feasible and, moreover, have proved effective in many cases. For more details, see (96), (97).

D_2. Mutation

Unlike the recombination (crossover) operators, which are binary operators (unless the multi-parent case), the mutation operators are unary operators, inspired by the biological mutation (i.e., an alteration in the genetic material of a cell of a living organism that is more or less permanent, and that can be transmitted to the cell's descendants), occupying the second place in the hierarchy of GAs variation operators. They apply to a single chromosome, resulting in a slightly modified individual, the 'mutant', also called offspring. A mutation operator is always stochastic, since the mutant depends on the outcomes obtained by random choices. Its role in the GA context (different from its role in other research areas met in evolutionary computing) is to prevent the situation that the population of chromosomes becomes too uniform (too similar chromosomes), implying thereby a slowing or even stopping of the evolution process, goal achieved by providing the so-called 'fresh blood' in the genes pool. From a technical point of view, the mutation operator alters one or more genes with a certain probability -the *mutation rate*.

It is worth noting that some types of EAs consider only mutation operators (without the use of crossover), while other algorithms may combine mutation and recombination.

Like in the case of recombination operators, mutation depends on the choice of the encoding used in the GAs. In the following, we briefly review the most popular mutation operators.

- *Mutation operator for binary representation* (*bitwise mutation*) considers, in principle, each gene separately, and allows each bit to flip from 0 to 1 or *vice versa*, with a small probability p_m. If the encoding length is l, then, because each bit has the same probability p_m of mutation, on average $l \cdot p_m$ values will be changed. An example of bitwise mutation is the following one. Consider the chromosome $(0,1,1,1,1)$ representing the binary encoding of the number 15. Assume that, generating five random numbers (in fact, *pseudo-random* numbers), one for each position, the 3rd and 5th random values are less than the mutation rate p_m. Then, $(0,1,\mathbf{1},1,\mathbf{1}) \rightarrow (0,1,\mathbf{0},1,\mathbf{0})$. The problem with this type of mutation is how to choose the mutation rate p_m in order to obtain a good result.

A concrete example of this mutation is the following one, (260). Thus, consider again the problem of maximizing the function $f : [-1,2] \rightarrow \mathbf{R}$, $f(x) = x \cdot \sin(10 \cdot \pi \cdot x) + 1$, by using GAs. Consider the chromosome $\mathbf{x} = 1.627888 \sim$ $(111000000|\mathbf{0}|111111000101)$, in which we wish to replace the 10-th gene using the bit-flipping mutation, and assume that we obtained for it a random number less than the mutation rate p_m, so we must proceed with the mutation. The corresponding value after mutation is $\mathbf{x}' = (111000000|\mathbf{1}|111111000101) \sim$ 1.630818, and, since $f(\mathbf{x}) = 2.250650$ is less than $f(\mathbf{x}') = 2.343555$, we conclude that we obtained an improvement over the initial situation. Note that if we considered another mutation, i.e., replacing the 5-th gene (0) with 1, then the corresponding value of f is -0.082257, so a worse result would be obtained.

- *Mutation operators for integer representation.* For integer encoding we mention two types of mutations.

  - *Random resetting* (*random choice*), which simply extends the 'bit-flipping' mutation described above, assuming that, with probability p_m, a new value is randomly chosen from the set of permissible values in each position. It is worth mentioning that this type of mutation is successfully applied to problems when the genes represent cardinal attributes.
  - *Creep mutation*, which applies mostly to ordinal attributes, consisting in the addition of a small (positive or negative) value to each gene with probability p. In this way, the value of each gene will be altered, with a certain probability, up or down, with values sampled at random from a distribution, generally symmetric about zero. These changes are usually small, and strongly depend on the distribution parameters. Thus, the problem of finding the appropriate values for these parameters has to be solved and sometimes two mutations are used in tandem.

- *Mutation operators for floating-point representation* change at random the values of each gene within its range of values (domain) after the scheme $(x_1, x_2, ..., x_l) \rightarrow (x_1', x_2', ..., x_l')$, $x_i, x_i' \in D_i$, where D_i is the range of values for the ith gene. We present below two common types of such operators.

  - *Uniform mutation*, which is based on the uniform choice of the values for the gene x_i, i.e., the value x_i' being selected using a continuous uniform distribution $U(a_i, b_i)$ (here, $D_i = (a_i, b_i)$). This is equivalent to the bitwise mutation and random resetting, respectively.
  - *Non-uniform mutation* (*normally distributed mutation*) is based, like the creep mutation for the integer representation, on the 'deformation' of the value of

each gene, by adding a certain amount obtained at random from a Normal (Gaussian) distribution with mean zero and arbitrary standard deviation (SD), user-specified, depending on the specific problem. Naturally, the new value thus obtained must belong to the gene's range of values. Technically speaking, since for the Gaussian distribution with mean zero, 95% of values are found in the confidence interval $I_{95\%}$, defined by $(-1.96 \times SD, 1.96 \times SD)$, we can conclude that most of the changes will be small, for a suitable value chosen for the standard deviation. Since the distribution underlying this mutation is determined from the beginning, this type of mutation is known as *non-uniform mutation with fixed distribution*. Note, for instance, in this context, a variant of the non-uniform mutation operator, (260), given by:

$$x_i' = \begin{cases} x_i + \triangle(t, b_i - x_i), & if \ a \ random \ digit \ is \ 0 \\ x_i - \triangle(t, x_i - a_i), & if \ a \ random \ digit \ is \ 1, \end{cases}$$

and aiming to reduce the disadvantage of random mutation. Because this mutation leads to problems if the optimum is near the feasible region's boundaries, an adaptive non-uniform mutation has been proposed, (275).

- *Mutation operators for permutation representation* can no longer act as above, that is considering each gene independently. We mention, in short, four of the most known such operators.

  – *Swap mutation*, in which one randomly picks two positions in the chromosome sequence and swaps their values. For instance, starting from the chromosome $(1, 2, 3, |4|, 5, 6, |7|, 8, 9)$ and choosing the 4-th and the 7-th position, one obtains the mutant $(1, 2, 3, |7|, 5, 6, |4|, 8, 9)$.
  – *Insert mutation*, in which one randomly picks two genes and moves one, so that it will be next to the other. Thus, starting from the above example, we get $(1, 2, 3, |4|, 5, 6, |7|, 8, 9) \rightarrow (1, 2, 3, |4, 7|, 5, 6, 8, 9)$.
  – *Scramble mutation*, which works as follows: one considers either the entire sequence of the chromosome or a subset of it, randomly chosen, then the corresponding genes' positions are mixed (in 'cooking language' -*making scrambled eggs*). If, for instance, we consider again the chromosome $(1, 2, 3, |4, 5, 6, 7|, 8, 9)$ and choose the sequence $(4, 5, 6, 7)$, we get the mutants $(1, 2, 3, |5, 7, 6, 4|, 8, 9)$ or $(1, 2, 3, |6, 7, 4, 5|, 8, 9)$.
  – *Inversion mutation*, which works as follows: one chooses at random two positions in the sequence, and reverses the order in which the values appear between the two points. Thus, in the above example, for the positions 4 and 7, we get $(1, 2, 3, |4, 5, 6, 7|, 8, 9) \rightarrow (1, 2, 3, |7, 6, 5, 4|, 8, 9)$.

E. *Algorithm parameters*

A genetic algorithm involves, as noted above, a set of parameters that appear in its building process, used to control it: population topology/size, probabilities concerning the applications of the variation operators (crossover and mutation probabilities), the total number of generations, etc. It is a notorious fact that the problem of setting (choosing) these parameters is crucial for a good performance of the algorithm. Thus, when building a genetic algorithm, one specifies, for instance, the following choices: the binary representation, the uniform crossover, the bit-flipping mutation, the tournament selection, and the generational model for the population. Obviously, when using a floating-point representation, or a permutation representation, and have to choose between multiple options, we have to handle much more parameters. Consequently, to be able to concretely build such an algorithm, we have to specify the appropriate parameters' values (e.g., population size, crossover rate p_c, mutation rate p_m, and tournament size k). Two approaches are known, in principle, to choose the best parameters' values, that is the best strategy parameters, in order to make the algorithm work efficiently:

- *Parameter tuning* means the finding of good values for the parameters, experimenting with different values, and then selecting the ones that give the best results on the testing problems. The basic idea is to find these 'good' values before the 'official' run of the algorithm, hence the name of *parameter tuning*. It is worth mentioning that the values selected for the system parameters remain fixed during the run (i.e., static parameters), being thus in contrast with the spirit of adaptability, underlying the evolutionary metaphor.
- *Parameter control*, on the other hand, represents the alternative, involving starting a run of the program with initial parameter values, and changing them during the run.

Although the parameter tuning method is a customary approach to the GAs design, there are many technical drawbacks related to it, regarding the fact that parameters are not independent, the process is time consuming (the larger the number of parameters and corresponding values, the longer the running time), etc. It is worth mentioning that, in addition, a long experience in this field has proved that specific problems require specific GAs parameters setups to work properly. In order to overcome all these limitations and many others, the idea of using dynamic parameters, instead of static parameters emerged. Technically, one replaces the fixed (constant) parameter p by a function $p(t)$, depending on the generation t, although finding this function is not just so simple. Another approach is based on the idea that led to the development of GAs, namely the optimization of different processes. Thus, it is natural to use GAs for tuning a GA to a particular problem (concretely, one for problem solving, and another for tuning the first one). Yet another approach is based on the presence of a human-designed feedback mechanism, utilizing actual information about the search process for determining new parameter values.

For a deeper insight in this matter, see (95), (97), (98), (388).

5.9.2 Architecture of GAs

In the previous sub-section we reviewed the components of a genetic algorithm, insisting where we thought it fit on certain technical details. Now the time has come to 'assemble' the constituent elements of such an algorithm into a coherent structure, illustrating its specific operating mode. Schematically, the architecture (flow chart) of any genetic algorithm (evolutionary algorithms, in general) can be represented as in (Fig. 5.59). This simple architecture can be implemented using the following structure, (259), (260).

```
BEGIN
    t ← 0
    initialize P(t)
    evaluate P(t)
    WHILE (NOT termination condition) DO
        BEGIN
            t ← t + 1
            select P(t) from P(t - 1)
            alter P(t)
            evaluate P(t)
        END
END
```

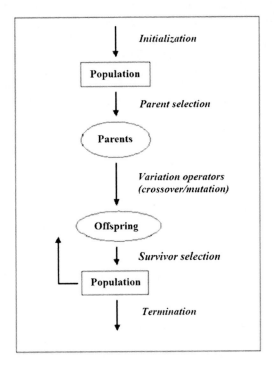

Fig. 5.59 Flow chart of a GA

Starting from this structure, we present below the classical scheme of a fundamental/canonical GA, (90).

Canonical GA scheme

   1. Choose a selection mechanism and initialize population $P(0)$.

   2. Assign $t \leftarrow 0$.

   3. Evaluate the chromosomes of the population $P(t)$ and retain the best of them.

   4. Apply the selection operator n times (n is the initial population size). The selected chromosomes form an intermediate population P_1 of the same size.

   5. Apply the recombination operator (crossover) to the population P_1. The offspring thus obtained will form the population P_2 together with the chromosomes from P_1, remained after the removal of the parents chosen for crossover.

   6. Apply the mutation operator to the population P_2, thus resulting a new generation of chromosomes $P(t+1)$.

   7. Assign $t \leftarrow t+1$.
If $t \leq N$, where N is the maximum number of generations, then return to step 3, otherwise the best performing chromosome is the sought solution for the problem that needs to be solved, and STOP.

Remark 5.31. 1) The first step in implementing any GA is represented by the process of generating an initial population. In the classic case, each member of this population is a binary sequence, called either "genotype", (188), or "chromosome", (333).
2) The implementation of a GA involves two stages. First, starting from the current population, a selection is applied in order to create the intermediate population. Secondly, the variation operators (recombination and mutation) are applied to the intermediate population in order to create the next population.
3) After applying the selection, recombination and mutation, the next population is then evaluated. Thus, the shift process from the current population to the next population, based on evaluation, selection, recombination, and mutation represents a generation in the algorithm implementation.
4. Both between the initialization step and the evaluation step (first phase), and between the alteration step and evaluation step (second phase), one can consider a local optimization procedure of the population $P(t)$, obtaining a modified GA scheme.

Note. The evolutionary algorithms, in general, and the genetic algorithm, particularly, have a general feature regarding the way they work. Thus, in the first phase, after initialization, the individuals are randomly spread over the whole search space, each of them trying to become the optimal solution. After a few generations, because of the changes obtained by using the variation operators, the new individuals abandon the low-fitness regions (the "valleys") looking for the high-fitness regions, in other words "climbing the hills". Finally, close to the end of the searching process, all individuals 'crowd' on some 'hills peaks', and, if these peaks represent local maxima and not global maximum, it is possible that the whole population "climbs a wrong hill". Starting from this suggestive comparison, we define a major problem concerning the searching process: *"the balance between the exploitation of the best available individuals at a time, and the robust exploration of the search space"*. Although there are no universally accepted definitions for these two aspects corresponding to the local search and global search, we can say that:

- *Exploration* represents the generation of new individuals in unknown regions (i.e., white spots on the map) of the search space, thus discovering potentially promising search areas and gaining, therefore, valuable information about the problem that needs to be solved;
- *Exploitation* represents the concentration of the search in the neighborhood of known good solutions, optimizing thus an area known as being promising, using the available information.

In this context, selection is commonly seen as the source of exploitation, while exploration is attributed to the operators mutation and recombination. Another widely spread opinion is that exploration and exploitation are opposite forces, (94).

5.9.3 Applications

It is obviously difficult to present a list of applications of GAs (and EAs, in general) as complete as possible. A quick Internet 'surfing' will provide a multitude of applications in very different and, say, unrelated fields, such as: engineering, art, economics, biology, genetics, operations research, robotics, social sciences, physics, chemistry, computing, etc.

A list of some of the most popular applications in real-world is shown below, although it is almost impossible to imagine an overview of the field of possible applications:

- Automated design in different fields: computer-automated design, design of mechatronic systems, design of industrial equipment, design of sophisticated trading systems in the financial sector, composite material design, multi-objective design of automotive components for crashworthiness, etc.

- Chemical kinetics, molecular structure optimization (chemistry). Computer-aided molecular design.

- Code-breaking (searching large solution spaces of ciphers for decryption).

- Robotics (learning robot behavior).

- Linguistic analysis, including grammar induction and other aspects of natural language processing.
- Game theory (equilibrium concepts).
- Mobile communications infrastructure optimization.
- Software engineering.
- TSP (Traveling salesman problem).
- Scheduling applications, including job-shop scheduling.
- Economics. Business optimization.
- Data mining (clustering, feature selection, classification, prediction, rule-discovery, etc.).
- Artificial creativity.
- Artificial Intelligence (NNs design, GAs optimization, etc.).
- Computer-automated design.
- Computer-aided (medical) diagnosis.
- Genetics (e.g., gene expression profiling).
- Multiple criteria production scheduling.
- Quality control.
- Biomimetics (e.g., biomimetic robotics).
- Computer gaming.
- Etc.

The GAs presentation was primarily due to their utility within the data mining process, since they represent an important tool in this area. Besides this motif, GAs represent a fascinating research field, their dynamics mimicking the natural evolutionary processes, and, therefore, we thought that our attempt to acquaint the reader with the basic knowledge about GAs is useful, along with the presentation within this book of the other research field of natural origin, namely NNs. Next, we present, for illustration, three ways GAs (EAs in general) can be applied to clustering problems, data classification and NNs optimization.

- *Application in clustering*, (157). Assume we have to divide the available data into three groups. A possible approach to this issue may be structured as follows.

  - *Step* 1: We can start grouping data at random using GAs. The fitness function within the algorithm will determine if a data sequence is a match for one of the three clusters. Thus, this fitness function could be anything that identifies some data sequences as "better fits" than other, e.g., a function that determines the level of similarity between data sequences within a group.
  - *Step* 2: One uses now the variation operators within GAs. Practically, if a data sequence in a dataset is found to be a good fit by the fitness function, then it will survive, and will be copied into a cluster. On the other hand, if the data sequence is not a good fit, then it can be mated with another sequence (crossover), to create o better offspring and so on.

- *Applications to classification problems*. The evolutionary classifiers that are classically available are very intricate and difficult to apply, due to their use of complicated credit assignment systems that penalize or reward the classification rules as well as of the complex schemes of their entire working mechanism. In this context, an important task would thus be to develop evolutionary engines capable of simultaneously finding simpler rules and of competitive accuracy. One of the recent approaches takes the viewpoint of the *Michigan school* regarding the representation - each individual encodes a single rule in an *"if... then..."* conjunctive form and the entire population constitutes the rule set. The credit assignment system is replaced by a mechanism that aims to create and maintain variable sized subpopulations that are connected to different optima in the search space and is called *genetic chromodynamics*. This classifier, (366) had been applied for spam detection, while extended research implied the further development of new algorithms within the genetic chromodynamics framework. The most performant of these, following validation on a suite of artificial test functions, had been utilized for classification, where the real-world problem was represented by diabetes mellitus diagnosis. The latest approach on evolutionary classifiers is a technique based on cooperative *coevolution*, (367). This method has been validated on three practical tasks: diabetes mellitus diagnosis, *Iris* plants classification and spam detection.

- *Application in NNs' design.* As previously mentioned, GAs can be used in designing many data mining 'tools' (e.g., NNs, even GAs, etc.), in order to optimize the way they solve the addressed problems. Since, for instance, finding the best set of synaptic weights in NNs is obviously an optimization task, the use of GAs is self-understood. The fact that GAs perform a global search, avoiding the chances of becoming caught in a local minimum, on the one hand, and, on the other hand, since they require no derivatives like the BP algorithm (gradient descent) does, recommends them as very efficient 'specialists' for this 'job'. From a historic point of view, the idea of evolving NNs using GAs can be found in Holland's classic book, (188), where a formal framework for studying adaptivity is presented, showing how the genetic processes can be generalized to be useful in many different disciplines. However, most of the studies in this domain are much more recent. Generally, the idea was to use GAs to evolve NNs in three main ways: setting the weights in a fixed architecture, learning NN topologies, and selecting training data as well as interpreting the output behavior of NNs, separately or simultaneously. Here we consider the first aspect only, briefly presenting some examples of such *hybrid systems*, applied in different real-world problems. Thus, there have been several studies illustrating the usefulness of using a GAs approach to make NNs more competitive in solving various concrete problems, by employing GAs to optimize the network weights. For instance, in the processing of passive sonar data from arrays of underwater acoustic receivers, (270), a multi-layered feedforward NN was trained, using a number of different genetic operators (mutation and crossover specifically tailored to weight optimization) in a set of experiments. Classic feedforward network tests (XOR problem, 424-encoder, two-bit adder) have been approached by training NNs using GAs, (401), by emphasizing mutation. In actuarial science, to forecast financial distress in life insurers, a NN optimized with a GA was used, (196). To model a three-dimensional ultrasonic positioning system, a NN model with random multipoint crossover and entire chromosome mutation was employed, (387). The search of the smoothing parameter in a probabilistic neural network, used for hepatic cancer detection, was performed using a GA approach, with whole arithmetic crossover and uniform mutation, (137). To improve the training of dynamic recurrent NNs, evolutionary algorithms were proposed and compared for Elman recurrent NNs in time-series prediction problems, (81). GA-based artificial NNs were compared with the log maximum-likelihood gradient ascent and the root-mean-square error minimizing gradient descent in the case of binary classification problems, (297). A hybrid neural network/genetic algorithm, with whole arithmetic crossover and non-uniform mutation, where the GA routine is used to set the MLP weights, was used for cancer detection and recurrence, (28). For an overview concerning the combination of NNs and GAs and their applications, see (334), (404), (268), (229).

The presence of GAs on the Internet is undeniable today. Starting from websites devoted to their presentation and implementation in different real-world situations, such as:

- Illinois Genetic Algorithms Laboratory -ILLiGAL
  (http://www-illigal.ge.uiuc.edu/about.html),
- Kanpur Genetic Algorithms Laboratory -KanGAL
  (http://www.iitk.ac.in/kangal/index.shtml),
- Evolutionary Computation Laboratory (George Mason University) -EClab
  (http://cs.gmu.edu/ ~eclab/),
- The MathWorks-Genetic Algorithm and Direct Search Toolbox
  (http://www.mathworks.com/products/gads/),

there are websites devoted to the evolutionary computing field, such as:

- International Society for Genetic and Evolutionary Computation -ISGEC
  (http://www.isgec.org/);
- Special Interest Group for Genetic and Evolutionary Computation -SIGEVO
  (http://www.sigevo.org/);
- Genetic Algorithms Research at Colorado State University -The GENITOR
  Group (http://www.cs.colostate.edu/~genitor/);
- Machine Learning and Evolutionary Computation Group -M&EC University of
  Torino (http://www.di.unito.it/~mluser/).

Besides the bibliographic references in this book or the specialized literature, we can also find on the Internet a variety of websites presenting tutorials, lecture notes, scientific papers and, in general, a lot of contributions in many various areas of interest, all showing the strong vitality of this area of research.

We conclude our brief journey into the world of GAs (EAs, in general), recalling an interesting remark on the usefulness of EAs in real-world applications, (104), (259): "...Evolutionary algorithms are much like a Swiss Army knife: a handy set of tools that can be used to address a variety of tasks. If you've ever used such a device, you know that this versatility comes at cost. For each application there is usually *a better tool devised specifically for that task. If you have to pry a nail out of a board, the claw of a hammer will probably do a better job than anything you'll find in a Swiss Army knife. Similarly, if you have to remove a splinter from your finger, a carefully crafted set of tweezers will usually do a better job than the makeshift version you'll find in the Swiss Army knife. Yet, if you don't know exactly what task you'll be faced with, the flexible nature of the Swiss Army knife comes in very handy. Imagine trying to remove a nail with tweezers, or to extract a splinter with the claw of a hammer! Having the Swiss Army knife provides you with the ability to address a wide variety of problems quickly and effectively, even though there might be a better tool for the job, and you don't have to carry a hammer, tweezers, etc.".* After these words, it seems that there is nothing else more to add, so, let us use with confidence the EAs -*Swiss Army knife.*

Chapter 6
Classification Performance Evaluation

Abstract. A great part of this book presented the fundamentals of the classification process, a crucial field in data mining. It is now the time to deal with certain aspects of the way in which we can evaluate the performance of different classification (and decision) models. The problem of comparing classifiers is not at all an easy task. There is no single classifier that works best on all given problems, phenomenon related to the *"No-free-lunch"* metaphor, i.e., each classifier ('restaurant') provides a specific technique associated with the corresponding costs ('menu' and 'price' for it). It is hence up to us, using the information and knowledge at hand, to find the optimal trade-off.

6.1 Costs and Classification Accuracy

In a classification problem, it is often important to specify the costs associated with correct or incorrect classifications. Doing so can be valuable when the cost of different misclassifications varies significantly. Classification accuracy is also a (statistical) measure showing how well the classifier correctly identifies the objects. To be as suggestive as possible, let us synthetically present these two issues, using the next two figures (matrices), which summarize the results of a classification process.

First, let us recall that evaluation of the performance of a classification model is based on the counts of (testing) objects correctly and incorrectly predicted. These counts are tabulated in the so-called *confusion matrix* -see Fig. 6.1. Briefly, the confusion matrix gives a detailed breakdown of misclassifications. The predicted class is displayed at the top of the matrix, and the observed class down the left side. Each cell contains a number showing how many cases that were actually of the given observed class were assigned by the model to the given predicted class.

Secondly, we can similarly create a *cost matrix* to bias the model to minimize cost or maximize benefit -see Fig. 6.2. The cost/benefit matrix is taken into consideration when the model is scored. It is important to point out that sometimes a more accurate classification is desired for some classes than others for reasons unrelated to relative class sizes (e.g., contagious disease vs. non-contagious disease, malignant tumors

F. Gorunescu: Data Mining: Concepts, Models and Techniques, ISRL 12, pp. 319–330.
springerlink.com   © Springer-Verlag Berlin Heidelberg 2011

CLASSIFICATION	PREDICTED CLASS		
		Class = YES	Class = NO
OBSERVED CLASS	Class = YES	a (*true positive*-TP)	b (*false negative* -FN)
	Class = NO	c (*false positive*-FP)	d (*true negative*-TN)

Fig. 6.1 Confusion matrix for a 2-class model

COST MATRIX	PREDICTED CLASS		
		Class = YES	Class = NO
OBSERVED CLASS	Class = YES	p	q
	Class = NO	r	s

Fig. 6.2 Cost matrix for a 2-class model

vs. benign tumors, etc.). Then, the cost in the cost matrix might be considered as a misclassification cost, being chosen accordingly.

The formulas related to the computation of cost and accuracy are the following:

$$Cost = p \times a + q \times b + r \times c + s \times d,$$

$$Accuracy = \frac{a+d}{a+b+c+d} = \frac{TP+TN}{TP+TN+FP+FN},$$

Example 6.1. Let us consider a classification process whose performance is summarized in the three following figures. Thus, the first figure refers to the costs assigned to the correct/incorrect predicted cases, costs that have been previously (*a priori*) established. The following two figures summarize the classification results, obtained by applying two different classifiers (model M_1 and model M_2) in a medical case of diagnosing tumors malignancy. Let us outline the fact that the way to estimate the costs corresponding to each category remains an open and sensitive problem.

COST MATRIX	PREDICTED CLASS		
		Class = malignant	Class = benign
OBSERVED CLASS	Class = malignant	-1	100
	Class = benign	1	0

Fig. 6.3 Cost matrix

CLASSIFICATION Model M_1	PREDICTED CLASS		
		Class = malignant	Class = benign
OBSERVED CLASS	Class = malignant	150	40
	Class = benign	60	250

Fig. 6.4 Confusion matrix for model M_1

CLASSIFICATION Model M_2	PREDICTED CLASS		
		Class = malignant	Class = benign
OBSERVED CLASS	Class = malignant	250	45
	Class = benign	5	200

Fig. 6.5 Confusion matrix for model M_2

Once calculations done, we obtain the following parameters measuring the performance of two classifiers:

$$Cost_{M_1} = -1 \times 150 + 100 \times 40 + 1 \times 60 + 0 = 3910,$$

$$Cost_{M_2} = -1 \times 250 + 100 \times 45 + 1 \times 5 + 0 = 4255,$$

$$Accuracy_{M_1} = \frac{400}{500} = 80\%,$$

$$Accuracy_{M_2} = \frac{450}{500} = 90\%.$$

Here comes the dilemma: what is the best model out of the two above? To answer this question we must first solve the dilemma: either a cancer patient would be incorrectly diagnosed as being healthy, or a healthy individual would be incorrectly diagnosed as having cancer.

As we saw from the above example, the issue which we have to handle after computing the two above pairs of parameters, seen as the measure of performance for two or more models, refers to the way in which we can use them in choosing the best performing model (see also subsection 6.3). This remains an open question, depending on the specific conditions in which the classification is used. In principle, we have to establish, however, a balance between cost and accuracy, that is an acceptable compromise for choosing the optimal classifier for the problem at hand.

In the same context regarding both the investigation of the classification performance of a particular model, and the comparison of several classifiers, in order to choose the most efficient one, we introduce another two important concepts. Thus, we will consider the *sensitivity* and the *specificity* as statistical measures of the performance of a binary classification. While sensitivity measures the proportion of true 'positives' that are correctly identified as such, specificity measures the proportion of true 'negatives' that are correctly identified. Basically, classification should be both sensitive and specific as much as possible, neither one is less important than the other. Moreover, we need to know the probability that the classifier will give the correct diagnosis and, unfortunately, the sensitivity and specificity do not give us this information. Thus, in addition, we also considered both the *positive predictive value* (PPV) and the *negative predictive value* (NPV). Recall that while PPV is the proportion of cases with 'positive' test results that are correctly diagnosed, NPV is the proportion of cases with 'negative' test results that are correctly diagnosed, (11), (12), (36). Technically, the formulas for the four above statistical measures of the performance are the following:

$$Sensitivity = \frac{Number\ of\ 'True\ Positives'}{Number\ of\ 'True\ Positives' + Number\ of\ 'False\ Negatives'},$$

$$Specificity = \frac{Number\ of\ 'True\ Negatives'}{Number\ of\ 'True\ Negatives' + Number\ of\ 'False\ Positives'},$$

$$PPV = \frac{Number\ of\ 'True\ Positives'}{Number\ of\ 'True\ Positives' + Number\ of\ 'False\ Positives'},$$

$$NPV = \frac{Number\ of\ 'True\ Negatives'}{Number\ of\ 'True\ Negatives' + Number\ of\ 'False\ Negatives'}.$$

Remark 6.1. 1) Sensitivity is also called *true positive rate* (TP *rate*), or *recall*. A sensitivity of 100% means that the classifier recognizes all observed positive cases - for instance, all people having cancer (malignant tumors) are recognized as being ill.

2) On the other hand, a specificity of 100% means that the classifier recognizes all observed negative cases - for instance, all healthy people (benign tumors) will be recognized as healthy.

3) Theoretically, the optimal classification model can achieve 100% sensitivity and 100% specificity, which is, practically, impossible.

We conclude this short presentation, by mentioning other indicators taken into consideration when evaluating a classifier:

- *False positive rate* (FP *rate*) = FP / (FP + TN) = 1 - *specificity*;
- *False negative rate* (FN *rate*) = FN / (TP + FN) = 1 - *sensitivity*;
- *Likelihood ratio positive* (LR+) = *sensitivity/*(1 - *specificity*);
- *Likelihood ratio negative* (LR-) = (1 - *sensitivity*)*/specificity*.

Remark 6.2. As we stated above, in medical diagnosis, the cost of misclassification is not equal (see the cancer detection case). Misclassifying patients as non-patients can cause more dangerous consequences than the other case. Consequently, patients may not have the necessary treatments and the consequence can be as serious as death, (298). Therefore, minimizing the false negative rate is one of the most important objectives for medical diagnostic rules. Usually, medical practitioners prefer high false positive rate to high false negative rate.

Note. The field in which these classification (prediction) performance indicators are often used is represented by the medical research (e.g., evidence-based medicine, see, for instance, (125)). Concretely, sensitivity and specificity are used to determine whether a test result usefully changes the probability that a condition exists, while the likelihood ratios are used to assess the value of performing a diagnostic test.

6.2 ROC (Receiver Operating Characteristic) Curve

We now present some basic notions about the ROC (*Receiver Operating Characteristic*) *curves*, widely used in assessing the results of predictions (forecasts). We schematically present the most important aspects regarding the ROC curves, just to familiarize the reader with this technique.

- The ROC curves were first developed by electrical and radar engineers during *World War II* for detecting enemy objects in battlefields (e.g., differentiating enemy aircrafts from flocks of birds -see, for instance, the story of Pearl Harbor attack in 1941, or problems of British radar receiver operators).
- ROC curves have long been used in *signal detection theory*, (156), to depict the trade-off between hit rates (signal) and false alarm rates (noise) of classifiers, (99), (375). Soon after that, they have been introduced in psychology to account for perceptual detection of signals, (374).
- ROC curves are commonly used in *medical research*, (424), (298), (230), (207), (321), (409), (423).
- ROC curves are also usually used in *machine learning* and *data mining* research. One of the earliest adopters of ROC curves in machine learning was Spackman, (354), who demonstrated the value of ROC curves in evaluating and comparing algorithms. Recent years have seen an increase in the use of ROC curves in the machine learning community, (306), (307).
- In *classification problems*, the ROC curve is a technique for visualizing, organizing and selecting classifiers, based on their performance.

In a classification problem using two decision classes (binary classification), each object is mapped to one element of the set of pairs (P, N), i.e., *positive* or *negative*. While some classification models (e.g., decision trees) produce a discrete class label (indicating only the predicted class of the object), other classifiers (e.g., naive Bayes, neural networks) produce a continuous output, to which different thresholds may be applied to predict class membership.

Technically, ROC curves, also known as ROC *graphs*, are two-dimensional graphs in which the TP rate is plotted on the Y-axis and the FP rate is plotted on the X-axis. In this way, a ROC graph depicts relative trade-offs between benefits ('true positives') and costs ('false positives'). We displayed in the figure below two types of ROC curves (discrete and continuous).

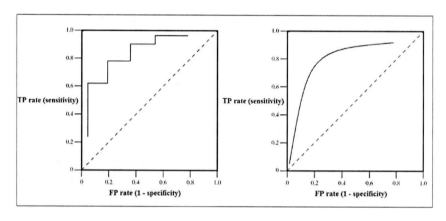

Fig. 6.6 ROC graph (discrete/continuous case)

Several points in the ROC space are important to note. The lower left point $(0,0)$ represents the strategy of never issuing a positive classification (i.e., no 'false positive' errors, but also no 'true positives'). The opposite strategy, of unconditionally issuing positive classifications, is represented by the upper right point $(1,1)$. The point $(0,1)$ represents the *perfect classification* (i.e., no FN and no FP). A completely random guess would give a point along a diagonal line (*line of no-discrimination*) from the left bottom to the top right corner. This diagonal line divides the ROC space as follows: (*a*) points above the diagonal represent good classification results, and (*b*) points below the diagonal line represent poor classification results. To conclude, informally, one point in ROC space is better than another if it is to the 'northwest' of the square (i.e., TP rate is higher, FP rate is lower, or both).

As we outlined above, a ROC curve is a two-dimensional "tool" used to assess classification performances. Accordingly, this "tool" is often used for classification models comparison (e.g., the machine learning community). But how can we use a graphical representation to decide which classifier is better (especially in ambiguous cases like the one represented in the figure below)?

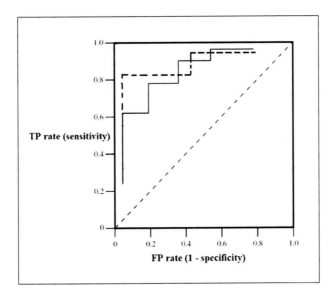

Fig. 6.7 Comparing classifiers (two ROC curves)

To address this issue we may want to reduce the ROC performance to a single scalar value representing expected performance. A common method is to calculate the *area under the ROC curve*, abbreviated AUC, (170) (320), (107). Since the AUC is a portion of the area of the unit square, its value will always be between 0.0 and 1.0. However, only the values over 0.5 are interesting, since random guessing produces the diagonal line between $(0,0)$ and $(1,1)$, which has an area of 0.5. AUC can be interpreted as the probability that, when we randomly pick one positive and one negative example, the classifier will assign a higher score to the positive example than to the negative (equivalent to the Wilcoxon test of ranks and Mann-Whitney *U* test). Therefore, a higher AUC value implies better classification performance, making it as a maximization objective. An algorithm used to compute AUC is to be found in (107). However, any attempt to summarize the ROC curve into a single number loses information about the pattern of trade-offs of the particular discriminator algorithm.

Remark 6.3. A rough guide for classifying the accuracy of a diagnostic test using AUC is the traditional system, presented below:

- 0.90 - 1.00 = excellent classification;
- 0.80 - 0.90 = good classification;

- 0.70 - 0.80 = fair classification;
- 0.60 - 0.70 = poor classification;
- 0.50 - 0.60 = failure.

Example 6.2. 1) Let us consider the hypothyroidism data regarding patients with suspected hypothyroidism reported by Goldstein and Mushlin, (132). Technically, they measured the T4 and TSH values in ambulatory patients with suspected hypothyroidism, and used the TSH values as a *gold standard* for determining which patients were truly hypothyroid (see table below).

Table 6.1 T4 values

T4 value	Hypothyroid	Euthyroid
≤ 5	18	1
5.1 - 7	7	17
7.1 - 9	4	36
≥ 9	3	39
Total	32	93

Under these circumstances, the ROC curve is used to illustrate how sensitivity and specificity change depending on the choice of the T4 level that defines hypothyroidism. Thus, starting from the above table, both the sensitivity and specificity values are computed, together with the TP and FP values, needed to represent the corresponding ROC curve (Table 6.2)

Table 6.2 Sensitivity and specificity values

Cut-point	Sensitivity	Specificity	True Positives	False Positives
5	0.56	0.99	0.56	0.01
7	0.78	0.81	0.78	0.19
9	0.91	0.42	0.91	0.58

The corresponding ROC curve is displayed in Fig. 6.8

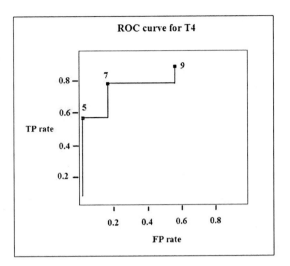

Fig. 6.8 The ROC curve for T4 value (hypothyroidism data)

In this case, AUC equals 0.86, so T4 would be considered to be "good" at separating hypothyroid from euthyroid patients.

Example 6.3. Consider now the use of the ROC curves for classifiers comparison. The two classifiers in competition are a MLP with two hidden layers and a RBF network. The database on which they have been applied is the Wisconsin Prognostic Breast Cancer (http://archive.ics.uci.edu/ml/machine-learning-databases/breast-cancer-wisconsin/), consisting of 569 cases, with two decision classes: (*a*) *benign* -357 instances (62.74%), and (*b*) *malign* -212 instances (37.25%), and with 30 numerical attributes. Thus, while MLP provided 97% training performance and 95% testing performance, RBF provided the same training performance, but 96% testing performance. Now, "*returning to our muttons*", AUC for MLP equals 0.98, less that AUC for RBF, which equals 0.994. Although both AUCs are very close to 1.0, implying excellent classification, RBF performs better on this occasion.

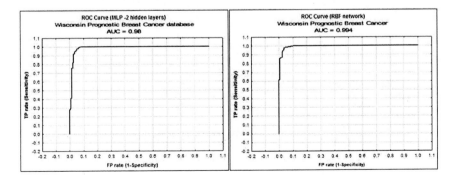

Fig. 6.9 The ROC curve for MLP and RBF (Wisconsin Prognostic Breast Cancer)

6.3   Statistical Methods for Comparing Classifiers

We saw above how we compare two or more classifiers using ROC curves. Besides visualization and the AUC, the method cannot be considered sufficient for a professional comparison. It is thus the aim of this subsection to briefly review some methods used for comparing classifiers, considering here only statistical tests, (378). Note that in this comparison only the testing accuracy is considered.

Let M_1 and M_2 be two classifiers, and suppose that M_1 achieves $p_1 = 85\%$ accuracy when evaluated on $N_1 = 30$ unknown objects, while M_2 achieves $p_2 = 75\%$ accuracy on $N_2 = 5000$ different objects. Based on this information, containing unbalanced testing datasets, we have to decide which is the best model. As it is easy to see, although M_1 has a higher accuracy (+10%), it was tested on a much smaller set (167 times smaller). So, how credible is the better performance of M_1 in comparison with M_2 in such circumstances?

To answer this question properly, we need statistical tools, namely the "*difference between two proportions*" test (z-test), used for computing the level of significance of the difference between two proportions (ratios). We present below the corresponding algorithm.

Algorithm for performance comparison (z-test)

Input:

- Enter the proportion p_1 (i.e., classification accuracy) corresponding to the first sample (model M_1);
- Enter the proportion p_2 (i.e., classification accuracy) corresponding to the second sample (model M_2);
- Enter the sample size (number of samples $\sim$ *testing dataset*) N_1 of the first sample;
- Enter the sample size N_2 of the second sample.

The corresponding p-level is computed based on the t-value for the respective comparison, given by:

$$|t| = \sqrt{\frac{N_1 \cdot N_2}{N_1 + N_2} \cdot \frac{|p_1 - p_2|}{\sqrt{p \cdot q}}},$$

where $p = \dfrac{p_1 \cdot N_1 + p_2 \cdot N_2}{N_1 + N_2}$, $q = 1 - p$, and there are $N_1 + N_2$ - 2 degrees of freedom.

Output: the significance level for the difference between the accuracies of the two models.

Thus, we can answer the question: *"How much confidence can we place on the larger accuracy?"*

Applying this algorithm to the above example, we obtain p-level (two-sided) = 0.21, which indicates that the two models are comparable in terms of prediction accuracy (no statistically significant difference, since $p > 0.05$).

Remark 6.4. If we consider in the above example that the first testing dataset N_1 (model M_1) contains 100 instances instead of just 30 (so, a better testing), then the corresponding significance level is $p = 0.02$ (< 0.05), so the first model is significantly more efficient than the second one.

Next, in order to determine the confidence interval for accuracy, (378), we first need to identify the accuracy's distribution, the accuracy measure being seen as a random variable. To do this, we can use the binomial distribution $B(N, p)$, based on multiple and independent Bernoulli trials, (287). Thus, given a testing dataset that contains N objects, let us consider X as the number of objects correctly classified by a model, and p as its *true accuracy*. Under these circumstances, X (seen as a random variable) has a binomial distribution with mean equaling $N \cdot p$ and variance equaling $N \cdot p \cdot (1 - p)$. Statistically speaking, the *empirical accuracy* is given by $acc = X/N$, governed also by the binomial distribution, with mean p and variance $p \cdot (1 - p)/N$. It is worth mentioning that the binomial distribution can be approximated by the Normal (Gaussian) distribution when N is large enough, (114), so we can consider the Normal distribution to estimate the confidence interval for the empirical accuracy, in this case. Accordingly, based on the Normal distribution, we can derive the following form for the confidence interval for acc:

$$P\left(-Z_{\alpha/2} \leq \frac{acc - p}{\sqrt{p(1-p)/N}} \leq Z_{1-\alpha/2}\right) = 1 - \alpha,$$

where $Z_{\alpha/2}$ and $Z_{1-\alpha/2}$ represents the upper and lower bounds obtained from the standard Normal distribution $N(0,1)$ at a confidence level α, with $\alpha \in (0,1)$. Since in this case $Z_{\alpha/2} = Z_{1-\alpha/2}$, the confidence interval (level α) for acc is given by:

$$\frac{2 \times N \times acc + Z_{\alpha/2}^2 \pm Z_{\alpha/2}\sqrt{Z_{\alpha/2}^2 + 4 \times N \times acc - 4 \times N \times acc^2}}{2(N + Z_{\alpha/2}^2)}.$$

Note that, usually, the 95% confidence interval is used ($\alpha = 0.05, Z_{\alpha/2} = 1.96$), (10). If, for instance, a model has an accuracy equaling 80%, evaluated on a testing dataset with $N = 100$ objects, then the 95% confidence interval is given by (71.1%, 86.7%).

Next, we are interested in evaluating the confidence interval for the difference between the error rates of two classification models, (378). Thus, let us again consider two classification models M_1 and M_2, with the corresponding testing datasets N_1 and N_2. The error rate of the first model is e_1, and the error rate of the second model is e_2. We have to answer the question: *"Is the difference $d = e_1 - e_2$ statistically significant?"*. As we mentioned above, for large enough testing datasets size, the difference d is Normally distributed, with mean d_t and variance σ_d^2. Under these circumstances, the $(1 - \alpha)$ confidence interval for the difference d is given by:

$$d_t = d \pm z_{\alpha/2} \cdot \widehat{\sigma}_d,$$

where:

$$\sigma_d^2 \simeq \widehat{\sigma}_d^2 = \frac{e_1(1 - e_1)}{N_1} + \frac{e_2(1 - e_2)}{N_2}.$$

Thus, if for instance, $e_1 = 15\%$ and $e_2 = 25\%$, with $N_1 = 30$, and $N_2 = 5000$, then $|d| = 0.1$, having the confidence interval given by (0.1 ± 0.128).
For more details concerning comparison tests, see, for instance (10).

Remark 6.5. All statistical tests used above relate to the branch of *Statistical inference*, based on the use of samples to estimate the true (theoretical) parameters of statistical populations (e.g., accuracy). Under these conditions, the confidence interval represents the range of values which we can be confident (with confidence level α) that includes the true value of the parameter. For instance, the 95% confidence interval for a certain parameter is usually interpreted as a range of values which contains the true value of the parameter, with probability 95%. Recall that all values in this interval are equally likely to occur, usually choosing the midpoint of the interval (*idem*).

References

[1] Abonyi, J., Feil, B.: Cluster analysis for data mining and system identification. Birkhäuser, Basel (2007)

[2] Ackley, D.: A connectionist machine for genetic hillclimbing. Kluwer Academic Publishers, Dordrecht (1987)

[3] Adamo, J.M.: Data Mining for association rules and sequential patterns: sequential and parallel algorithms. Springer, Heidelberg (2001)

[4] Agrawal, R., Imielinski, T., Swami, A.: Mining association rules between sets of items in large databases. In: Proc. 1993 ACM SIGMOD Conference, Washington, pp. 207–216 (1993)

[5] Agrawal, R., Srikant, R.: Fast Algorithms for Mining Association Rules in Large Databases. In: Proc. 20th International Conference on Very Large Data Bases, pp. 487–499. Morgan Kaufmann, San Francisco (1994)

[6] Agrawal, R., Srikant, R.: Mining sequential patterns. In: Proc. 11th ICDE 1995, Taipei, Taiwan, pp. 3–14 (March 1995)

[7] Agrawal, R., Srikant, R.: Mining sequential patterns: Generalizations and performance improvements. In: Apers, P.M.G., Bouzeghoub, M., Gardarin, G. (eds.) EDBT 1996. LNCS, vol. 1057, pp. 3–17. Springer, Heidelberg (1996)

[8] Aldrich, J.H., Nelson, F.D.: Linear probability, logit and probit models. Sage University Papers, Thousand Oaks (1985)

[9] Allen, D.M.: Mean square error of prediction as a criterion for selecting variables. Technometrics 13, 469–475 (1971)

[10] Altman, D.G.: Practical statistics for medical research. Chapman and Hall, Boca Raton (1991)

[11] Altman, D.G., Bland, J.M.: Diagnostic tests 1: Sensitivity and specificity. BMJ 308, 1552 (1994)

[12] Altman, D.G., Bland, J.M.: Diagnostic tests 2: Predictive values. BMJ 309, 102 (1994)

[13] Andersen, R.: Modern Methods for Robust Regression. Sage University Paper Series on Quantitative Applications in the Social Sciences, 07-152. Sage, Thousand Oaks (2008)

[14] An, L., Tong, L.: Similarity, Boolean Reasoning and Rule Induction. In: Proc. Second International Symposium on Intelligent Information Technology Application, vol. 1, pp. 7–12 (2008)

[15] Anderson, J.A.: An introduction to Neural Networks. MIT Press, Cambridge (1995)

[16] Aris, R.: Mathematical modelling techniques. Pitman (1978)

[17] Aspray, W., Burks, A.: Papers of John von Neumann on computing and computer theory. Charles Babbage Institute Reprint Series for the history of computing, vol. 12. MIT Press, Cambridge (1986)

[18] Back, T.: Self-adaptation. In: Back, T., Fogel, D.B., Michalewicz, Z. (eds.) Evolutionary computation 2: advanced algorithms and operators, ch. 21, pp. 188–211. Institute of Physics Publishing (2000)

[19] Back, T., Fogel, D.B., Michalewicz, Z. (eds.): Evolutionary computation 1: basic algorithms and operators. Institute of Physics Publishing, Bristol (2000)

[20] Baker, J.E.: Reducing bias and inefficiency in the selection algorithm. In: Grefenstette, J.J. (ed.) Proc. 2nd International Conference on Genetic Algorithms and Their Applications, pp. 14–21. Lawrence Erlbaum, Hillsdale (1987)

[21] Banzhaf, W., Nordin, P., Keller, R.E., Francone, F.D.: Genetic Programming: An Introduction: On the Automatic Evolution of Computer Programs and Its Applications. Morgan Kaufmann, San Francisco (1998)

[22] Barricelli, N.A.: Esempi numerici di processi di evoluzione. Methodos, 45–68 (1954)

[23] Barto, A.G., Sutton, R.S., Anderson, C.W.: Neuronlike adaptive elements that can solve difficult learning control problems. IEEE Transactions on Systems, Man, and Cybernetics SMC-13, 834–846 (1983)

[24] Bates, D.M., Watts, D.G.: Nonlinear Regression Analysis and Its Applications. Wiley Series in Probability and Statistics. Wiley, Chichester (1988)

[25] Bayes, T.: An Essay towards solving a Problem in the Doctrine of Chances. Philosophical Transactions of the Royal Society of London 53, 370–418 (1763)

[26] Beachley, N.H., Harrison, H.L.: Introduction to dynamic systems analysis. Harper and Row, New York (1978)

[27] Belciug, S., Gorunescu, F., Gorunescu, M., Salem, A.B.: Assessing Performances of Unsupervised and Supervised Neural Networks in Breast Cancer Detection. In: Proc. 7th International Conference on INFOrmatics and Systems (INFOS 2010), ADEM, pp. 80–87 (2010)

[28] Belciug, S., Gorunescu, F.: A hybrid neural network/genetic algorithm applied to the breast cancer detection and recurrence. Expert Systems. Willey& Blackwell (2011) (in press)

[29] Bender, E.A.: An introduction to mathematical modeling. John Wiley, Chichester (1978)

[30] Berkhin, P.: Survey of clustering data mining techniques. Technical Report, Accrue Software, San Jose, California,
http://citeseer.nj.nec.com/berkhin02survey.html

[31] Berry, M., Linoff, G.: Mastering Data Mining. John Wiley and Sons, Chichester (2000)

[32] Bezdek, J.C.: Pattern recognition with fuzzy objective function algorithms. Plenum Press, New York (1981)

[33] Beyer, H.G.: The theory of evolution strategies. Springer, Heidelberg (2010)

[34] Bishop, C.: Neural networks for pattern recognition. Oxford University Press, Oxford (1995)

[35] Blumer, A., Ehrenfeucht, A., Haussler, D., Warmuth, M.K.: Occam's razor. Information Processing Letters 24, 377–380 (1987)

[36] Boring, S.: Sensitivity, specificity, and predictive value. In: Walker, H.K., Hall, W.D., Hurst, J.W. (eds.) Clinical Methods: The History, Physical, and Laboratory Examinations, ch. 6, 3rd edn. Butterworths, London (1990)

[37] Borji, A., Hamidi, M.: Evolving a fuzzy rule-base for image segmentation. World Academy of Science, Engineering and Technology 28, 4–9 (2007)

[38] Boser, B., Guyon, I., Vapnik, V.N.: A training algorithm for optimal margin classifiers. In: Proc. 5th Annual Workshop on Computational Learning Theory, San Matero, California, USA, pp. 144–152 (1992)

[39] Box, G.E.P.: Evolutionary operation: a method of increasing industrial productivity. Applied Statistics 6, 81–101 (1957)

[40] Box, G.E.P., Jenkins, G.M.: Time series analysis: Forecasting and control. Holden-Day, San Francisco (1976)

[41] Boyce, B.R., Meadow, C.T., Kraft, D.H.: Measurement in information science. Academic Press, London (1994)

[42] Brause, R., Langsdorf, T., Hepp, M.: Neural Data Mining for Credit Card Fraud Detection. In: Proc. 11th IEEE International Conference on Tools with Artificial Intelligence, vol. 103, pp. 103–106 (1999)

[43] Breiman, L., Friedman, J.H., Olshen, R.A., Stone, C.J.: Classification and regression trees. Wadsworth and Brooks/Cole Advanced Books and Software, Monterey (1984)

[44] Bremermann, H.J.: Optimization through evolution and recombination. In: Yovits, M.C., Jacobi, G.T., Goldstein, G.D. (eds.) Self-Organizing Systems, pp. 93–106. Spartan Books, Washington (1962)

[45] Brown, M.P., Grundy, W.N., Lin, D., Cristianini, N., Sugnet, C.W., Furey, T.S., Ares Jr., M., Haussler, D.: Knowledge-based Analysis of Microarray Gene Expression Data by Using Support Vector Machines. Proc. of the National Academy of Sciences USA 97(1), 262–267 (2000)

[46] Bull, L., Kovacs, T. (eds.): Foundations of Learning Classifier Systems: An Introduction. STUDFUZZ, vol. 183. Springer, Heidelberg (2005)

[47] Bull, L., Bernardo-Mansilla, E., Holmes, J. (eds.): Learning Classifier Systems in Data Mining: An Introduction. SCI, vol. 125. Springer, Heidelberg (2008)

[48] Burbidge, R., Trotter, M., Buxton, B., Holden, S.: Drug Design by Machine Learning: Support Vector Machines for Pharmaceutical Data Analysis. Computers and Chemistry 26(1), 5–14 (2001)

[49] Burges, C.: A tutorial on Support Vector Machines for pattern recognition. Data Mining and Knowledge Discovery 2, 121–167 (1998)

[50] Cabena, P., Hadjinian, P., Stadler, R., Verhees, J., Zanasi, A.: Discovering Data Mining from Concept to Implementation. Prentice-Hall, Englewood Cliffs (1997)

[51] Cacoulos, T.: Estimation of a multivariate density. Ann. Inst. Stat. Math (Tokyo) 18, 179–189 (1966)

[52] Cazzaniga, M., Salerno, F., Borroni, G., Ceriani, R., Stucchi, G., Guerzoni, P., Casiraghi, M.A., Tommasini, M.: Prediction of asymptomatic cirrhosis in chronic hepatitis C patients: accuracy of artificial neural networks compared with logistic regression models. Eur. J. Gastroenterol. Hepatol. 21(6), 681–687 (2009)

[53] Chaudhuri, S., Dayal, U.: Data Warehousing and OLAP for Decision Support. In: ACM SIGMOD Int. Conf. on Management of Data, vol. 26(2), pp. 507–508. ACM Press, Tucson (1997)

[54] Chen, H., Wang, X.F., Ma, D.Q., Ma, B.R.: Neural network-based computer-aided diagnosis in distinguishing malignant from benign solitary pulmonary nodules by computed tomography. Chin. Med. J. (Engl.) 120(14), 1211–1215 (2007)

[55] Cheng, C.-H., Chen Jr., S.: Diagnosing cardiovascular disease using an enhanced rough sets approach. Applied Artificial Intelligence 23(6), 487–499 (2009)

[56] Cheng, J.-H., Chen, H.-P., Cheng, K.-L.: Business failure prediction model based on grey prediction and rough set theory. WSEAS Transactions on Information Science and Applications 6(2), 329–339 (2009)

[57] Cherkassky, V., Mulier, F.: Learning from data: concepts, theory and methods. Wiley, Chichester (1998)

[58] Chi, C.L., Street, W.N., Wolberg, W.H.: Application of artificial neural network-based survival analysis on two breast cancer datasets. In: AMIA Annu. Symp. Proc., vol. 11, pp. 130–134 (2007)

[59] Chiu, J.S., Wang, Y.F., Su, Y.C., Wei, L.H., Liao, J.G., Li, Y.C.: Artificial neural network to predict skeletal metastasis in patients with prostate cancer. J. Med. Syst. 33(2), 91–100 (2009)

[60] Christensen, T., Neuberger, J., Crowe, J., et al.: Beneficial effect of azathioprine and prediction of prognosis in primary biliary cirrhosis: final results of an international trial. Gastroenterology 89, 1084–1091 (1985)

[61] Clark, P., Niblett, T.: The CN2 Induction Algorithm. Machine Learning 3(4), 261–283 (1989)

[62] Cliff, N., Krus, D.J.: Interpretation of canonical variate analysis: Rotated vs. unrotated solutions. Psychometrika 41(1), 35–42 (1976)

[63] Cochran, W.G.: Sampling Techniques. John Wiley and Sons, Chichester (1977)

[64] De Cock, M., Cornelis, C., Kerre, E.E.: Fuzzy Rough Sets: Beyond the Obvious. In: Proc. FUZZIEEE 2004 (2004 IEEE International Conference on Fuzzy Systems), vol. 1, pp. 103–108 (2004)

[65] Codd, E.F., Codd, S.B., Salley, C.T.: Providing OLAP (On-line Analytical Processing) to User-Analysts: An IT Mandate. Codd and Date Inc. (1993)

[66] Cohen, W.: Fast Effective Rule Induction. In: Proc. 12th Intl. Conf. on Machine Learning, Tahoe City, CA, USA, pp. 115–123 (1995)

[67] Cooley, W.W., Lohnes, P.R.: Multivariate data analysis. Wiley, New York (1971)

[68] Cortes, C., Vapnik, V.N.: Support vector networks. Machine Learning 20, 273–297 (1995)

[69] Cost, S., Salzberg, S.: A Weighted Nearest Neighbor Algorithm for Learning with Symbolic Features. Machine Learning 10(1), 57–78 (1993)

[70] Cover, T.M.: eometrical and statistical properties of systems of linear inequalities with applications in pattern recognition. IEEE Transactions on Electronic Computers EC-14, 326–334 (1965)

[71] Cox, D.R., Miller, D.: The theory of stochastic processes. Methuen (1965)

[72] Cox, D.R.: Regression models and life tables. J. Roy. Statist. Soc. [B] 34, 187–202 (1972)

[73] Cristianini, N., Shawe-Taylor, J.: An introduction to Support Vector Machines and other kernel-based learning methods. Cambridge University Press, Cambridge (2000)

[74] Darlington, R.B.: Regression and linear models. McGraw-Hill, New York (1990)

[75] Darwiche, A.: Modeling and reasoning with Bayesian Networks. Cambridge University Press, Cambridge (2009)

[76] Darwin, C.: On the Origin of Species by Means of Natural Selection, or the Preservation of Favoured Races in the Struggle for Life, 1st edn. John Murray, London (1859)

[77] Dasarathy, B.V. (ed.): Nearest Neighbour Norms: Nearest Neighbour Pattern Classification Techniques. IEEE Computer Society Press, Los Alamitos (1991)

[78] Davis, L.: Handbook of Genetic Algorithms. Van Nostrand Reinhold, New York (1991)

[79] De Jong, K.: An analysis of the behaviour of a class of genetic adaptive systems. PhD thesis, University of Michigan (1975)

[80] De Jong, K.: Evolutionary computation. MIT Press, Cambridge (2002)

[81] Delgado, M., Pegalajar, M.C., Cuellar, M.P.: Memetic evolutionary training for recurrent neural networks: an application to time-series prediction. Expert Systems 23(2), 99–115 (2006)

[82] Domingos, P.: The Role of Occam's Razor in Knowledge Discovery. Data Mining and Knowledge Discovery 3, 409–425 (1999)

[83] Dos Santos, W.P., de Souza, R.E., dos Santos Filho, P.B.: Evaluation of Alzheimer's disease by analysis of MR images using multilayer perceptrons and Kohonen SOM classifiers as an alternative to the ADC maps. In: Conf. Proc. IEEE Eng. Med. Biol. Soc., pp. 2118–2121 (2007)

[84] Draper, N.R., Smith, H.: Applied regression analysis, 2nd edn. John Wiley, Chichester (1981)

[85] Drucker, H., Burges, C.J.C., Kaufman, L., Smola, A., Vapnik, V.: Support Vector Regression Machines. In: Mozer, M.C., Jordan, M.I., Petsche, T. (eds.) Advances in Neural Information Processing Systems, vol. 9, pp. 155–161. MIT Press, Cambridge (1997)

[86] Dubois, D., Prade, H.: Rough fuzzy sets and fuzzy rough sets. International Journal of General Systems 17, 191–209 (1990)

[87] Dubois, D., Prade, H.: Putting rough sets and fuzzy sets together. In: Slowinski, R. (ed.) Intelligent Decision Support-Handbook of Applications and Advances of the Rough Sets Theory, pp. 203–232. Kluwer Academic Publishers, Dordrecht (1992)

[88] Duda, R.O., Hart, P.E.: Pattern classification and scene analysis. John Wiley, Chichester (1973)

[89] Duda, R.O., Hart, P.E., Stork, D.H.: Pattern Classification, 2nd edn. Wiley Interscience, Hoboken (2001)

[90] Dumitrescu, D.: Genetic algorithms and evolutionary strategies applications in Artificial Intelligence and conexes domains. Blue Publishing House, Cluj-Napoca (2000)

[91] Dunteman, G.H.: Principal Components Analysis. Series: Quantitative applications in the social science. Sage University Paper 69 (1989)

[92] Dym, C.L., Ivey, E.S.: Principles of mathematical modeling. Academic Press, London (1980)

[93] Dzubera, J., Whitley, D.: Advanced correlation analysis of operators for the Traveling Salesman Problem. In: Davidor, Y., Männer, R., Schwefel, H.-P. (eds.) PPSN 1994. LNCS, vol. 866, pp. 68–77. Springer, Heidelberg (1994)

[94] Eiben, A.E., Schippers, C.A.: On evolutionary exploration and exploitation. Fundamenta Informaticae 35(1-4), 35–50 (1998)

[95] Eiben, A.E., Hinterding, R., Michalewicz, Z.: Parameter control in evolutionary algorithms. IEEE Transactions on Evolutionary Computation 3(2), 124–141 (1999)

[96] Eiben, A.E.: Multiparent recombination in evolutionary computing. In: Gosh, A., Tsutsui, S. (eds.) Advances in Evolutionary Computation: Theory and Applications, pp. 175–192. Springer, Heidelberg (2003)

[97] Eiben, A.E., Smith, J.E.: Introduction to Evolutionary Computing. Springer, Heidelberg (2003)

[98] Eiben, A.E., Michalewicz, Z., Schoenauer, M., Smith, J.E.: Parameter Control in Evolutionary Algorithms. SCI, vol. 54. Springer, Heidelberg (2007)

[99] Egan, J.P.: Signal detection theory and ROC analysis. Series in Cognition and Perception. Academic Press, New York (1975)

[100] El-Darzi, E., Gorunescu, M., Gorunescu, F.: A non-homogeneous kinetic model in the patients flow modeling. In: Gorunescu, F., El-Darzi, E., Gorunescu, M. (eds.) Proc. First East European Conference on Health Care Modelling and Computation-HCMC 2005, Craiova, Romania, pp. 99–106. Medical University Press (2005)

[101] El-Darzi, E., Abbi, R., Vasilakis, C., Gorunescu, F., Gorunescu, M., Millard, P.: Length of stay-based clustering methods for patient grouping. In: McClean, S., Millard, P., El-Darzi, E., Nugent, C. (eds.) Intelligent Patient Management. Studies in Computational Intelligence, vol. 189, pp. 39–56. Springer, Heidelberg (2009)

[102] El-Gamal, M.A., Mohamed, M.D.: Ensembles of Neural Networks for Fault Diagnosis in Analog Circuits. Journal of Electronic Testing: Theory and Applications 23(4), 323–339 (2007)

[103] Elsner, U.: Static and Dynamic Graph Partitioning. A comparative study of existing algorithms. Logos (2002)

[104] English, T.M.: Evaluation of evolutionary and genetic optimizers: no free lunch. In: Fogel, L.J., Angeline, P.J., Back, T. (eds.) Proc. 5th Annual Conference on Evolutionary Programming, pp. 163–169. MIT Press, Cambridge (1996)

[105] Everitt, B., Landau, S., Leese, M.: Cluster analysis. Wiley, Chichester (2001)

[106] Fausett, L.V.: Fundamentals of Neural Networks: architectures, algorithms and applications. Prentice-Hall, Englewood Cliffs (1993)

[107] Fawcett, T.: An introduction to ROC analysis. Pattern Recognition Letters 27, 861–874 (2006)

[108] Fayyad, U., Piatestky-Shapiro, G., Smyth, P., Uthurusamy, R.: Advances in Knowledge Discovery and Data Mining. AAAI/MIT Press (1996)

[109] Fibak, J., Pawlak, Z., Slowinski, K., Slowinski, R.: Rough sets based decision algorithm for treatment of duodenal ulcer by HSV. Bull. Polish Acad. Sci. Biological Sci. 34(10-12), 227–246 (1986)

[110] Fisher, D.: Improving Inference through Conceptual Clustering. In: Proc. of 1987 AAAI Conferences, Seattle, Washington, pp. 461–465 (1987)

[111] Fisher, D.: Knowledge acquisition via incremental conceptual clustering. Machine Learning 2, 139–172 (1987)

[112] Fisher, R.A.: On the mathematical foundations of theoretical statistics. Philosophical Transactions of the Royal Society A222, 309–368 (1922)

[113] Fisher, R.A.: The use of multiple measurements in taxonomic problems. Annual Eugenics 7(Part II), 179–188 (1936)

[114] Foucart, T., Bensaber, A., Garnier, R.: Methodes pratiques de la Statistique. Masson (1987)

[115] Fraser, A.S.: Simulation of Genetic Systems by Automatic Digital Computers. I. Introduction. Australian Journal of Biological Sciences 10, 484–491 (1957)

[116] Frawley, W., Piatetsky-Shapiro, G., Matheus, C.: Knowledge Discovery in Databases: An Overview. AI Magazine (1992)

[117] French, M.N., Krajewski, W.F., Cuykendall, R.R.: Rainfall forecasting in space and time using a neural network. Journal of Hydrlogy 137(1-4), 1–31 (1992)

[118] Freund, J.: Modern elementary statistics, 11th edn. Pearson/Prentice Hall (2004)

[119] Fukunaga, K.: Introduction to statistical pattern recognition. Academic Press, London (1990)

[120] Gao, X., Zhao, R., Qi, C., Shi, Z.: A rough set method to treat ECG signals for predicting coronary heart disease. Journal of Biomedical Engineering 25(5), 1025–1028 (2008)

[121] Gan, G., Ma, C., Wu, J.: Data clustering: Theory, algorithms, and applications. SIAM, Philadelphia (2007)

[122] Garcia-Orellana, C.J., Gallardo-Caballero, R., Macias-Macias, M., Gonzalez-Velasco, H.: SVM and neural networks comparison in mammographic CAD. In: Conf. Proc. IEEE Eng. Med. Bio. Soc., pp. 3204–3207 (2007)

[123] Garcia-Laencina, P.J., Sancho-Gomez, J.L., Figueiras-Vidal, A.R., Verleysen, M.: K-nearest neighbours with mutual information for simultaneous classification and missing data imputation. Neurocomputing 72(7-9), 1483–1493 (2009)

[124] Gilchrist, W.: Statistical modeling. John Wiley, Chichester (1984)

[125] Glasziou, P., Del Mar, C., Salisbury, J.: Evidence-based medicine workbook. BMJ Publishing Group (2003)

[126] Glesner, M., Pochmuller, W.: Neurocomputers: An Overview of Neural Networks in VLSI. Chapman and Hall, Boca Raton (1994)

[127] Gluck, R., Klimov, A.V.: Occam's Razor in Metacomputation: the Notion of a Perfect Process Tree. In: Cousot, P., Filé, G., Falaschi, M., Rauzy, A. (eds.) WSA 1993. LNCS, vol. 724, pp. 112–123. Springer, Heidelberg (1993)

[128] Goh, C., Law, R., Mok, H.: Analyzing and forecasting tourism demand: A rough sets approach. Journal of Travel Research 46(3), 327–338 (2008)

[129] Goldberg, D.E.: Genetic Algorithms in Search, Optimization, and Machine Learning. Addison-Wesley, Reading (1989)

[130] Goldberg, D.E., Lingle, R.: Alleles, loci, and the travelling salesman problem. In: Grefenstette, J.J. (ed.) Proceedings 1st International Conference on Genetic Algorithms and Their Applications, pp. 154–159. Lawrence Erlbaum, Hillsdale (1985)

[131] Goldberg, D.E.: Sizing Populations for Serial and Parallel Genetic Algorithms. Addison-Wesley, Reading (1989)

[132] Goldstein, B.J., Mushlin, A.I.: Use of single thyroxine test to evaluate ambulatory medical patients for suspected hypothyroidism. J. Gen. Intern. Med. 2, 20–24 (1987)

[133] Gordon, A.D.: Classification, 2nd edn. Chapman and Hall, Boca Raton (1999)

[134] Gorsuch, R.L.: Factor analysis, 2nd edn. Lawrence Erlbaum Assoc. Inc., Hillsdale (1983)

[135] Gorunescu, F.: k-means clustering: a heuristic approach to improve the treatments effectiveness. Craiova Medical Journal 5(3), 421–423 (2003)

[136] Gorunescu, F.: Architecture of Probabilistic Neural Networks: estimating the adjustable smoothing factor. In: Proc. 4th International Conference on Artificial Intelligence and Digital Communications. Research Notes in Artificial Intelligence and Digital Communications (Subseries of Research Notes in Computer Science), vol. 104, pp. 56–62 (2004)

[137] Gorunescu, F., Gorunescu, M., El-Darzi, E., Gorunescu, S.: An Evolutionary Computational Approach to Probabilistic Neural Network with Application to Hepatic Cancer Diagnosis. In: Tsymbal, A., Cunningham, P. (eds.) Proc. 18th IEEE International Symposium on Computer-Based Medical Systems -IEEE CBMS 2005, pp. 461–466. IEEE Computer Science Press, Los Alamitos (2005)

[138] Gorunescu, F., Gorunescu, M., El-Darzi, E., Ene, M., Gorunescu, S.: Performance enhancement approach for Probabilistic Neural Networks. In: Gorunescu, F., El-Darzi, E., Gorunescu, M. (eds.) Proc. First East European Conference on Health Care Modelling and Computation-HCMC 2005, pp. 142–148. Medical University Press (2005)

[139] Gorunescu, F., Gorunescu, M., El-Darzi, E., Ene, M., Gorunescu, S.: Statistical Comparison of a Probabilistic Neural Network Approach in Hepatic Cancer Diagnosis. In: Proc. Eurocon2005 -IEEE International Conference on "Computer as a tool", pp. 237–240. IEEE Press, Los Alamitos (2005)

[140] Gorunescu, M., Gorunescu, F., Ene, M., El-Darzi, E.: A heuristic approach in hepatic cancer diagnosis using a probabilistic neural network-based model. In: Janssen, J., Lenca, P. (eds.) Proc. 11th Applied Stochastic Models and Data Analysis - ASMDA 2005, pp. 1016–1025 (2005)

[141] Gorunescu, F., Gorunescu, M., El-Darzi, E., Gorunescu, S., Revett, K., Khan, A.: A Cancer Diagnosis System Based on Rough Sets and Probabilistic Neural Networks. In: Gorunescu, F., El-Darzi, E., Gorunescu, M. (eds.) Proc. First East European Conference on Health Care Modelling and Computation - HCMC 2005, pp. 149–159. Medical University Press (2005)

[142] Gorunescu, F., Gorunescu, M., El-Darzi, E., Ene, M.: Comparing classifiers: Probabilistic Neural Networks vs. k-means clustering. In: Proc. 5th International Conference on Artificial Intelligence and Digital Communications. Research Notes in Artificial Intelligence and Digital Communications (Subseries of Research Notes in Computer Science), Annals University of Craiova, vol. 105, pp. 6–11 (2005)

[143] Gorunescu, F., Gorunescu, M.: A heuristic approach in cancer research using evolutionary computation. International Journal of Computer Research (Nova Science) 12(4), 59–65 (2005)

[144] Gorunescu, F.: Benchmarking Probabilistic Neural Network algorithms. In: Proc. 6th International Conference on Artificial Intelligence and Digital Communications. Research Notes in Artificial Intelligence and Digital Communications (Subseries of Research Notes in Computer Science), vol. 106, pp. 1–7 (2006)

[145] Gorunescu, M., Gorunescu, F., Revett, K.: A Neural Computing-based approach for the Early Detection of Hepatocellular Carcinoma. International Journal of Biomedical Sciences, Transactions on Engineering, Computing and Technology (Enformatika) 17, 65–68 (2006)

[146] Gorunescu, M., Revett, K.: A novel noninvasive cancer diagnosis using neural networks. In: Proc. 7th International Conference on Artificial Intelligence and Digital Communications. Research Notes in Artificial Intelligence and Digital Communications (Subseries of Research Notes in Computer Science), vol. 107, pp. 16–20 (2007)

[147] Gorunescu, F., Gorunescu, M., Revett, K., Ene, M.: A hybrid incremental/Monte Carlo searching technique for the "smoothing" parameter of Probabilistic Neural Networks. In: Proc. International Conference on Knowledge Engineering, Principles and Techniques -KEPT 2007, pp. 107–114. Cluj University Press (2007)

[148] Gorunescu, M., Gorunescu, F., Revett, K.: Investigating a Breast Cancer Dataset Using a Combined Approach: Probabilistic Neural Networks and Rough Sets. In: Proc. 3rd ACM International Conference on Intelligent Computing and Information Systems, ICICIS 2007, pp. 246–249. Police Press, Cairo (2007)

[149] Gorunescu, F., Gorunescu, M., Gorunescu, S., Saftoiu, A., Vilmann, P.: Neural Computing: Application in Non-invasive Cancer Detection. Case Studies in Business, Industry and Government Statistics (CSBIGS) 2(1), 38–46 (2008)

[150] Gorunescu, F., Saftoiu, A., Vilmann, P.: Neural Networks-based dynamic medical diagnosis. In: Proc. 4th ACM International Conference on Intelligent Computing and Information Systems, ICICIS 2009, pp. 30–35 (2009)

[151] Gorunescu, F., Gorunescu, M., Saftoiu, A., Vilmann, P., Belciug, S.: Competitive/Collaborative Neural Computing System for Medical Diagnosis in Pancreatic Cancer Detection. Expert Systems. Willey&Blackwell (2010), doi:10.1111/j.1468-0394.2010.00540.x

[152] Gorunescu, F., El-Darzi, E., Belciug, S., Gorunescu, M.: Patient grouping optimization using a hybrid Self-Organizing Map and Gaussian Mixture Model for length of stay-based clustering system. In: Proc. IEEE International Conference on Intelligent Systems, IS 2010, pp. 173–178 (2010)

[153] Gorunescu, F., Gorunescu, M., El-Darzi, E., Gorunescu, S.: A statistical framework for evaluating neural networks to predict recurrent events in breast cancer. International Journal of General Systems 39(5), 471–488 (2010)

[154] Gourieroux, C., Monfort, A.: Time series and dynamic models. Cambridge University Press, Cambridge (1997)

[155] Gower, J.C., Legendre, P.: Metric and Euclidean properties of dissimilarity coefficients. Journal of Classification 3, 5–48 (1986)

[156] Green, D.M., Swets, J.M.: Signal detection theory and psychophysics. John Wiley and Sons Inc., New York (1966)

[157] Groth, R.: Data Mining. A hands-on approach for business professionals. Prentice-Hall, Englewood Cliffs (1998)

[158] Grubbs, F.: Procedures for Detecting Outlying Observations in Samples. Technometrics 11(1), 1–21 (1969)

[159] Grunwald, P.: The Minimum Description Length principle. MIT Press, Cambridge (2007)

[160] Grzymala, J., Siddhave, S.: Rough set Approach to Rule Induction from Incomplete Data. In: Proc. IPMU 2004, 10th International Conference on information Processing and Management of Uncertainty in Knowledge-Based System (2004)

[161] Gurney, K.: An introduction to Neural Networks. CRC Press, Boca Raton (1997)

[162] Guyon, I., Weston, J., Barnhill, S., Vapnik, V.: Gene Selection for Cancer Classification using Support Vector Machines. Machine Learning 46(1-3), 389–422 (2002)

[163] Guyon, I., Elisseeff, A.: An introduction to variable and feature selection. Journal of Machine Learning Research 3, 1157–1182 (2003)

[164] Halkidi, M., Batistakis, Y., Vazirgiannis, M.: On clustering validation techniques. Journal of Intelligent Information Systems 17(2/3), 107–145 (2001)

[165] Hammer, B.: Learning with Recurrent Neural Networks. LNCIS, vol. 254. Springer, Heidelberg (2000)

[166] Han, J., Kamber, M.: Data Mining Concepts and Techniques, 2nd edn. Morgan Kaufmann Publishers, San Francisco (2006)

[167] Han, J., Pei, J., Yin, Y., Mao, R.: Mining frequent patterns without candidate generation. Data Mining and Knowledge Discovery 8, 53–87 (2004)

[168] Hand, D., Mannila, H., Smyth, P.: Principles of Data Mining. MIT Press, Cambridge (2001)

[169] Hand, D., Yu, K.: Idiot's Bayes - not so stupid after all? International Statistical Review 69(3), 385–399 (2001)

[170] Hanley, J.A., McNeil, B.J.: The meaning and use of the area under a receiver operating characteristic (ROC) curve. Radiology 143, 29–36 (1982)

[171] Hansen, L.P., Sargent, T.J.: Robustness. Princeton University Press, Princeton (2007)

[172] Harris, S.: Total Information Awareness official responds to criticism. In: Government Executive.com (January 31, 2003),
http://www.govexec.com/dailyfed/0103/013103h1.htm

[173] Hassan, M.R., Hossain, M.M., Bailey, J., Ramamohanarao, K.: Improving *k*-Nearest Neighbour Classification with Distance Functions Based on Receiver Operating Characteristics. In: Daelemans, W., Goethals, B., Morik, K. (eds.) ECML PKDD 2008, Part I. LNCS (LNAI), vol. 5211, pp. 489–504. Springer, Heidelberg (2008)

[174] Hassanien, A., Ali, J.: Image classification and retrieval algorithm based on rough set theory. S. Afr. Comput. J. 30, 9–16 (2003)

[175] Hassanien, A.: Rough set approach for attribute reduction and rule generation: A case of patients with suspected breast cancer. Journal of the American Society for Information Science and Technology 55(11), 954–962 (2004)

[176] Hassanien, A., Abraham, A., Peters, J., Schaefer, G., Henry, C.: Rough sets and near sets in medical imaging: a review. IEEE Transactions on Information Technology in Biomedicine 13(6), 955–968 (2009)

[177] Hastie, T., Tibshirani, R., Friedman, J.: The Elements of Statistical Learning: Data Mining, Inference, and Prediction. Springer, Heidelberg (2001)

[178] Hastie, T., Tibshirani, R.: Generalized additive models (with discussion). Statistical Science 1, 297–318 (1986)

[179] Hastie, T., Tibshirani, R.: Generalized additive models. Chapman and Hall, Boca Raton (1990)

[180] Haupt, R., Haupt, S.E.: Practical genetic algorithms. Wiley, Chichester (2004)

[181] Haykin, S.: Neural Networks, 2nd edn. A comprehensive foundation. Prentice-Hall, Englewood Cliffs (1999)

[182] Hertz, J., Krogh, A., Palmer, R.G.: Introduction to the theory of neural computation. Addison Wesley, Redwood City (1991)

[183] Heyer, L.J., Kruglyak, S., Yooseph, S.: Exploring expression data: identification and analysis of coexpressed genes. Genome Res. 9(11), 1106–1115 (1999)

[184] Hill, T., Lewicki, P.: Statistics, methods and applications. A comprehensive reference for science, industry, and data mining. StatSoft (2006)

[185] Hincapie, J.G., Kirsch, R.F.: Feasibility of EMG-based neural network controller for an upper extremity neuroprosthesis. IEEE Trans. Neural. Syst. Rehabil. Eng. 17(1), 80–90 (2009)

[186] Hinton, G., Sejnowski, T.J. (eds.): Unsupervised Learning: Foundations of Neural Computation. MIT Press, Cambridge (1999)

[187] Hoffmann, A.: Paradigms of artificial intelligence. Springer, Heidelberg (1998)

[188] Holland, J.H.: Adaptation. In: Rosen, R., Snell, F.M. (eds.) Progress in Theoretical Biology, vol. 4, Plenum, New York (1976)

[189] Holland, J.H.: Adaptation in Natural and Artificial Systems, 1st edn. University of Michigan Press, Ann Arbor (1975); MIT Press, Cambridge (1992)

[190] Holte, R.C.: Very Simple Classification Rules Perform Well on Most Commonly Used Datasets. Machine Learning 11, 63–90 (1993)

[191] Hopfield, J.J.: Neural networks and physical systems with emergent collective computational abilities. Proc. National Academy of Sciences of the USA 79(8), 2554–2558 (1982)

[192] Hoppner, F., Klawonn, F., Kruse, R., Runkler, T.: Fuzzy Cluster Analysis. Wiley, Chichester (1999)

[193] Hoskinson, A.: Creating the ultimate research assistant. Computer 38(11), 97–99 (2005)

[194] Hotelling, H.: Relations between two sets of variates. Biometrika 28, 321–377 (1936)

[195] Hua, S., Sun, Z.: Support Vector Machine Approach for Protein Subcellular Localization Prediction. Bioinformatics 17(8), 721–728 (2001)

[196] Huang, C., Dorsey, E., Boose, M.A.: Life Insurer Financial Distress Prediction: A Neural Network Model. Journal of Insurance Regulation 13(2), 131–167 (Winter 1994)

[197] Huber, P.J.: Robust Statistics. John Wiley, Chichester (1981)

[198] Huet, S., Bouvier, A., Poursat, M.-A., Jolivet, E.: Statistical tools for nonlinear regression, 2nd edn. Springer Series in Statistics. Springer, Heidelberg (2004)

[199] Hunt, E.B., Marin, J., Stone, P.T.: Experiments in Induction. Academic Press, New York (1966)

[200] Hwang, J.-N., Choi, J., Oh, S., Marks, R.: Query-based learning applied to partially trained multilayer perceptron. IEEE Transactions on Neural Networks 2(1), 131–136 (1991)

[201] Ince, E.A., Ali, S.A.: Rule based segmentation and subject identification using fiducial features and subspace projection methods. Journal of Computers 2(4), 68–75 (2007)

[202] Jackson, J.E.: A User's Guide to Principal Components. Wiley series in probability and mathematical statistics. Wiley-IEEE (2003)

[203] Jain, A.K., Dubes, R.C.: Algorithms for Clustering Data. Prentice-Hall, New Jersey (1988)

[204] Jambu, M.: Exploratory and multivariate data analysis. Academic Press, London (1991)

[205] Jensen, F.: Bayesian networks and decision graphs. Statistics for Engineering and Information Science. Springer, Heidelberg (2001)

[206] Jiang, Z., Yamauchi, K., Yoshioka, K., Aoki, K., Kuroyanagi, S., Iwata, A., Yang, J., Wang, K.: Support Vector Machine-Based Feature Selection for Classification of Liver Fibrosis Grade in Chronic Hepatitis C. Journal of Medical Systems 20, 389–394 (2006)

[207] Jin, H., Lu, Y.: A non-inferiority test of areas under two parametric ROC curves. Contemporary Clinical Trials 30(4), 375–379 (2009)

[208] Joachims, T.: Learning to Classify Text Using Support Vector Machines: Methods, Theory and Algorithms. Springer, Heidelberg (2002)

[209] Jolliffe, I.T.: Principal component analysis, 2nd edn. Springer series in statistics. Springer, Heidelberg (2002)

[210] Kahn, B., Strong, D., Wang, R.: Information Quality Benchmarks: Product and Service Performance. Communications of the ACM, 184–192 (April 2002)

[211] Kalamatianos, D., Liatsis, P., Wellstead, P.E.: Near-infrared spectroscopic measurements of blood analytes using multi-layer perceptron neural networks. In: Conf. Proc. IEEE Eng. Med. Biol. Soc., pp. 3541–3544 (2006)

[212] Kaufman, L., Rousseeuw, P.J.: Finding groups in data: an introduction to cluster analysis. John Wiley and Sons, Chichester (2005)

[213] Khan, A.U., Bandopadhyaya, T.K., Sharma, S.: Classification of Stocks Using Self Organizing Map. International Journal of Soft Computing Applications 4, 19–24 (2009)

[214] Kim, J.-O., Mueller, C.: Introduction to factor analysis. Series 07-013. Sage Publications, Beverly Hills (1978)

[215] Kim, J.-O., Mueller, C.: Factor Analysis. Statistical Methods and Practical Issues. Series 07-014. Sage Publications, Beverly Hills (1978)

[216] Kitajima, M., Hirai, T., Katsuragawa, S., Okuda, T., Fukuoka, H., Sasao, A., Akter, M., Awai, K., Nakayama, Y., Ikeda, R., Yamashita, Y., Yano, S., Kuratsu, J., Doi, K.: Differentiation of common large sellar-suprasellar masses effect of artificial neural network on radiologists' diagnosis performance. Acad. Radiol. 16(3), 313–320 (2009)

[217] Kohonen, T.: An introduction to neural computing. Neural Networks 1, 3–16 (1988)

[218] Kohonen, T.: Self-Organizing Maps (3rd extended edn.). Springer Series in Information Sciences, vol. 30. Springer, Heidelberg (2001)

[219] Komorowski, J., Polkovski, L., Skowron, A.: Rough Sets: A tutorial. In: 11th European Summer School in Language, Logic and Information. Lecture Notes for ESSLLI 1999, Utrecht, Holland (1999)

[220] Komorowski, J., Ohrn, A., Skowron, A.: The ROSETTA Rough Set Software System. In: Klosgen, W., Zytkow, J. (eds.) Handbook of Data Mining and Knowledge Discovery, ch. D.2.3. Oxford University Press, Oxford (2002)

[221] Koski, T., Noble, J.: Bayesian Networks: An Introduction. Wiley Series in Probability and Statistics (2009)

[222] Koza, J.: Genetic Programming: On the Programming of Computers by Means of Natural Selection. MIT Press, Cambridge (1992)

[223] Kowalczyk, W., Piasta, Z.: Rough sets inspired approach to knowledge discovery in business databases. In: Wu, X., Kotagiri, R., Korb, K.B. (eds.) PAKDD 1998. LNCS, vol. 1394, Springer, Heidelberg (1998)

[224] Kumar, A., Agrawal, D.P., Joshi, S.D.: Multiscale rough set data analysis with application to stock performance modeling. Intelligent Data Analysis 8(2), 197–209 (2004)

[225] Kuo, S.J., Hsiao, Y.H., Huang, Y.L., Chen, D.R.: Classification of benign and malignant breast tumors using neural networks and three-dimensional power Doppler ultrasound. Ultrasound Obstet. Gynecol. 32(1), 97–102 (2008)

[226] Lai, K.C., Chiang, H.C., Chen, W.C., Tsai, F.J., Jeng, L.B.: Artificial neural network-based study can predict gastric cancer staging. Hepatogastroenterology 55(86-87), 1859–1863 (2008)

[227] Langdon, W., Poli, R.: Foundations of genetic programming. Springer, Heidelberg (2002)

[228] Larose, D.: Discovering Knowledge in Data: An Introduction to Data Mining. John Wiley, Chichester (2005)

[229] Larose, D.: Data mining methods and models. Wiley Interscience, Hoboken (2006)

[230] Lasko, T.A., Bhagwat, J.G., Zou, K.H., Ohno-Machado, L.: The use of receiver operating characteristic curves in biomedical informatics. Journal of Biomedical Informatics 38(5), 404–415 (2005)

[231] Lebowitz, M.: Experiments with incremental concept formation. Machine Learning 2, 103–138 (1987)

[232] Levine, M.: Canonical analysis and factor comparison. Sage Publications, Beverly Hills (1977)

[233] Liang, K.H., Krus, D.J., Webb, J.M.: K-fold crossvalidation in canonical analysis. Multivariate Behavioral Research 30, 539–545 (1995)

[234] Lin, T.Y., Cercone, N. (eds.): Rough sets and data mining. Analysis of imprecise data. Kluwer Academic Publishers, Dordrecht (1997)

[235] Lin, T.Y., Tremba, J.: Attribute Transformations on Numerical Databases. In: Terano, T., Chen, A.L.P. (eds.) PAKDD 2000. LNCS, vol. 1805. Springer, Heidelberg (2000)

[236] Lin, T.Y.: Attribute transformations for data mining. I: Theoretical explorations. International Journal of Intelligent Systems 17(2), 213–222 (2002)

[237] Lin, T.Y., Yiyu, Y., Yao, Y.Y., Zadeh, L. (eds.): Data Mining, Rough Sets and Granular Computing. Physica-Verlag, Heidelberg (2002)

[238] Lindeman, R.H., Merenda, P.F., Gold, R.: Introduction to bivariate and multivariate analysis. Scott, Foresman and Co., New York (1980)

[239] Lippmann, R.P.: An introduction to computing with neural nets. IEEE ASSP Magazine 4, 4–22 (1987)

[240] Liu, H., Hussain, F., Tan, C.L., Dash, M.: Discretization: An Enabling Technique. Data Mining and Knowledge Discovery 6(4), 393–423 (2002)

[241] Liu, H., Hiroshi, M.: Feature Selection for Knowledge Discovery and Data Mining. Kluwer Academic Publishers, Dordrecht (1998)

[242] Liu, H., Hiroshi, M. (eds.): Feature Extraction, Construction and Selection: A Data Mining Perspective. Springer International Series in Engineering and Computer Science (1998)

[243] Loeve, M.: Probability Theory I, 4th edn. Springer, Heidelberg (1977)

[244] Luo, F.-L., Unbehauen, R.: Applied Neural Networks for Signal Processing. Cambridge University Press, Cambridge (1999)

[245] Maglogiannis, I., Sarimveis, H., Kiranoudis, C.T., Chatziioannou, A.A., Oikonomou, N., Aidinis, V.: Radial basis function neural networks classification for the recognition of idiopathic pulmonary fibrosis in microscopic images. IEEE Trans. Inf. Technol. Biomed. 12(1), 42–54 (2008)

[246] Maqsood, I., Abraham, A.: Weather analysis using ensemble of connectionist learning paradigms. Applied Soft Computing 7(3), 995–1004 (2007)

[247] Mandic, D., Chambers, J.: Recurrent Neural Networks for Prediction: Learning Algorithms, Architectures and Stability. Wiley, Chichester (2001)

[248] Mathias, K., Whitley, D.: Genetic Operators, the Fitness Landscape and the Traveling Salesman Problem. In: Manner, R., Manderick, B. (eds.) Parallel Problem Solving from Nature, pp. 219–228. Elsevier, Amsterdam (1992)

[249] Maung, Z.M., Mikami, Y.: A rule-based syllable segmentation of Myanmar Text. In: Proc. IJCNLP 2008 Workshop on NLP for Less Privileged Languages, pp. 51–58 (2008)

[250] McLachlan, G.J., Peel, D.: Finite Mixture Models. John Wiley and Sons, New York (2000)

[251] Martinez, W.L., Martinez, A.L.: Exploratory Data Analysis with MATLAB. CRC Press, Boca Raton (2004)

[252] Martinez, L.C., da Hora, D.N., Palotti, J.R., Meira, W., Pappa, G.: From an artificial neural network to a stock market day-trading system: a case study on the BM and F BOVESPA. In: Proc. 2009 International Joint Conference on Neural Networks, pp. 3251–3258 (2009)

[253] McCulloch, W.S., Pitts, W.A.: A logical calculus of the ideas immanent in nervous activity. Bulletin of Mathematics and Biophysics 5, 115–133 (1943)

[254] McLachlan, G., Krishnan, T.: The EM Algorithm and extensions, 2nd edn. Wiley, Chichester (2008)

[255] McLaren, C.E., Chen, W.P., Nie, K., Su, M.Y.: Prediction of malignant breast lesions from MRI features: a comparison of artificial neural network and logistic regression techniques. Acad. Radiol. 16(7), 842–851 (2009)

[256] Meinel, L.A., Stolpen, A.H., Berbaum, K.S., Fajardo, L.L., Reinhardt, J.M.: Breast MRI lesion classification: improved performance of human readers with a backpropagation neural network computer-aided diagnosis (CAD) system. J. Magn. Reson. Imaging 25(1), 89–95 (2007)

[257] Menzies, T., Hu, Y.: Data Mining For Very Busy People. IEEE Computer 36(11), 22–29 (2003)

[258] Metz, C.E.: Basic principles of ROC analysis. Sem. Nuc. Med. 8, 283–298 (1978)

[259] Michalewicz, Z., Fogel, D.B.: How to solve it: Modern Heuristics. Springer, Heidelberg (2004)

[260] Michalewicz, Z.: Genetic Algorithms + Data Structures = Evolution Programs, 3rd edn. Springer, Heidelberg (1996)

[261] Michalski, R.S.: Knowledge acquisition through conceptual clustering: A theoretical framework and an algorithm for partitioning data into conjunctive concepts. International Journal of Policy Analysis and Information Systems 4, 219–244 (1980)

[262] Michalski, R.S., Stepp, R.E.: Learning from observation: Conceptual clustering. In: Michalski, R.S., Carbonell, J.G., Mitchell, T.M. (eds.) Machine Learning: An Artificial Intelligence Approach, pp. 331–363. Tioga, Palo Alto (1983)

[263] Michel, A.N., Liu, D.: Qualitative analysis and synthesis of recurrent neural networks. Pure and Applied Mathematics. CRC Press, Boca Raton (2002)

[264] Miller, B.L., Goldberg, D.E.: Genetic Algorithms, Tournament Selection and the Effects of Noise. Complex Systems 9, 193–212 (1996)

[265] Minsky, M.L., Papert, S.A.: Perceptrons. MIT Press, Cambridge (1969) (expanded edn., 1987)

[266] Mirzaaghazadeh, A., Motameni, H., Karshenas, M., Nematzadeh, H.: Learning flexible neural networks for pattern recognition. World Academy of Science, Engineering and Technology 33, 88–91 (2007)

[267] Mitchell, T.M.: Machine learning. McGraw-Hill, New York (1997)

[268] Mitchell, M.: An introduction to genetic algorithms (complex adaptive systems). MIT Press, Cambridge (1998)

[269] Moore, D.S., McCabe, G.P.: Introduction to the Practice of Statistics, 3rd edn. W.H. Freeman, New York (1999)

[270] Montana, D., Davis, L.: Training feedforward networks using genetic algorithms. In: Proc. International Joint Conference on Artificial Intelligence. Morgan Kaufmann, San Francisco (1989)

[271] Mrozek, A., Plonka, L., Rough, L.: sets in image analysis. Foundations of Computing Decision Sciences 18(3-4), 259–273 (1993)

[272] Mulaik, S.A.: Foundations of Factor Analysis, 2nd edn. Chapman and Hall/CRC Statistics in the Social and Behavioral Scie (2009)

[273] Muller, K.R., Smola, A.J., Ratsch, G., Scholkopf, B., Kohlmorgen, J., Vapnik, V.: Predicting Time Series with Support Vector Machines. In: Gerstner, W., Hasler, M., Germond, A., Nicoud, J.-D. (eds.) ICANN 1997. LNCS, vol. 1327, pp. 999–1004. Springer, Heidelberg (1997)

[274] Mustapha, N., Jalali, M., Jalali, M.: Expectation Maximization clustering algorithm for user modeling in Web usage mining systems. European Journal of Scientific Research 32(4), 467–476 (2009)

[275] Neubauer, A.: Adaptive non-uniform mutation for genetic algorithms. In: Computational Intelligence Theory and Applications. LNCS, Springer, Heidelberg (1997)

[276] Nguyen, H.S., Nguyen, S.H.: Pattern extraction from data. Fundamenta Informaticae 34, 129–144 (1998)

[277] Nguyen, H.D., Yoshihara, I., Yasunaga, M.: Modified edge recombination operators of genetic algorithms for the traveling salesman problem. In: 26th Annual Conference of the IEEE Industrial Electronics Society, IECON 2000, vol. 4, pp. 2815–2820 (2000)

[278] Nigsch, F., Bender, A., van Buuren, B., Tissen, J., Nigsch, E., Mitchell, J.B.O.: Melting Point Prediction Employing k-nearest Neighbor Algorithms and Genetic Parameter Optimization. Journal of Chemical Information and Modeling 46(6), 2412–2422 (2006)

[279] Norton, P.G., Dunn, E.V.: Snoring as a risk factor for disease: an epidemiological survey. Br. Med. J. 291, 630–632 (1985)

[280] Obe, O.O., Shangodoyin, D.K.: Artificial neural network based model for forecasting sugar cane production. Journal of Computer Science 6(4), 439–445 (2010)

[281] Ogulata, S.N., Sahin, C., Erol, R.: Neural network-based computer-aided diagnosis in classification of primary generalized epilepsy by EEG signals. J. Med. Syst. 33(2), 107–112 (2009)

[282] Ohrn, A., Komorowski, J.: ROSETTA: A Rough Set Toolkit for Analysis of Data. In: Proc. Third International Joint Conference on Information Sciences, Fifth International Workshop on Rough Sets and Soft Computing (RSSC 1997), vol. 3, pp. 403–407 (1997)

[283] Ohrn, A.: ROSETTA Technical Reference Manual, Department of Computer and Information Science, Norwegian University of Science and Technology (NTNU), Trondheim, Norway (2000)

[284] Oliver, J., Wallace, C.S.: Inferring decision graphs. In: Proc. of the 1992 Aust. Joint Conf. on Artificial Intelligence, pp. 361–367 (1992)

[285] Oliver, I.M., Smith, D.J., Holland, J.: A study of permutation crossover operators on the travelling salesman problem. In: Grefenstette, J.J. (ed.) Proc. 2nd International Conference on Genetic Algorithms and Their Applications, pp. 224–230. Lawrence Erlbaum, Hillsdale (1987)

[286] O'Neill, S., Leahy, F., Pasterkamp, H., Tal, A.: The effects of chronic hyperinflation, nutritional status, and posture on respiratory muscle strength in cystic fibrosis. Am. Rev. Respir. Dis. 128, 1051–1054 (1983)

[287] Osaki, S.: Applied stochastic system modeling. Springer, Heidelberg (1992)

[288] Osuna, E., Freund, R., Girosi, F.: Training Support Vector Machines: An Application to Face Detection. In: Proc. of the IEEE Conference on Computer Vision and Pattern Recognition, pp. 130–136 (1997)

[289] Park, L.-J.: Learning of Neural Networks for Fraud Detection Based on a Partial Area Under Curve. In: Wang, J., Liao, X.-F., Yi, Z. (eds.) ISNN 2005. LNCS, vol. 3497, pp. 922–927. Springer, Heidelberg (2005)

[290] Parzen, E.: On estimation of a probability density function and mode. Ann. Math. Stat. 33, 1065–1076 (1962)

[291] Pawlak, Z.: Rough Sets. International Journal of Computer and Information Sciences 11, 341–356 (1982)

[292] Pawlak, Z.: Rough Sets: Theoretical Aspects of Reasoning About Data. Kluwer Academic Publishing, Dordrecht (1991)

[293] Pearson, K.: Contributions to the mathematical theory of evolution II: Skew variation in homogeneous material. Philosophical Transactions of the Royal Society of London 186, 343–414 (1895)

[294] Pearson, K.: Das Fehlergesetz und seine Verallgemeinerungen durch Fechner und Pearson. A Rejoinder. A Rejoinder. Biometrika 4, 169–212 (1905)

[295] Pearson, K.: On the generalized probable error in multiple normal correlation. Biometrika 6, 59–68 (1908)

[296] Pedrycz, W., Skowron, A.: Fuzzy and rough sets. Handbook of data mining and knowledge discovery. Oxford University Press, Inc., New York (2002)

[297] Pendaharkar, P.C.: A comparison of gradient ascent, gradient descent and genetic-algorithm-based artificial neural networks for the binary classification problem. Expert Systems 24(2), 65–86 (2007)

[298] Pepe, M.S.: The statistical evaluation of medical tests for classification and prediction. Oxford University Press, Oxford (2004)

[299] Piatetsky-Shapiro, G.: Discovery, analysis, and presentation of strong rules. In: Piatetsky-Shapiro, G., Frawley, W. (eds.) Knowledge Discovery in Databases, AAAI/MIT Press, Cambridge/MA (1991)

[300] Polkowski, L., Skowron, A. (eds.): Rough Sets in Knowledge Discovery 1: Methodology and Applications. Physica-Verlag, Heidelberg (1998)

[301] Polkowski, L., Skowron, A. (eds.): Rough Sets in Knowledge Discovery 2: Applications, Case Studies and Software Systems. Physica-Verlag, Heidelberg (1998)

[302] Polkowski, L., Tsumoto, S., Lin, T.Y. (eds.): Rough Set Methods and Applications: New Developments in Knowledge Discovery in Information Systems. STUDFUZZ. Physica-Verlag, Heidelberg (2000)

[303] Polkowski, L.: Rough Sets. Physica-Verlag, Heidelberg (2002)

[304] Pontil, M., Verri, A.: Support Vector Machines for 3D Object Recognition. IEEE Transactions on Pattern Analysis and Machine Intelligence 20(6), 637–646 (1998)

[305] Price, R., Shanks, G.: A Semiotic Information Quality Framework. In: Proc. IFIP International Conference on Decision Support Systems (DSS2004): Decision Support in an Uncertain and Complex World, Prato (2004)

[306] Provost, F., Fawcett, T.: Analysis and visualization of classifier performance: Comparison under imprecise class and cost distributions. In: Proc. 3rd International Conference on Knowledge Discovery and Data Mining (KDD 1997), pp. 43–48. AAAI Press, Menlo Park (1997)

[307] Provost, F., Fawcett, T., Kohavi, R.: The case against accuracy estimation for comparing induction algorithms. In: Shavlik, J. (ed.) Proc. ICML 1998, pp. 445–453. Morgan Kaufmann, San Francisco (1998)

[308] Quinlan, J.R.: Induction of decision trees. Machine Learning 1, 81–106 (1986)

[309] Quinlan, J.R., Rivest, R.L.: Inferring decision trees using the minimum description length principle. Information and Computation 80(3), 227–248 (1989)

[310] Quinlan, J.R.: C4.5: Programs for Machine Learning. Morgan Kaufmann, San Francisco (1993)

[311] Radzikowska, A.M., Kerre, E.E.: A comparative study of fuzzy rough sets. Fuzzy Sets and Systems 126, 137–156 (2002)

[312] Raj, V.S.: Better performance of neural networks using functional graph for weather forecasting. In: Proc. 12th WSEAS International Conference on Computers, pp. 826–831 (2008)

[313] Rechenberg, I.: Evolutionsstrategie Optimierung technischer Systeme nach Prinzipien der biologischen Evolution. PhD thesis (1971); reprinted by Fromman-Holzboog (1973)

[314] Revett, K., Gorunescu, F., Gorunescu, M., El-Darzi, E., Ene, M.: A Breast Cancer Diagnosis System: A Combined Approach Using Rough Sets and Probabilistic Neural Networks. In: Proc. Eurocon 2005-IEEE International Conference on Computer as a tool, Belgrade, Serbia, pp. 1124–1127. IEEE Computer Society Press, Los Alamitos (2005)

[315] Revett, K., Gorunescu, F., Gorunescu, M., Ene, M.: Mining A Primary Biliary Cirrhosis Dataset Using Rough Sets and a Probabilistic Neural Network. In: Proc. of the IEEE Intelligent Systems 2006 -IS06, Westminster, London, pp. 284–289. IEEE Press, Los Alamitos (2006)

[316] Revett, K., Gorunescu, F., Salem, A.B.: Feature Selection in Parkinson's Disease: A Rough Sets Approach. In: Proc. International Multiconference on Computer Science and Information Technology-IMCSIT 2009, Mrgowo, Poland, vol. 4, pp. 425–428 (2009)

[317] Ripley, B.D.: Pattern Recognition and Neural Networks. Cambridge University Press, Cambridge (1996)

[318] Rissanen, J.: Modeling by the shortest data description. Automatica 14, 465–471 (1978)

[319] Rissanen, J.: Information and Complexity in Statistical Modeling. Springer, Heidelberg (2007)

[320] *** ROC graphs: Notes and practical considerations for researchers. HP Labs, Tech. Rep. HPL-2003-4 (2003)

[321] Roemer, V.M., Walden, R.: Sensitivity, specificity, receiver-operating characteristic (ROC) curves and likelihood ratios for electronic foetal heart rate monitoring using new evaluation techniques. Z. Geburtshilfe Neonatol. 214(3), 108–118 (2010)

[322] Rokach, L., Maimon, O.: Data mining with decision trees. Series in Machine Perception and Artificial Intelligence, vol. 69. World Scientific, Singapore (2007)

[323] Romesburg, H.C.: Cluster Analysis for Researchers, 2nd edn. Krieger Pub. Co. (2004)

[324] Rosenblatt, F.: The Perceptron: A probabilistic model for information storage and organization in the brain. Psychoanalytic Review 65, 386–408 (1958)

[325] Rosenblatt, F.: Principles of neurodynamics: Perceptrons and the theory of brain mechanisms. Spartan, Washington (1962)

[326] Ryan, T.: Modern Regression Methods, Solutions Manual, 2nd edn. Wiley Series in Probability and Statistics. Wiley, Chichester (2009)

[327] Saad, E.W., Choi, J.J., Vian, J.L., Wunsch, D.C.: Query-based learning for aerospace applications. IEEE Trans. Neural Netw. 14(6), 1437–1448 (2003)

[328] Sandage, A.: EDWIN HUBBLE. JRASC 83(6), 1889–1953 (1989)

[329] Saftoiu, A., Gorunescu, F., Rogoveanu, I., Ciurea, T., Ciurea, P.: A Dynamic Forecasting Model to Predict the Long-Term Response During Interferon Therapy in Chronic Hepatitis C. World Congresses of Gastroenterology 59(3), 349 (1998)

[330] Saftoiu, A., Ciurea, T., Gorunescu, F., Rogoveanu, I., Gorunescu, M., Georgescu, C.: Stochastic modeling of the tumor volume assessment and growth patterns in hepatocellular carcinoma. Journal of Oncology/Bulletin du Cancer 91(6), 162–166 (2004)

[331] Saftoiu, A., Vilmann, P., Gorunescu, F., Gheonea, D.I., Gorunescu, M., Ciurea, T., Popescu, G.L., Iordache, A., Hassan, H., Iordache, S.: Neural network analysis of dynamic sequences of EUS elastography used for the differential diagnosis of chronic pancreatitis and pancreatic cancer. Gastrointestinal Endoscopy 68(6), 1086–1094 (2008)

[332] Sanzogni, L., Kerr, D.: Milk production estimates using feed forward artificial neural networks. Computer and Electronics in Agriculture 32(1), 21–30 (2001)

[333] Schaffer, J.D.: Some Effects of Selection Procedures on Hyperplane Sampling by Genetic Algorithms. In: Davis, L. (ed.) Genetic Algorithms and Simulated Annealing. Pitman (1987)

[334] Schaffer, J.D., Whitley, D., Eshelman, L.: Combination of Genetic Algorithms and Neural Networks: The state of the art. In: Combination of Genetic Algorithms and Neural Networks. IEEE Computer Society, Los Alamitos (1992)

[335] Schervish, M.J.: P-Values: What They Are and What They Are Not. The American Statistician 50(3), 203–206 (1996)

[336] Schlkopf, B., Smola, A.: Learning with Kernels: Support Vector Machines, Regularization, Optimization, and Beyond (Adaptive Computation and Machine Learning). MIT Press, Cambridge (2001)

[337] Schwefel, H.P.: Numerische Optimierung von Computer-Modellen. PhD thesis (1974); reprinted by Birkhäuser (1977)

[338] Seber, G.A.F., Wild, C.J.: Nonlinear Regression. Wiley Series in Probability and Statistics. Wiley, Chichester (1989)

[339] Senthilkumaran, N., Rajesh, R.: A Study on Rough Set Theory for Medical Image Segmentation. International Journal of Recent Trends in Engineering 2(2), 236–238 (2009)

[340] Shafer, J., Agrawal, R., Mehta, M.: SPRINT: A scalable parallel classifier for Data Mining. In: Proc. 22nd VLDB Conference Mumbai (Bombay), India, pp. 544–555 (1996)

[341] Shakhnarovish, G., Darrell, T., Indyk, P. (eds.): Nearest-Neighbor Methods in Learning and Vision. MIT Press, Cambridge (2005)

[342] Shank, D.B., McClendon, R.W., Paz, J., Hoogenboom, G.: Ensemble artificial neural networks for prediction of dew point temperature. Applied Artificial Intelligence 22(6), 523–542 (2008)

[343] Shvaytser, H., Peleg, S.: Fuzzy and probability vectors as elements of vector space. Information Sciences 36, 231–247 (1985)

[344] Shu, C., Burn, D.H.: Artificial neural network ensembles and their application in pooled flood frequency analysis. Water Resources Research 40(9), W09301,1–10 (Abstract) (2004), doi: 10.1029/2003WR002816

[345] Shuai, J.-J., Li, H.-L.: Using rough set and worst practice DEA in business failure prediction. In: Ślęzak, D., Yao, J., Peters, J.F., Ziarko, W.P., Hu, X. (eds.) RSFDGrC 2005. LNCS (LNAI), vol. 3642, pp. 503–510. Springer, Heidelberg (2005)

[346] Skowron, A.: Boolean reasoning for decision rules generation. In: Komorowski, J., Raś, Z.W. (eds.) ISMIS 1993. LNCS, vol. 689, pp. 295–305. Springer, Heidelberg (1993)

[347] Slezak, D.: Approximate Entropy Reducts. Fundamenta Informaticae 53(3,4), 365–390 (2002)

[348] Slowinski, R., Zopounidis, C.: Applications of the rough set approach to evaluation of bankruptcy risk. International J. Intelligent Systems in Accounting, Finance and Management 4(1), 27–41 (1995)

[349] Smith, T.F.: Occam's razor. Nature 285(5767), 620 (1980)

[350] Smith, J.E., Fogarty, T.C.: Operator and parameter adaptation in genetic algorithms. Soft Computing 1(2), 81–87 (1997)

[351] Song, T., Jamshidi, M., Lee, R.R., Hung, M.: A modified probabilistic neural network for partial volume segmentation in brain MR image. IEEE Trans. Neural Networks 18(5), 1424–1432 (2007)

[352] Song, J.H., Venkatesh, S.S., Conant, E.A., Arger, P.H., Sehgal, C.M.: Comparative analysis of logistic regression and artificial neural network for computer-aided diagnosis of breast masses. Acad. Radiol. 12(4), 487–495 (2005)

[353] Sorjamaa, A., Hao, J., Lendasse, A.: Mutual Information and k-Nearest Neighbors Approximator for Time Series Prediction. In: Duch, W., Kacprzyk, J., Oja, E., Zadrożny, S. (eds.) ICANN 2005. LNCS, vol. 3697, pp. 553–558. Springer, Heidelberg (2005)

[354] Spackman, K.A.: Signal detection theory: Valuable tools for evaluating inductive learning. In: Proc. 6th International Workshop on Machine Learning, pp. 160–163. Morgan Kaufman, San Mateo (1989)

[355] Spearman, C.: General intelligence objectively determined and measured. American Journal of Psychology 15, 201–293 (1904)

[356] Spearman, C.: The nature of intelligence and the principles of cognition. Macmillan, London (1923)

[357] Spearman, C.: The abilities of man. Macmillan, London (1927)

[358] Specht, D.F.: Probabilistic neural networks for classification mapping or associative memory. In: Proc. IEEE International Conference on Neural Networks, vol. 1, pp. 525–532 (1988)

[359] Specht, D.F.: Probabilistic neural networks. Neural Networks 3, 110–118 (1990)

[360] Specht, D.F.: Probabilistic neural networks and the polynomial adaline as complementary techniques for classification. IEEE Trans. on Neural Networks 1(1), 111–121 (1990)

[361] Specht, D.F.: A Generalized Regression Neural Network. IEEE Transactions on Neural Networks 2(6), 568–576 (1991)

[362] Srisawat, A., Phienthrakul, T., Kijsirikul, B.: SV-kNNC: An Algorithm for Improving the Efficiency of k-Nearest Neighbor. In: Yang, Q., Webb, G. (eds.) PRICAI 2006. LNCS (LNAI), vol. 4099, pp. 975–979. Springer, Heidelberg (2006)

[363] Stefansky, W.: Rejecting Outliers in Factorial Designs. Technometrics 14, 469–479 (1972)

[364] Steinwart, I., Christmann, A.: Support Vector Machines. Information Science and Statistics. Springer, Heidelberg (2008)

[365] Stevens, J.: Applied multivariate statistics for the social sciences. Erlbaum, Hillsdale (1986)

[366] Stoean, C., Preuss, M., Gorunescu, R., Dumitrescu, D.: Elitist generational genetic chromodynamics - a new radii-based evolutionary algorithm for multimodal optimization. In: 2005 IEEE Congress on Evolutionary Computation (CEC 2005), pp. 1839–1846. IEEE Computer Society Press, Los Alamitos (2005)

[367] Stoean, C., Preuss, M., Dumitrescu, D., Stoean, R.: Cooperative Evolution of Rules for Classification. In: IEEE Post-proceedings Symbolic and Numeric Algorithms for Scientific Computing (SYNASC 2006), pp. 317–322. IEEE Press, Los Alamitos (2006)

[368] Stoean, R., Preuss, M., Stoean, C., El-Darzi, E., Dumitrescu, D.: An Evolutionary Approximation for the Coefficients of Decision Functions within a Support Vector Machine Learning Strategy. Foundations on Computational Intelligence 1, 83–114 (2009)

[369] Stoean, R., Preuss, M., Stoean, C., El-Darzi, E., Dumitrescu, D.: Support Vector Machine Learning with an Evolutionary Engine. Journal of the Operational Research Society, Special Issue: Data Mining and Operational Research: Techniques and Applications 60(8), 1116–1122 (2009)

[370] Stoean, R., Stoean, C., Lupsor, M., Stefanescu, H., Badea, R.: Evolutionary-Driven Support Vector Machines for Determining the Degree of Liver Fibrosis in Chronic Hepatitis C. Artificial Intelligence in Medicine. Elsevier (2010) (in press)

[371] Sunay, A.S., Cunedioglu, U., Ylmaz, B.: Feasibility of probabilistic neural networks, Kohonen self-organizing maps and fuzzy clustering for source localization of ventricular focal arrhythmias from intravenous catheter measurements. Expert System 26(1), 70–81 (2009)

[372] Sutton, R.S., Barto, A.G.: Reinforcement learning. MIT Press, Cambridge (1999)

[373] Suykens, J.A.K., Vandewalle, J.P.L., De Moor, B.L.R.: Artificial neural networks for modelling and control of non-linear systems. Kluwer Academic Publishers, Dordrecht (1996)

[374] Swets, J.A.: The relative operating characteristic in psychology. A technique for isolating effects of response bias finds wide use in the study of perception and cognition. Science 182(4116), 990–1000 (1973)

[375] Swets, J.A., Dawes, R.M., Monahan, J.: Better decisions through science. Scientific American 283, 82–87 (2000)

[376] Świniarski, R.W.: Rough Sets and Bayesian Methods Applied to Cancer Detection. In: Polkowski, L., Skowron, A. (eds.) RSCTC 1998. LNCS (LNAI), vol. 1424, pp. 609–616. Springer, Heidelberg (1998)

[377] Syswerda, G.: Uniform crossover in genetic algorithms. In: Scaffer, J.D. (ed.) Proc. 3rd International Conference on Genetic Algorithms, pp. 2–9. Morgan Kaufmann, San Francisco (1989)

[378] Tan, P.-N., Steinbach, M., Kumar, V.: Introduction to Data Mining. Addison-Wesley, Reading (2005)

[379] Tay, F., Shen, L.: Economic and financial prediction using rough sets model. European Journal of Operational Research 141(3), 641–659 (2002)

[380] Thomson, S.K.: Sampling, 2nd edn. Series in Probability and Statistics. Wiley, Chichester (2002)

[381] Thuraisingham, B.: Data Mining: Technologies, Techniques, Tools, and Trends. CRC Press, Boca Raton (1999)

[382] Thurstone, L.L.: Multiple factor analysis. Psychological Review 38, 406–427 (1931)

[383] Titterington, D.M., Smith, A.F.M., Makov, U.E.: Statistical Analysis of Finite Mixture Distributions. John Wiley and Sons, New York (1985)

[384] Tong, S., Chang, E.: Support Vector Machine Active Learning for Image Retrieval. In: Proc. of the 9th ACM international conference on Multimedia, vol. 9, pp. 107–118 (2001)

[385] Toussaint, G.T.: Geometric proximity graphs for improving nearest neighbor methods in instance-based learning and data mining. International Journal of Computational Geometry and Applications 15(2), 101–150 (2005)

[386] Tukey, J.: Exploratory Data Analysis. Addison-Wesley, Reading (1977)

[387] Unnikrishnan, N., Mahajan, A., Chu, T.: Intelligent system modeling of a three-dimensional ultrasonic positioning system using neural networks and genetic algorithms. Proc. Institution of Mechanical Engineers, Part I: Journal of Systems and Control Engineering 217(5), 367–377 (2003)

[388] Vajda, P., Eiben, A.E., Hordijk, W.: Parameter control methods for selection operators in genetic algorithms. In: Rudolph, G., Jansen, T., Lucas, S., Poloni, C., Beume, N. (eds.) PPSN 2008. LNCS, vol. 5199, pp. 620–630. Springer, Heidelberg (2008)

[389] Vapnik, V.N., Chervonenkis, Y.: On the uniform convergence of relative frequencies of events to their probabilities. Theoretical Probability and its applications 17, 264–280 (1971)

[390] Vapnik, V.N.: Estimation of dependencies based on empirical data (translated by S. Kotz). Springer, Heidelberg (1982)

[391] Vapnik, V.N.: The nature of statistical learning theory. Springer, Heidelberg (1995)

[392] Vapnik, V.N., Golowich, S.E., Smola, A.J.: Support Vector Method for Function Approximation, Regression Estimation and Signal Processing. In: NIPS 1996, Denver, USA, pp. 281–287 (1996)

[393] Vapnik, V.N.: Statistical learning theory. Wiley, Chichester (1998)

[394] Ward, J.H.: Hierarchical grouping to optimize an objective function. Journal of the American Statistical Association 58(301), 236–244 (1963)

[395] Velleman, P., Hoaglin, D.: The ABC's of EDA: Applications, Basics, and Computing of Exploratory Data Analysis. Duxbury, Boston (1981)

[396] Wand, Y., Wang, R.: Anchoring Data Quality Dimensions in Ontological Foundations. Communications of the, pp. 86–95. ACM Press, New York (1996)

[397] Wang, J., Neskovic, P., Cooper, L.N.: Improving nearest neighbor rule with a simple adaptive distance measure. Pattern Recognition Letters 28, 207–213 (2007)

[398] Westphal, C., Blaxton, T.: Data Mining Solutions. John Wiley, Chichester (1998)

[399] Whitley, D., Kauth, J.: Genitor: A different genetic algorithm. In: Proc. of the Rocky Mountain Conference on Artificial Intelligence, pp. 118–130 (1988)

[400] Whitley, D.: The GENITOR algorithm and selection pressure: Why rank-based allocation of reproductive trials is best. In: Proc. of the Third International Conference on Genetic Algorithms, pp. 116–123. Morgan Kaufmann, San Francisco (1989)

[401] Whitley, D., Hanson, T.: Optimizing neural networks using faster more accurate genetic search. In: Proc. 3rd International Conference on Genetic Algorithms, pp. 391–396. Morgan Kaufmann, San Francisco (1989)

[402] Whitley, D., Starkweather, T., Fuquay, D.'.A.: Scheduling problems and traveling salesman: The genetic edge recombination operator. In: Proc. 3rd International Conference on Genetic Algorithms, pp. 133–140. Morgan Kaufmann, San Francisco (1989)

[403] Whitley, D., Starkweather, T., Shaner, D.: Traveling Salesman and Sequence Scheduling: Quality Solutions Using Genetic Edge Recombination. Handbook of Genetic Algorithms. Van Nostrand (1990)

[404] Whitley, D.: Genetic Algorithms in Engineering and Computer Science. In: Periaux, J., Winter, G. (eds.) John Wiley and Sons Ltd., Chichester (1995)

[405] Whitley, D.: Permutations. In: Back, T., Fogel, D.B., Michalewicz, Z. (eds.) Evolutionary computation 1: basic algorithms and operators, ch. 33.3, pp. 274–284. Institute of Physics Publishing, Bristol (2000)

[406] Widrow, B., Hoff, M.E.: Adaptive switching circuits. IRE WESCON Convention Record 4, 96–104 (1960); reprinted in Anderson and Rosenfeld (1988)

[407] Widrow, B., Lehr, M.A.: 30 years of adaptive neural networks: perceptron, madaline, and backpropagation. Proc. of the IEEE 78(9), 1415–1442 (1990)

[408] Wikinews: U.S. Army intelligence had detected 9/11 terrorists year before (August 18, 2005),
http://en.wikinews.org/w/index.php?title=U.S.
_Army_intelligence_had_detected_9/11_terrorists
_year_before%2C_says_officer&oldid=130741

[409] Wishart, G.C., Warwick, J., Pitsinis, V., Duffy, S., Britton, P.D.: Measuring performance in clinical breast examination. Br. J. Surg. 7(8), 1246–1252 (2010)

[410] Witten, I.H., Eibe, F.: Data Mining: Practical Machine Learning Tools and Techniques, 2nd edn. Morgan Kaufmann, San Francisco (2005)

[411] Wood, S.: Generalized additive models. An introduction with R. Chapman and Hall/CRC (2006)

[412] Yao, J., Herbert, J.: Financial time-series analysis with rough sets. Applied Soft Computing 9(3), 1000–1007 (2009)

[413] Ye, N., Farley, T.: A scientific approach to cyberattack detection. Computer 38(11), 55–61 (2005)

[414] Ye, N., Chen, Q.: Computer intrusion detection through EWMA for autocorrelated and uncorrelated data. IEEE Trans. Reliability 52(1), 73–82 (2003)

[415] Ye, N.: Mining computer and network security data. In: Ye, N. (ed.) The Handbook of Data Mining, pp. 617–636. Lawrence Erlbaum Assoc., Mahwah (2003)

[416] Yeh, W., Huang, S.W., Li, P.C.: Liver Fibrosis Grade Classification with B-mode Ultrasound. Ultrasound in Medicine and Biology 29(9), 1229–1235 (2003)

[417] Zadeh, L.: Fuzzy sets. Information and Control 8, 338–353 (1965)

[418] Zaki, M.J., Parthasarathy, S., Ogihara, M., Li, W.: New algorithms for fast discovery of association rules. Technical Report TR651 (1997)

[419] Zaknich, A.: Neural networks for intelligent signal processing. Series in Innovative Intelligence, vol. 4. World Scientific, Singapore (2003)

[420] Zhang, C., Zhang, S.: Association Rule Mining. Models and Algorithms. LNCS (LNAI), vol. 2307. Springer, Heidelberg (2002)

[421] Zhong, N.: Using Rough Sets with Heuristics for Feature Selection. Journal of Intelligent Information Systems 16, 199–214 (2001)

[422] Ziarko, W.: Rough Sets, Fuzzy Sets and Knowledge Discovery. Springer, Heidelberg (1994)

[423] Zur, R.M., Pesce, L.L., Jiang, Y.: The effect of two priors on Bayesian estimation of "Proper" binormal ROC curves from common and degenerate datasets. Acad. Radiol. 17(8), 969–979 (2010)

[424] Zweig, M.H., Campbell, G.: Receiver-operating characteristic (ROC) plots: a fundamental evaluation tool in clinical medicine. Clinical Chemistry 39(8), 561–577 (1993)

Index

CPSIA information can be obtained at www.ICGtesting.com
Printed in the USA
LVOW090303100112

263143LV00006B/73/P